ROADSIDE GEOLOGY

of WASHINGTON

Second Edition

MARLI B. MILLER
AND DARREL S. COWAN

Photographs by Marli B. Miller

2017
Mountain Press Publishing Company
Missoula, Montana

First Printing, October 2017
All rights reserved

Geological maps for the road guides are based primarily on
the *Geologic Map of Washington,* 1:250,000 series, by the
Washington Division of Geology and Earth Resources

Roadside Geology is a registered trademark
of Mountain Press Publishing Company

Library of Congress Cataloging-in-Publication Data

Names: Miller, Marli Bryant, 1960- author. | Cowan, Darrel S., 1945- author.
Title: Roadside geology of Washington / Marli B. Miller and Darrel S. Cowan ;
  photographs by Marli B. Miller.
Description: Second edition. | Missoula, Montana : Mountain Press Publishing
  Company, 2017. | Series: Roadside geology series | Includes
  bibliographical references and index.
Identifiers: LCCN 2017028083 | ISBN 9780878426775 (pbk. : alk. paper)
Subjects: LCSH: Geology—Washington (State)—Guidebooks. | Washington
  (State)—Guidebooks.
Classification: LCC QE175 .A48 2017 | DDC 557.97—dc23
LC record available at https://lccn.loc.gov/2017028083

Printed in Canada

MP Mountain Press
PUBLISHING COMPANY
P.O. Box 2399 • Missoula, MT 59806 • 406-728-1900
800-234-5308 • info@mtnpress.com
www.mountain-press.com

To my brother, Lloyd,
for his love and inspiration.

To my aunt, Jane Miller Bigsby Kettering,
and her wonderful family,
for making Washington such a special place.

–Marli Miller

To all of my colleagues and students
at the University of Washington,
my academic home in my adopted state.

–Darrel Cowan

Acknowledgments

What a privilege and joy it's been to write this book! Having gone to the University of Washington for graduate school from the mid-1980s until the early 1990s, I benefited from the first edition of this book by David Alt and Don Hyndman and had a long-standing interest in the region's geology. It wasn't until this project, however, that I took the time to put it all together. And Washington's geology is truly amazing.

Most importantly, I was able to write this book with my former thesis advisor and close friend Darrel Cowan. His wide-ranging perspective on the geologic framework of Washington proved invaluable over and over again. Darrel wrote the introductions to chapters 2 through 7, and together we drove thousands of miles of Washington roads.

Our project benefited immeasurably from Jennifer Carey and Jessica Czajkowski. Jennifer, our editor, made herself available to answer our many questions and made our sometimes jargon-plagued text readable. She helped sort through the too-many photographs, pointed out important places we missed, and wrote the book's index. Jessica, a geologist and editor at the Washington State Department of Natural Resources (DNR), provided easy access to publications, maps, and diagrams of the DNR, introduced me to other geologists at the DNR, helped me with the geologic maps, took me into the field, reviewed one of the road guides, and put me up at her house! She is also the portal manager for the DNR's online interactive geologic map, which proved to be hands-down the best resource available on Washington geology for this book.

Many other people helped, too. Andy Buddington (Spokane Community College), Trevor Contreras (Washington DNR), Ralph Dawes (Wenatchee Valley College), and Michael Polenz (Washington DNR) took me into the field and reviewed some of the road guides. Pat Pringle (Centralia College) reviewed the road guides for the South Cascades, Donna Whitney (University of Minnesota) reviewed the guide for the North Cascades Highway, and Nick Zentner (Central Washington University) reviewed several of the guides for central Washington. Mitch Allen (Washington DNR), Cheryl Dawes (Wenatchee), and David Lewis (Bellingham) took me into the field.

I benefited greatly from discussions with colleagues at the University of Oregon, most notably Becky Dorsey, Ted Fremd, Gene Humphreys, Greg Retallack, Paul Wallace, and Ray Weldon. In addition, Dave Tucker of Western Washington University and Nick Zentner of Central Washington University gave helpful pointers. My friends and colleagues Barb Nash of the University of Utah and Ellen Morris Bishop of Joseph, Oregon, provided ideas and much-appreciated inspiration. At the University of Washington, Darrel engaged Jody Bourgeois, Harvey Greenberg, John Stone, and Kathy Troost. He also benefited from discussions with Ted Doughty of Spokane, Ralph Haugerud and Rowland Tabor of the US Geological Survey, and Jeff Tepper at the University of Puget Sound.

Thanks also to Tom and Lorna Bigsby, Lisa Bryce and David, Margot and Cameron Lewis, Andy and Teresa Buddington, Ralph and Cheryl Dawes, Joe and Alyse Gass, Doug Norseth and Bruce Hegna, and Elaine Sachter and Michael Newman, who invited me into their homes while I traveled. My daughters, Lindsay and Megan, and my friends, Maria Gibson and Corey Jarrett, accompanied me on some road trips.

Finally, we both thank Jeannie Painter for the wonderful design and layout of this book and Chelsea Feeney for editing the many maps and diagrams.

—Marli Miller

Moonrise over Mt. Shuksan from near Artist Point at the end of WA 542 in the North Cascades.

modified from Washington Division of Geology and Earth Resources

UNCONSOLIDATED DEPOSITS

Qs — Quaternary sediments, dominantly nonglacial; includes alluvium and volcaniclastic, glacial outburst flood, eolian, landslide, and coastal deposits

Qg — Quaternary sediments, dominantly glacial drift; includes alluvium

SEDIMENTARY ROCKS

Tls — late Tertiary (Pliocene-Miocene)

Tes — early Tertiary (Oligocene-Paleocene)

Mzs — Mesozoic

MzPzs — Mesozoic-Paleozoic

Pzs — Paleozoic

p€s — Precambrian

VOLCANIC ROCKS

Qv — Quaternary

QPv — Quaternary-Pliocene

Tlv — late Tertiary (Pliocene-Miocene)

Tlvc — late Tertiary Columbia River Basalt Group

Tev — early Tertiary (Oligocene-Paleocene); not of Siletz terrane

Ev — Eocene Siletz terrane

Mzv — Mesozoic

INTRUSIVE IGNEOUS ROCKS

Ti — Tertiary

Mzi — Mesozoic

Pzi — Paleozoic

METAMORPHIC ROCKS

pTm — pre-Tertiary

pKm — pre-Cretaceous

Pzm — Paleozoic

p€m — Precambrian

u — ultramafic rocks

——— fault

——▲— thrust fault; teeth on upper plate

——■— detachment fault; blocks on upper plate

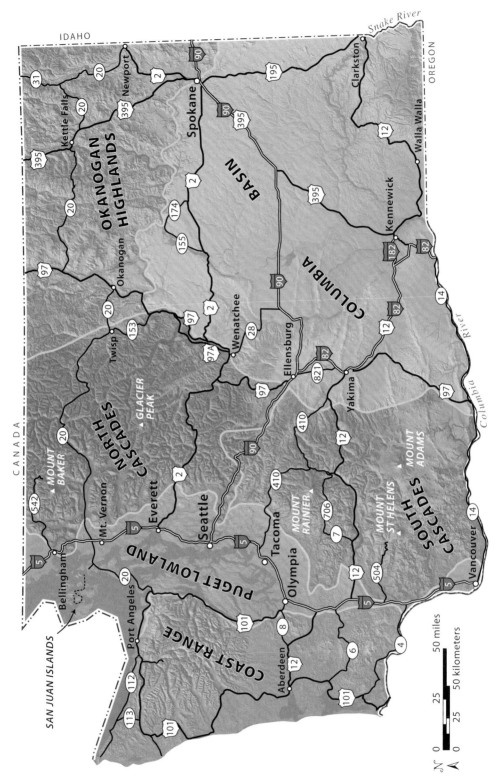

Roads and regions of Roadside Geology of Washington.

Contents

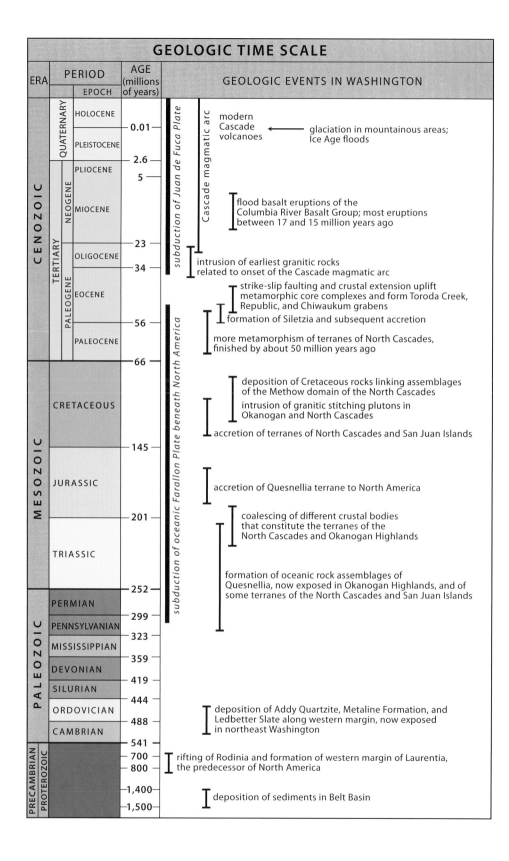

GEOLOGIC TIME SCALE

ERA	PERIOD	EPOCH	AGE (millions of years)	GEOLOGIC EVENTS IN WASHINGTON

CENOZOIC

QUATERNARY — HOLOCENE — 0.01
PLEISTOCENE — 2.6

TERTIARY / NEOGENE — PLIOCENE — 5
MIOCENE — 23

TERTIARY / PALEOGENE — OLIGOCENE — 34
EOCENE — 56
PALEOCENE — 66

MESOZOIC

CRETACEOUS — 145
JURASSIC — 201
TRIASSIC — 252

PALEOZOIC

PERMIAN — 299
PENNSYLVANIAN — 323
MISSISSIPPIAN — 359
DEVONIAN — 419
SILURIAN — 444
ORDOVICIAN — 488
CAMBRIAN — 541

PRECAMBRIAN / PROTEROZOIC

700 — 800 — 1,400 — 1,500

subduction of Juan de Fuca Plate

Cascade magmatic arc

subduction of oceanic Farallon Plate beneath North America

modern Cascade volcanoes ← glaciation in mountainous areas; Ice Age floods

flood basalt eruptions of the Columbia River Basalt Group; most eruptions between 17 and 15 million years ago

intrusion of earliest granitic rocks related to onset of the Cascade magmatic arc

strike-slip faulting and crustal extension uplift metamorphic core complexes and form Toroda Creek, Republic, and Chiwaukum grabens

formation of Siletzia and subsequent accretion

more metamorphism of terranes of North Cascades, finished by about 50 million years ago

deposition of Cretaceous rocks linking assemblages of the Methow domain of the North Cascades

intrusion of granitic stitching plutons in Okanogan and North Cascades

accretion of terranes of North Cascades and San Juan Islands

accretion of Quesnellia terrane to North America

coalescing of different crustal bodies that constitute the terranes of the North Cascades and Okanogan Highlands

formation of oceanic rock assemblages of Quesnellia, now exposed in Okanogan Highlands, and of some terranes of the North Cascades and San Juan Islands

deposition of Addy Quartzite, Metaline Formation, and Ledbetter Slate along western margin, now exposed in northeast Washington

rifting of Rodinia and formation of western margin of Laurentia, the predecessor of North America

deposition of sediments in Belt Basin

INTRODUCTION

Although Washington is the smallest state in the American West, it displays some of the most diverse geology anywhere and brims with geologic superlatives. The Channeled Scabland was carved by some of the largest floods ever, the North Cascades is the most rugged and most heavily glaciated mountain range in the Lower Forty-Eight, and Mount St. Helens is the most active volcano in the conterminous United States. Washington's rocks, which consist of every major type, reflect an active geologic history that spans well over 1 billion years.

We divide Washington into six geologically distinct physiographic regions. The Coast Range, made primarily of young oceanic basalt and sedimentary rock, includes the Olympic Mountains to the north and Willapa Hills to the south. The Puget Lowland, largely covered by ice during the Pleistocene Epoch, encompasses all of Puget Sound and the San Juan Islands and includes the I-5 corridor to the southern border with Oregon. The North Cascades, rugged mountains of mostly metamorphic and igneous rock, extend northward from I-90 to Canada, whereas the volcanic South Cascades extend southward from I-90 to Oregon. In the northeastern corner of the state, the Okanogan Highlands area consists of a wide variety of igneous, metamorphic, and sedimentary rock and is probably Washington's most geologically diverse area. The Columbia Basin, covered largely by basaltic lava flows that were scoured by Ice Age floods, makes up Washington's southeastern quarter.

Before we get into the eventful geologic history that created the Evergreen State, we provide an introduction to the rocks you can expect to see here, as well as the plate tectonic settings in which they formed.

Washington's Rocks–A Primer

Igneous Rocks

Igneous rocks form through the cooling and crystallization of molten rock, a liquid state. Molten rock is called *magma* when it is underground and *lava* when it erupts onto Earth's surface. When magma cools underground, the resulting rock is termed *intrusive*. When lava cools on the surface, the resulting rock is called *extrusive* or *volcanic*. Volcanic rocks typically contain small crystals because they cool quickly on Earth's surface, whereas intrusive rocks typically display easily visible crystals because they cool more slowly, which gives crystals time to grow.

Both intrusive and extrusive rocks are further classified based largely on silica content. Granite and rhyolite are the intrusive and extrusive, silica-rich varieties, whereas gabbro and basalt are the intrusive and extrusive, silica-poor varieties. Important intermediate rocks are diorite (intrusive) and andesite (extrusive). Magmas can erupt violently when they contain a high proportion of water or gas, which expands explosively as the rising magma decompresses. Silica-rich magmas, being very viscous, can make for especially large eruptions because water and gas within the thick magma require more energy to escape. Explosive eruptions produce glass-rich material called *tephra*. Ash and pumice are varieties of tephra.

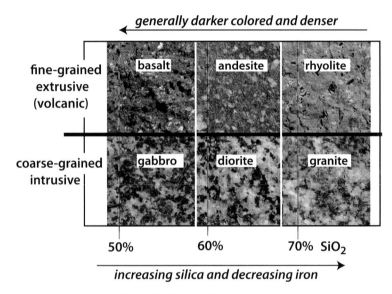

Six major types of igneous rocks.

From the lava flows of the Columbia Basin to the active volcanoes of the Cascade Range and the submarine lavas of the Coast Range, eruptions of molten rock have produced a variety of extrusive rocks in Washington. The modern Cascade Range contains all types of volcanic rocks, but the snow-clad volcanoes consist mostly of andesitic rock. Both the Columbia Basin and the underlying basement rock of the Coast Range consist mostly of basalt. The basalt of the Coast Range erupted under the ocean between about 55 to 50 million years ago. Much of it displays rounded "pillows" from the chilling effect of water on the lava. As blobs of lava erupted on the seafloor, their outer margins rapidly chilled into rock that insulated their interiors, which stayed soft and malleable. Settling onto the seafloor, they conformed to the underlying surface to look like a pile of pillows. Pillow basalt that formed on the ocean floor also shows up in the San Juan Islands. The basalt of the Columbia Basin, called the Columbia River Basalt Group, erupted from southeastern Washington and northeastern Oregon as enormous floods of basalt from 17 to about 6 million years ago. Many flows of the Columbia River Basalt Group also contain pillow lavas, but these formed as the lava flows poured into lakes.

Pillow basalt on Lopez Island.

Bodies of intrusive rocks, usually called *plutons*, are most prevalent in the Cascade Range and Okanogan Highlands, but narrow dikes that cut across older rocks can be found scattered throughout all the different regions. Most of the plutons consist of varieties of granite. In the North Cascades and Okanogan Highlands, many of these rocks are tonalite, a variety of granite that contains mostly plagioclase but very little orthoclase feldspar. In much of northern Washington, we see places where several plutons coalesced into large batholiths.

Metamorphic Rocks

Metamorphic rocks experienced a change, or metamorphism, because of high temperature and/or pressure, usually brought about by deep burial of the rock. Most metamorphic rocks display a layering, called *foliation*, which forms perpendicular to the direction of applied pressure. In general, the higher the temperature of metamorphism, the coarser grained the crystals are that define this layering. Some metamorphic rocks do not contain foliation because they formed solely from high temperatures. For example, they may have formed next to intrusive bodies where they were heated by the crystallizing magma but experienced no pressure. Metamorphic rocks can start out originally as sedimentary or igneous rocks.

The North Cascades, Okanogan Highlands, and Olympic Mountains contain the most extensive areas of metamorphic rocks. Probably the most common of these rocks are greenstone, phyllite, and gneiss. Greenstone forms when basalt is metamorphosed under fairly low temperatures. Phyllite typically originates when shale is heated by low metamorphic temperatures. Gneiss forms under high temperatures.

Washington also hosts mylonite and blueschist, unusual metamorphic rocks that tell us some important things about their geologic histories. Mylonites form in fault zones under high temperatures, such as in the middle crust, where the rock flows instead of breaks. These rocks tend to be strongly foliated, but unlike other high-temperature metamorphic rocks, mylonites are fine grained. They also typically display lineations on their surfaces, formed by elongate minerals. Blueschist is a blue-tinted, foliated metamorphic rock whose blue color is derived from certain blue minerals, most notably glaucophane, a sodium-rich amphibole. These minerals form only under the high-pressure and low-temperature conditions of subduction zones. The sinking cold material in subduction zones is subject to high pressures long before it can heat to very high temperatures.

Typical gneiss (left) is coarse grained, but mylonitic gneiss (right) is very fine grained because it recrystallized in a high-temperature fault zone. Note the metamorphic layering, or foliation.

Sedimentary Rocks

Sedimentary rocks consist of sediment, either broken particles of preexisting rock or material precipitated by either biological or chemical processes. The sediment accumulates in horizontal layers, called beds, in a variety of environments on Earth's surface, such as rivers, lakes, or oceans. In a sequence of sedimentary rocks or lava flows, the oldest lies at the bottom and the youngest on top, allowing us to trace the geologic history of an area.

Sedimentary rocks show up in all parts of Washington, even in parts that are dominated by other types of rock. The basalt-covered Columbia Basin, for example, also hosts the Ellensburg and Latah Formations, which were deposited in rivers and lakes during quiet intervals between volcanic eruptions. Parts of the North Cascades contain sedimentary rocks that were deposited on Earth's surface while intrusion and metamorphism were happening below. Sedimentary rocks resting on top the older Coast Range basalt typify the Coast Range of Washington.

Washington also hosts important sedimentary deposits, accumulations of loose or somewhat compacted sediment that has not yet been turned to rock. The most abundant of these deposits, glacial till and outwash, formed during the Pleistocene Epoch, when the northern and high-altitude parts of the state were covered by ice. Till, deposited directly by ice, consists of material of all sizes and shapes without any clear sedimentary layering. In contrast, glacial outwash is deposited by rivers and streams running off the front of the glacier and so typically shows layering and size sorting of material. In addition, coarse flood gravels and fine-grained lake deposits were deposited by the enormous Ice Age floods that formed the Channeled Scabland.

Glacial till (left), deposited directly from ice, contains material of all sizes, whereas outwash (right) is sorted and deposited in layers by running water. This outcrop of till is located along WA 20 in the Okanogan Highlands. The outwash can be found near the town of Concrete in the North Cascades.

Plate Tectonics

The theory of plate tectonics explains much of Washington's geology. It holds that the Earth's outer shell, called the *lithosphere*, is broken into large fragments, called *plates*, that move gradually through time. As these plates interact with each other along their edges, they change the shape of the plate margins, generate earthquakes and volcanoes, and add landmass to the continental plates. Washington, lying next to a plate boundary, has been subject to these processes since its oldest rocks formed over 1 billion years ago.

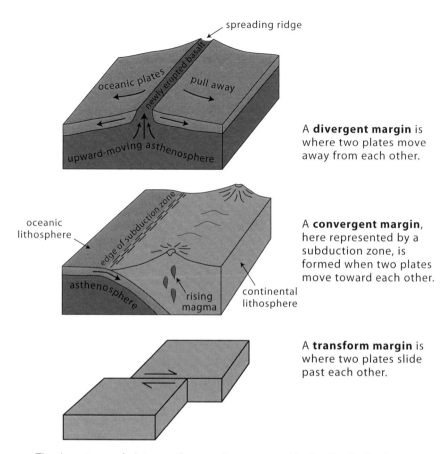

spreading ridge

A **divergent margin** is where two plates move away from each other.

A **convergent margin**, here represented by a subduction zone, is formed when two plates move toward each other.

A **transform margin** is where two plates slide past each other.

The three types of plate margins are all represented in the Pacific Northwest.
–Modified from Miller, 2014

Today, hidden beneath the ocean off the Pacific Northwest coast is every type of plate margin: *convergent*, where two plates move together; *divergent*, where two plates pull apart; and *transform*, where two plates slide side by side. About 50 miles (80 km) off Washington's coast, the oceanic Juan de Fuca Plate converges with the continental North American Plate. Being denser than continental lithosphere, the Juan de Fuca Plate sinks beneath the North American Plate at the Cascadia subduction zone. About another 100 miles (160 km) to the west lies a divergent margin, where the Juan de Fuca Plate spreads eastward away from the west-moving Pacific Plate. As the two plates pull apart, submarine volcanic eruptions of basalt form new oceanic lithosphere along the axis of the rift. The northeast-trending spreading center, or ridge, is offset in some places by transform faults that enable different sections of the spreading center to continue spreading even though they are not continuous with each other.

The subduction of the Juan de Fuca Plate beneath the North American Plate takes place up and down the coast of the Pacific Northwest, lending a consistency to the geology. The Cascades parallel the subduction zone, extending southward through Oregon to Lassen Peak in northern California and northward to Mt. Garibaldi in southern British Columbia. Each of these places hosts a coast range that was uplifted because of the converging plates. Each highlights various Cascade volcanoes and experiences periodic large-magnitude earthquakes. In addition, the subduction carries material toward the continent, where it may get accreted onto the continent's edge.

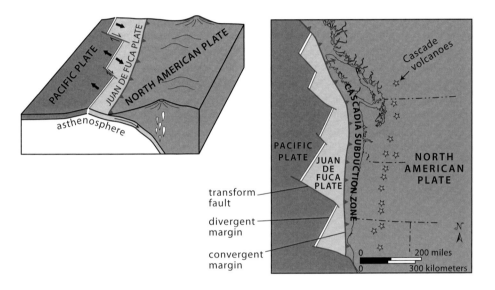

Plate margins of the Pacific Northwest. –From Miller, 2014

The Cascade volcanoes receive magma derived from the subduction. The subducting Juan de Fuca Plate heats up and releases water, which rises into the hot, dry, lower part of the overriding North American Plate. There, the added water causes localized melting of rock because it lowers the rock's melting point. The magma rises and erupts at volcanoes that form a mountain chain parallel to and east of the upper edge of the subduction zone. Because the Earth's surface is curved, the volcanoes form an arc, although given its short length the shape of the Cascade volcanic arc is not as apparent as those above longer subduction zones around the edge of the Pacific Ocean.

Not all the magma below a volcanic chain erupts. Much of it stays beneath the surface and crystallizes into intrusive igneous rock, which in many cases is granite or diorite in composition. If the area is later uplifted and eroded, we can see these exposed magma chambers on the surface as plutons. Some parts of today's Cascade Range display exposed plutons, such as the

Tatoosh Granite near Mt. Rainier and plutons of the Chilliwack Batholith in the North Cascades. Together, the intrusive and extrusive components of an arc are called a *magmatic arc*; those on continents are continental magmatic arcs, while those built on oceanic crust are island arcs.

Faults and Earthquakes

Washington is not without its faults—fractures in rock along which one side has moved relative to the other. They form because crustal stresses can exceed the strength of the rock, causing it to break and slide. Three main types of stress cause faulting: compression, tension, and horizontal shear.

Geologists identify fault types based on the type of slip, or relative movement, of the rock on either side. Dip-slip faults, in which the movement takes place parallel to the dip, or inclination, of the fault are either normal or thrust (or reverse) faults. If the block residing below the fault surface, called the *footwall*, rises with respect to the overlying *hanging wall*, then it's a normal fault, which results in crustal extension. If the footwall drops relative to the hanging wall, then it's a thrust fault, formed by crustal compression. Thrust faults that dip greater than about 45 degrees are typically called reverse faults.

Strike-slip faults are those in which motion is side-by-side, parallel to the fault's strike, most easily visualized as the direction of the length of the fault. These faults form by horizontal shear. In right-lateral faults, the side across

strike-slip faults (parallel to strike)

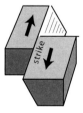

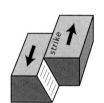

right-lateral
horizontal shear

left-lateral
horizontal shear

oblique-slip fault

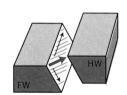

FW footwall
HW hanging wall

dip-slip faults (parallel to dip)

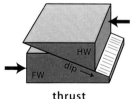

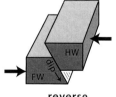

normal
horizontal extension

thrust
horizontal compression

reverse
horizontal compression

Different types of faults. –Modified from Miller, 2014

the fault from the viewer moves right; in left-lateral faults, the side across the fault from the viewer moves left. Faults with in-between directions of slip are oblique-slip faults.

Slip on faults causes earthquakes. In general, the greater the amount of slip along a fault and the larger its area of slippage, the larger the earthquake is. With its many modern and ancient faults, Washington has a long history of earthquake activity. In 2001, the Puget Sound area was rocked by the magnitude 6.8 Nisqually earthquake that caused more than $2 billion in damage. An earthquake of similar magnitude occurred near Wenatchee, just east of the Cascades, in 1872. In between those years, at least seven earthquakes between magnitude 4 and magnitude 7 occurred in Washington. Today's setting, adjacent to the plate margin, ensures that events like this will continue.

Earthquakes in Washington tend to occur in one of three places: on one of the myriad fault zones that break up the North American Plate above the subduction zone, deep on the subduction zone, or near the upper edge of the subduction zone. Every earthquake in the last three hundred years in Washington has occurred either above the zone or deep within it. The last one near the upper edge of the subduction zone took place on January 26, 1700, and it was a magnitude 9.

Brian Atwater of the US Geological Survey and his international team of colleagues were able to determine the exact time, date, and size of this last big subduction zone earthquake. They found sites up and down the coast that showed evidence for abrupt, earthquake-induced subsidence about three hundred years ago. After combing through historical records, they found records of a tsunami hitting Japan on January 27, 1700, and reasoned that the quake had occurred at about 9 p.m. the evening before. The size of the tsunami, which likely reached more than 15 feet (4.6 m) in height, combined with the length of coastline affected by the subsidence, suggested a magnitude 9 earthquake.

To get a sense of the size of a magnitude 9 earthquake, consider that the energy released during each successive step in the scale corresponds to an approximate increase of 32 times. That is, a magnitude 7 earthquake releases 32 times the energy of a magnitude 6 earthquake. A magnitude 8 earthquake therefore releases about 1,000 times (32 x 32) the energy of a magnitude 6 earthquake and a magnitude 9 releases more than 30,000 times (32 x 32 x 32) the energy! Given that the destructive Nisqually earthquake in 2001 was below a magnitude 7, the effects of a magnitude 9 could be catastrophic.

Continental Accretion and Mountain Building

If you were to look at the oldest, deepest-level rock, called the basement, at any given place in the western two-thirds of Washington, you would be looking at rock that originated somewhere other than North America and was later added to the continent. This process of continental accretion caused the entire western United States to grow westward through time. Accretion occurs at subduction zones where topographically high areas on the subducting plate, such as oceanic plateaus or island arcs, don't fully subduct and get scraped off

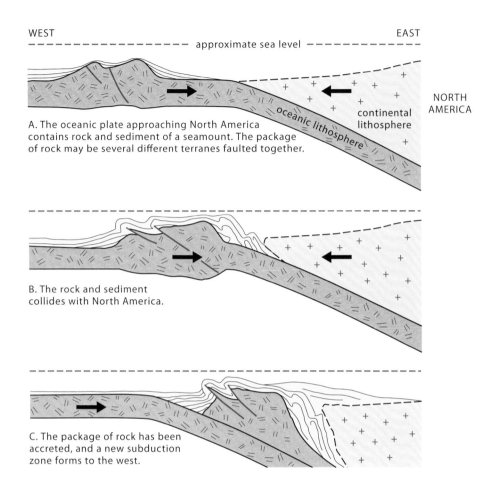

WEST   EAST
------------------------------ approximate sea level ------------------------------

A. The oceanic plate approaching North America
contains rock and sediment of a seamount. The package
of rock may be several different terranes faulted together.

NORTH
AMERICA

oceanic lithosphere

continental
lithosphere

B. The rock and sediment
collides with North America.

C. The package of rock has been
accreted, and a new subduction
zone forms to the west.

Schematic cross sections to illustrate the accretion of a terrane to the edge of North America. –Modified from Miller, 2014

onto the continent. Accretion can also take place when crustal blocks slide into place along strike-slip faults. Blocks of accreted material qualify as terranes if they are separated from other blocks by faults and have their own distinct geologic histories. Accreted terranes form such a disorganized patchwork that, in many cases, we can't tell how they relate to each other.

The compressive stresses generated by terranes colliding with the continent can build mountains. At least three major accretionary events have affected Washington since the beginning of the Mesozoic Era, each faulting and folding the basement rock. The timing of accretion becomes more recent toward the west, so if you go to Washington's eastern side, you can see evidence of the first major accretionary event. There, rock of the Quesnellia terrane was accreted onto preexisting North America. Farther to the west, you cross into where terranes of the North Cascades were accreted onto Quesnellia, and west of there, you cross onto the Siletz terrane, which was accreted onto the western edge of the terranes of the North Cascades.

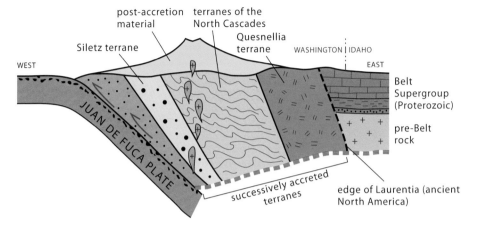

Schematic east-west cross section of Washington shows the subducting Juan de Fuca Plate, successively accreted terranes, and the edge of ancient North America. —Modified from Miller, 2014

Washington's Geologic History

PRECAMBRIAN ERA
(ending 541 million years ago)

Washington contains relatively few Precambrian rocks, nearly all of which are confined to the Okanogan Highlands in the northeastern corner of the state. These rocks point to the existence of a preexisting continent, called Rodinia, that broke up in late Precambrian time to form the western edge of today's North America.

Washington's geologic history begins with its oldest rocks, a thick sequence of sedimentary rocks called the Belt Supergroup, deposited from about 1.5 to 1.4 billion years ago. These rocks extend across northern Idaho and into western Montana, outlining a large basin that was probably a mostly shallow inland sea. The basin formed on a continent, called Rodinia, that preceded North America. Rocks of the Belt Supergroup abruptly end a short distance west of Spokane, where they encounter rocks of the accreted terranes. Somehow the basin was cut off, likely by a continental rift. The rifting occurred during late Precambrian time, sometime between 750 to 700 million years ago. The rifted western side of the basin now probably resides in either Australia, Antarctica, or Siberia. The rifted eastern side became the western edge of nascent North America, which

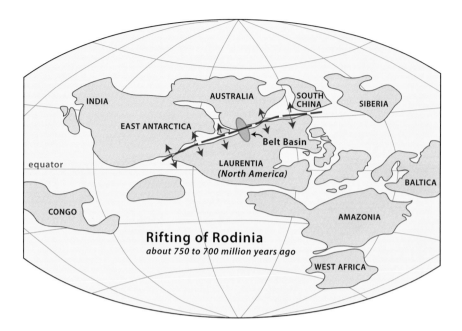

Map showing a proposed distribution of landmasses at the time Rodinia rifted, between about 750 to 700 million years ago. The Belt Basin, shown in red, was split in two by this rifting. –Modified from Goodge and others, 2008

included the remaining half of the Belt Basin. We have evidence for this rifting in the Windermere Supergroup, a series of now-metamorphosed sedimentary and volcanic rocks that sits on top of the Belt rocks. The Windermere Supergroup occupies much of the far northeast corner of Washington and extends into northern Idaho. The well-exposed parts of the Belt Supergroup along Washington's highways consist mostly of metamorphosed shale and sandstone of the Prichard Formation, which lies at the bottom of the sequence. In contrast to much of the younger Belt rocks, the Prichard likely accumulated in deep ocean water.

Another group of Precambrian rocks in Washington, those of the Yellow Aster Complex of the North Cascades, is part of an accreted terrane that was added to the rifted western edge of North America during the Mesozoic Era. Its origin bears an unknown relation to the early edge of North America.

PALEOZOIC ERA
(541–252 million years ago)

Washington has two Paleozoic histories, that of the early western margin of North America, preserved in the eastern Okanogan Highlands, and that of its accreted terranes, scattered throughout basement exposures of the western Okanogan Highlands, the North Cascades, and the San Juan Islands.

The rifting of Rodinia created an early version of the Pacific Ocean and a broad marine shelf along the western edge of Laurentia, the continental predecessor of North America. Throughout much of the Paleozoic, shallow marine sediments accumulated on this shelf. In northeastern Washington, we see the Addy Quartzite (also called the Gypsy Quartzite) as the base of the Paleozoic sequence, overlain by the Metaline Formation and Ledbetter Slate, each of which has been metamorphosed. The younger Paleozoic rocks of northeastern Washington have a volcanic component that remains enigmatic.

Elsewhere during the Paleozoic, Washington's early accreted terranes were forming in a variety of oceanic environments, including the deep ocean, subduction zones, and island arcs. Each environment hosted its own characteristic assemblage of rocks. For example, a deep-ocean assemblage might consist of deepwater shale or chert deposited on top of oceanic basalt, whereas an island arc assemblage might consist of andesite and basaltic lavas and tephras formed by the volcanoes, plus limestone deposited on nearby reefs. To complicate things, many of these rocks were metamorphosed during and subsequent to accretion.

One of the more intriguing aspects of Washington's accreted terrane story stems from fossils within outcrops of late Paleozoic (Permian) limestone on San Juan Island. These fossils, a variety of a single-celled marine organism called foraminifera, have no counterparts in North America. Instead, they seemed to come from the far western Pacific in what was then the ancient Tethys Sea. Elsewhere along the west coast of North America, other Tethyan fossils, including corals and radiolaria, have been found in similar-age limestone and chert in terranes. Their presence in North America implies that at least some of the accreted terranes traveled incredible distances.

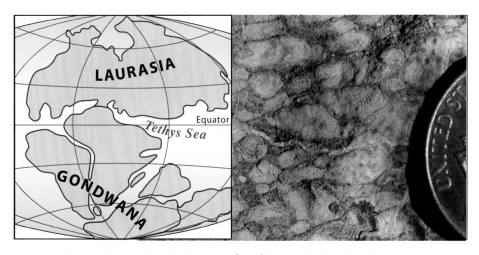

Tethyan foraminifera in limestone (right) formed in Permian time in the Tethys Sea (left). –Fossil from the collection of Gregory Retallack

Radiolarian chert, also called ribbon chert, is a fine-bedded rock of siliceous skeletons of radiolaria that settled to the quiet seafloor.

A mélange is a mixture of rocks that were caught between colliding plates.

Turbidites, which contain thin graded beds, are deposited on the seafloor of fore-arc basins.

— radiolarian (ribbon chert)
— mélange
— turbidites
⌁ pillow basalt
— blueschist

island arc   **magmatic arc**

seamount

fore-arc basin

fore-arc basin

oceanic lithosphere   oceanic lithosphere

continental lithosphere

Blueschist, a metamorphic rock, only forms in subduction zones.

Pillow basalt forms when magma erupts underwater, usually when new ocean crust forms at a spreading ridge.

Some important oceanic settings and their rocks. The Puget Lowland is a modern-day fore-arc basin, the low-lying area between the volcanic arc and its subduction zone. —Modified from Miller, 2014

MESOZOIC ERA

(252–66 million years ago)

Throughout the Mesozoic Era, Washington's diverse terranes were assembled into two larger composite terranes that were accreted sequentially onto the continental margin. The accretion grew the margin westward and caused two periods of mountain building, which left evidence in the form of faults, folds, and igneous intrusions.

Explaining when various terranes came together or were accreted to North America has long been a challenge to geologists who study accretionary tectonics. Two of the more direct ways of dating accretionary events comes from igneous and sedimentary rocks. Where intrusive bodies cut across terrane boundaries, the age of the intrusion must be younger than the merging of the terranes. These intrusions are called *stitching plutons* because they literally stitch the terranes together. Where sedimentary rocks overlie terrane boundaries or contain material derived from both terranes, they also must be younger than the merging of the terranes. These sedimentary rocks are called *linking assemblages*. If either the stitching pluton or overlying sedimentary rock has characteristics indicative of a continental origin, then it's likely that they formed after the terranes were accreted.

In Washington, the first major accretionary event occurred during the Jurassic Period and included the Quesnellia terrane as well as several others now found in Canada. The accretion caused mountain building, now evident in

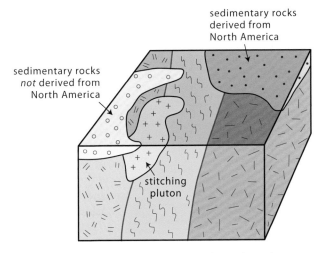

Linking assemblages and stitching plutons help determine when terranes came together. The yellow sedimentary rocks were deposited after the green and orange terranes came together, but they don't give information about when the terranes were accreted to North America. Similarly, the stitching pluton intruded after those terranes joined. The brown sedimentary rocks, being derived from North America, were deposited after the orange terrane was accreted to the continent. –Modified from Miller, 2014

the faulting and folding of both Quesnellia rocks and those of the North American margin. If you drive US 97 between Omak and Tonasket, for example, you can see highly deformed Triassic limestone of Quesnellia. Farther east, along WA 31 to Metaline Falls, you can see some of the deformation in the sedimentary rocks of ancient North America. On a geologic map, you can see that these sedimentary rocks form an arc-shaped band called the Kootenay arc.

The second major accretionary event involved the many terranes of the North Cascades and San Juan Islands. These terranes formed in a variety of oceanic settings over a period of several hundred million years of the Paleozoic and Mesozoic Eras before coming together as a large composite terrane. Over the last forty years or so, geologists have been able to piece together some seemingly disparate rock units, but how all the terranes fit together remains a subject of debate. Most researchers agree, however, that the timing of accretion was early in the Cretaceous Period. Their evidence comes partly from the Cretaceous sedimentary rocks of the Methow domain of the North Cascades, which contain material derived from both the newly accreted terranes and the North American continent.

CENOZOIC ERA
(66 million years ago to present)

Early in the Cenozoic Era, Washington's plate margin experienced its third accretionary event, which led to the subsequent formation of the modern Cascadia subduction zone and magmatic arc. During the latter part of the Cenozoic Era, beginning about 17 million years ago, southern Washington began to establish its present character. Among other things, it experienced multiple eruptions of the Columbia River Basalt Group; the development of the Yakima fold belt; and in the last 2 million years, a glacial period that greatly shaped the appearance of nearly the entire state.

Siletzia, a gigantic oceanic plateau, island, and seamount complex, was accreted about 50 million years ago in Cenozoic time. This terrane forms the basement of the Coast Range from southern Oregon to southwestern British Columbia. Its rocks consist mostly of brecciated, massive, and pillowed basalt flows that erupted underwater, and a minor amount of basalt-rich sandstone and even some limestone deposited during volcanically quiet times. Some of the basalt, especially that near Olympia, likely flowed overland because it exhibits the vertical cooling fractures called *colonnade*. In Washington, rocks of Siletzia are generally called the Crescent Formation, and in British Columbia they are called the Metchosin Volcanics.

With a volume of about 500,000 cubic miles (2 million km^3), Siletzia's size rivals that of the composite terranes of the Mesozoic Era. Not only did it add substantially to the edge of the plate margin, but when it drifted into the subduction zone, it managed to bring everything to a grinding halt. It literally stopped the subduction and shut off the arc! It wasn't until some 15 million years later that a new subduction zone formed immediately west of Siletzia,

Pillow basalt of Siletzia at Cape Disappointment, near the mouth of the Columbia River.

which gave rise to the modern Cascade volcanic arc. See the chapters on the North and South Cascades for information on the modern Cascade volcanoes.

Beginning about 60 million years ago and continuing for some time after the accretion of Siletzia, strike-slip faulting and crustal extension formed sedimentary basins and caused volcanism and intrusion of granites in much of northern Washington. The Golden Horn Batholith, for example, crystallized about 48 million years ago. The Straight Creek fault, which trends north-south through the center of the North Cascades, experienced more than 55 miles (90 km) of right-lateral slip soon after the accretion of Siletzia. Among other things, its movement broke apart a just-formed extensional basin. Today, the Chuckanut Formation, which forms the hills near Bellingham, represents the western side of the basin, whereas the Swauk Formation, in the mountains between Cle Elum and Leavenworth, represents the eastern side. The extension also caused the uplift of the metamorphic core complexes of the Okanogan Highlands.

Beginning 16.8 million years ago, lavas of the Columbia River Basalt Group erupted from long fissures in southeastern Washington and northeastern Oregon. The basaltic lavas covered more than 77,220 square miles (200,000 km²) of Washington and Oregon, including parts of western Idaho and northern Nevada, and had a volume that exceeded 52,800 cubic miles (220,000 km³). This volume is more than fifty times that estimated by the National Park Service for the volume of air filling the Grand Canyon in Arizona. It's even more amazing when you consider that 94 percent of the lava erupted before 14 million years ago after only 3 million years.

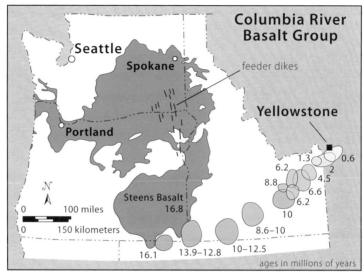

Distribution of Columbia River Basalt Group. Geologists think a plume from the mantle was responsible for both the flood basalts and the inception of the Yellowstone hot spot.

View of layers of flood basalts downstream from Palouse Falls.

Most of the lavas of the Columbia River Basalt Group lie within the Columbia Basin, but some of the flows poured down the Columbia River all the way to the Pacific Ocean. The flows form spectacular cliffs along the Columbia River and show up as far north as Hoquiam at the south end of the Olympics. Near the coast, the advancing lavas also managed to invade downward into the Astoria Formation to form large intrusive bodies. These intrusions don't cross any of Washington's major highways, but several lie along US 101 in northern Oregon.

Even during the first couple million years when the eruptions were most prolific, there were periods of relative quiet when layers of sediment were deposited in lakes and streams. In most of central and southern Washington, these deposits are called the Ellensburg Formation, whereas near Spokane and Grand Coulee, they're called the Latah Formation, and near the coast, they're called the Vantage Horizon.

During and after eruptions of the Columbia River Basalt Group, southern Washington and northern Oregon began rumpling into a series of anticlines and synclines called the Yakima fold belt. The belt trends roughly east-west from the Cascades to the Idaho Border, although many of the individual folds lie at some angle to the belt. Because basalt tends to be resistant to erosion, the folds dictate the modern landscape. Anticlines form high ridges across much of the Columbia Basin, such as Umtanum Ridge and the Horse Heaven Hills, whereas the synclines form broad valleys, such as the Kittitas Valley near Ellensburg. Many of the anticlines formed along thrust faults, some of which exhibit modern-day seismicity that suggests the folds are still growing.

Pleistocene Epoch

For all of its variability, Washington's landscape was more profoundly affected in the last 2 million years directly and indirectly by glaciation than by anything else. A continental ice sheet several thousands of feet thick advanced southward from Canada over much of northern Washington, while smaller alpine glaciers sculpted most high-elevation areas. Alpine glaciers still adorn many of the high peaks of the North Cascades and Olympics, as well as the Cascade volcanoes.

The glaciers deposited thick accumulations of glacial till over most of northern Washington, and rivers and streams running off the front of the glaciers deposited outwash. In the southern Puget Lowland, much of the till and outwash were streamlined by the ice into narrow ridges called *drumlins*. Along the margin of the lowland, much of the outwash was deposited in deltas, where the sediment-charged rivers emptied into lakes.

In a less direct way, the glaciers caused the enormous floods that tore through eastern and central Washington and created the Channeled Scabland. The floods, described in greater detail in the introduction to the Columbia Basin, formed when gigantic lakes, dammed by glacial ice, emptied catastrophically when the ice dams ruptured. Most of the floods came from Glacial Lake Missoula in Montana, which covered some 3,000 square miles (7,800 km²) and reached depths of more than 1,500 feet (460 m). However, some floods came from Glacial Lake Columbia, where an ice tongue dammed the Columbia

Glaciated peaks in the North Cascades.

Aerial view of part of the flood-torn Channeled Scabland about 25 miles (40 km) southeast of Ritzvi.

River, and from Glacial Lake Kootenay in southern British Columbia. Because of the variability of the flood sources, we call them the Ice Age floods instead of the Missoula floods, following the recommendation of Bruce Bjornstad, who has written the definitive books on the floods.

Glacial abrasion, where rock embedded in glacial ice wears away at bedrock, created much of the dust and fine silt that makes up the great loess deposits of Washington's Palouse Hills. Another source for the loess was the dust and silt included in the flood deposits. Winds, largely driven by the temperature contrast between ice-covered areas and barren land, transported the fine material and deposited it over much of southeastern Washington. In some places, the loess exceeds 200 feet (60 m) in thickness.

Landslides and Today's Landscape

Probably more than anything else, landslides shape Washington's landscape. From Grand Coulee to the Columbia River Gorge to the Olympic Mountains, they come in a variety of styles and sizes, but they all move bodies of rock and/ or soil down slopes because of gravity. They tend to be more frequent on steep slopes and after heavy rainfalls, so they are most prevalent in western Washington. Besides shaping the landscape, landslides can cause significant property damage and loss of life. Large landslides appear as bright yellow areas marked "Qls" on nearly every geologic map in this book.

Some of the more significant landslides occur in sequences of rock or sediment that contain a relatively impermeable layer, such as a clay-rich layer deposited in a lake or formed by chemical alteration. Groundwater tends to accumulate on the upper surfaces of these impermeable layers, to buoy the overlying rocks and facilitate sliding. In the Columbia River Gorge, for example, giant landslides form in the Columbia River basalts: water percolates down cracks in the basalt and underlying sediment and ponds on top of the clay-rich Ohanapecosh Formation. In Seattle, most landslides form in glacial till and stream sediment that slide off the top of clay-rich lake deposits.

The terrible slide of March 2014 at Oso, in the Cascade foothills, occurred in unconsolidated glacial sediments similar to those in the Seattle area. The sediments form steep hills rising some 600 feet (180 m) above the North Fork Stillaguamish River and have cut loose slides many times in the past. The Oso slide was especially destructive because the sediments rapidly disintegrated into a somewhat fluidized debris flow that poured 0.6 mile (1 km) to the other side of the valley. In the process, the slide destroyed a neighborhood of about thirty-five single-family homes and took the lives of forty-three people.

Sea stack visible through Hole in the Wall at Rialto Beach.

COAST RANGE

From the Columbia River in the south to the Olympic Mountains in the north, the Coast Range offers dramatic changes in topography and scenery. The tidal flats and estuaries of Willapa Bay and Grays Harbor extend from the coast into the rolling, forested Willapa Hills. To the north rises the majestic Olympic Peninsula, home to a luxuriant temperate rain forest, sandy beaches guarded by sea stacks, and active glaciers flanking the summit of 7,980-foot-high (2,432 m) Mt. Olympus. The signs marking tsunami evacuation routes along coastal roads remind us that an active subduction zone, where the Juan de Fuca oceanic plate descends beneath the continent, lies only 75 miles (120 km) offshore to the west.

Siletzia and Its Sedimentary Cover

The geology of the Coast Range records a 50-million-year history of subduction. The geologic basement in the Coast Range is the Siletz terrane, or Siletzia, a thick—perhaps as much as 9 miles (14 km)—sequence of basaltic volcanic rocks that erupted from about 56 to 49 million years ago. These volcanic rocks, along with some interbedded sedimentary rock, are called the Crescent Formation in Washington. Although geologists still debate the origin of Siletzia, it likely was an oceanic plateau that formed near the northwest coast of North America. A consensus holds that the terrane arrived at a subduction zone and was accreted—added—to western North America about 50 million years ago.

Oligocene-age crab fossil on the inside of a concretion formed in the Lincoln Creek Formation. –Rock sample from collection of Jessica Czajkowski

After accretion, up to 15,000 feet (4,600 m) of well-bedded sediments were deposited in the shallow ocean on the Siletzia basalts. This rock record is especially complete, although not particularly well exposed, in the Willapa Hills. There, the sedimentary rocks range in age from the Eocene to Oligocene Lincoln Creek Formation to the Miocene Montesano Formation.

A very different source of basalt also occurs in the Willapa Hills. Lava of the Columbia River Basalt Group, erupted 250 miles (400 km) to the east, flowed through the valley of the Columbia River all the way to the Pacific Ocean. These flows testify to not only the remarkable fluidity of basalt but also the antiquity of the Columbia River: the present course of the river must have existed 16 million years ago, and perhaps it formed even earlier.

Olympic Mountains and the Olympic Peninsula

A new subduction zone developed about 35 million years ago along the western edge of accreted Siletzia. The Olympic Subduction Complex consists of sediments eroded from the continent and then scraped off the subducting oceanic plate and fashioned into an accretionary wedge. The term *accretionary wedge* derives from the shape of the accreted sediments in a cross-sectional profile. Although the wedge is present along the entire length of the subduction zone from northern Vancouver Island to northernmost California, it is either entirely submerged offshore or buried beneath the hard rocks of Siletzia—except on the Olympic Peninsula.

A simplified geologic map of the Olympic Peninsula shows three distinct areas. The central part of the peninsula—the core—is composed of the Olympic

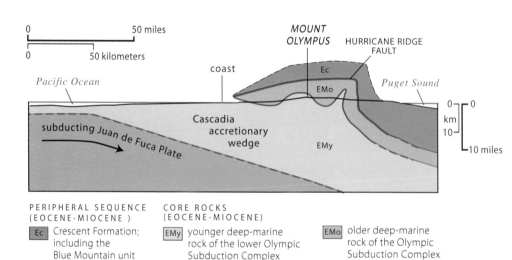

Cross section across the accretionary wedge through the Olympic core to Puget Sound. The core rocks and Crescent Formation shown above today's land surface have been eroded away along the line of the cross section but show up to the north. –Modified from Brandon and others, 1998

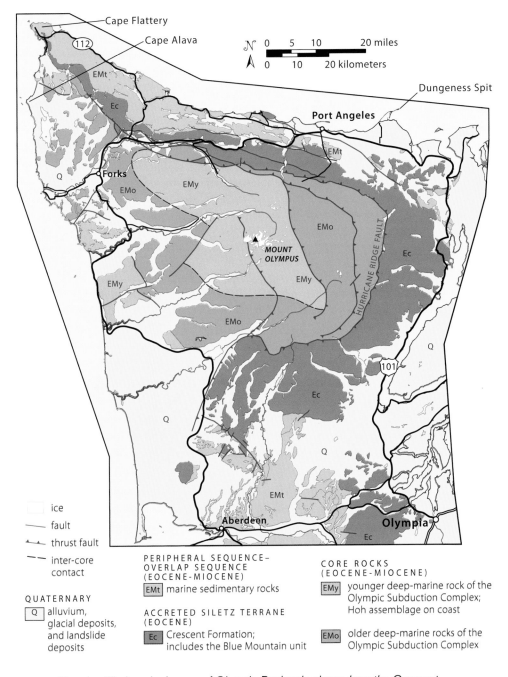

The simplified geologic map of Olympic Peninsula shows how the Crescent Formation forms a horseshoe around the interior of the Olympic Mountains.

Subduction Complex. Siletzia, composed of the Crescent Formation and inter-bedded sedimentary rocks, surrounds the core and is separated from it by the Hurricane Ridge fault, a major thrust. The Peripheral sequence, composed of Eocene and younger strata, overlies Siletzia. Both Siletzia and the Peripheral sequence form a ring, or horseshoe, around the core.

The horseshoe shape of the Siletz terrane resulted from folding. The lid of Siletzia has been domed upward and eroded away to reveal the Olympic Subduction Complex beneath. The core of the peninsula is one of the best opportunities on Earth to examine an accretionary wedge on land. The wedge consists of variably deformed but poorly dated sedimentary rocks that range in age from Eocene in the eastern core to Miocene at and near the coast.

The most interesting and accessible part of the Olympic Subduction Complex in the western part of the core is generally termed the Hoh assemblage. The Hoh is composed of sandstone, mudstone, and minor pillow basalt that cannot be assigned to a conventional well-bedded stratigraphic sequence. Moreover, a distinctive component of the assemblage is *mélange*—literally a mixture of blocks of diverse rock types immersed in a mudstone matrix. Mélanges the world over are characteristic components of accretionary wedges formed at subduction zones, but their origin is still fiercely debated. Many of the scenic headlands and sea stacks at Ruby Beach and Second and Third Beaches, all

Mélange at Third Beach, just south of La Push. This single outcrop is cut by numerous small fault zones and consists of sandstone, red mudstone, and some basalt.

accessible by easy hikes from roads in Olympic National Park, consist of blocks eroded from the mélange by waves.

The Olympic Peninsula is one of the few places in Washington where people have drilled for oil. In the early twentieth century, people were lured there by natural seeps of oil and gas along the Hoh River. Early publications refer to "smell muds," the term the Native Americans reputedly used for the hydrocarbon-charged shale in the Hoh assemblage. Exploration was active beginning in 1913 and again in the 1930s. The center of operations came to be known as Oil City, and the decaying remains of buildings and rusting drilling apparatuses are still scattered through the forest. The eleven wells that were drilled in the 1930s yielded a few barrels of oil, but overall the ventures were a commercial failure. Another well was drilled in 1948, again without success. Major energy companies, doubtlessly encouraged by the shows of hydrocarbons near Oil City, drilled six exploratory holes in the 1960s off the southern coast of Washington, but none was developed commercially.

The tiny glaciers on the summit massifs of the Olympic Mountains belie the role that glaciation played in the northern Coast Range, particularly on the Olympic Peninsula and in the Strait of Juan de Fuca. Prior to about 15,000 years ago, during the maximum extent of glaciation in North America, glaciers descended from the core of the range down the major river valleys, including the Hoh, Queets, Quinault, and Elwha. Moreover, part of the vast Cordilleran Ice Sheet to the north extended south into what we now know as the Strait of Georgia. As it encountered the mountainous Olympic Peninsula, it divided into a lobe that progressed southward into present-day Puget Sound and a lobe that progressed northwestward out the Strait of Juan de Fuca. In 1913, the geologist J Harlen Bretz, who later would become much better known for his documentation of the huge Ice Age floods that surged across the Columbia Basin, relied on his remarkable observations in the field to publish a report stating that the

The Olympic Mountains today host a fraction of the ice that dominated the landscape during the Pleistocene Epoch. This view is from near the top of Hurricane Ridge Road in Olympic National Park.

surface of the Juan de Fuca Lobe had been at elevations of about 3,000 feet (900 m) along the northeast and north coast of the Olympic Peninsula. His evidence was boulders of granite, a rock type unknown in the bedrock of the peninsula, that must have been carried from Canada and left as erratics—out-of-place strangers—in the glacial deposits. Imagine that at the maximum extent of glacial ice, you could have stood on the mountain slopes along the north side of the peninsula and seen a thick lobe of glacial ice extending northwestward in the strait as far as the present-day coastline at Cape Flattery.

Spits, Estuaries, and Sea Stacks

The Coast Range in Washington is bounded on its north and east sides by the Pacific Ocean and on its south by the Columbia River, so it is not surprising that a variety of landforms here owe their origin to the action of water on land. North of the mouth of the Columbia River, Willapa Bay and Grays Harbor are two prominent estuaries—shallow embayments protected from the open Pacific Ocean by long, low ridges covered with sand. The Long Beach Peninsula is a vacation destination famous for its nearly continuous sandy beaches on the Pacific Ocean side. The sand is supplied from sediment delivered to the Pacific by the Columbia River and carried northward by north-flowing ocean currents—a process coastal geologists call *longshore drift*. Although the peninsula protects Willapa Bay from active erosion by the Pacific Ocean, the influence of periodically rising and falling tides has created a network of marshes, wetlands, and mudflats that are especially well developed at the southern end of the bay. The Willapa Bay National Wildlife Refuge, established in 1937, is a scenic mecca of resident and migrating birds.

Tidal slough along US 101 near Willapa Bay. Left photo shows it at low tide; right photo shows it at high tide.

Dungeness Spit, which curves a little over 5 miles (8 km) northeastward from the coast of the Olympic Peninsula east of Port Angeles, is also encompassed by a national wildlife refuge—a home to hundreds of species of birds. The spit and bay it protects are probably most well known because they gave their name to the Dungeness crab, a species that has a much wider range, from Alaska to southern California. The general northeastward movement of sand along the spit is driven by waves from the Strait of Juan de Fuca.

Most of the 1,200-mile course (1,930 km) of the Columbia River, from its headwaters in British Columbia southward into Washington and then westward along the border of Washington and Oregon to the Pacific Coast, is in a deep valley. Within several miles of its mouth, however, the river widens and looks more like a lake than a slowly flowing river. The muddy strip on its northern bank, periodically inundated and exposed by rising and falling tides, indicates its proximity to the ocean.

The stunning headlands and sea stacks lending a wild beauty to the coast of the Olympic Peninsula testify to the relentless erosion of land by the Pacific Ocean. Cape Alava is the westernmost point in the contiguous United States, and Cape Flattery is the northwesternmost point. The headlands and sea stacks near Cape Alava consist of the Hoh assemblage. Many of these, strangely enough, contain basalt of the Crescent Formation of the Siletz terrane. The nearest present-day exposures of the Crescent basalt are about 10 miles (16 km)

Many of the sea stacks at Ruby Beach are easily accessible.

to the east. How these blocks of Eocene basalt became incorporated into the Miocene part of the subduction complex is an intriguing question that motivates geologists to continue their research in the Olympics.

GUIDES TO THE COAST RANGE

US 12
INTERSTATE 5—ABERDEEN
45 miles (72 km)

US 12 follows the Chehalis River to its mouth at Aberdeen, the same route that meltwater, flowing from the ice sheet in the Puget Lowland, took to the ocean. Along the way, the road passes some outstanding but widely separated bedrock exposures that represent each major rock unit of this part of the Willapa Hills.

For about the first 11 miles (18 km) to Oakville, you gain some nice views of the Chehalis River floodplain and surrounding hills. The road crosses several abandoned river channels west of Rochester, most noticeably 0.4 mile (640 m) east of milepost 41. Other ones are far subtler, best seen on an aerial image. The channels likely originated from the migration of the Black River, a tributary of the Chehalis that shares this part of the floodplain. The Black River rises to the north, below a low divide that separates it from Puget Sound. During late Pleistocene times, when the Puget Lowland was full of ice, much of its meltwater overtopped the divide and flowed down the Black River. This huge volume of water then spilled into the Chehalis River and flowed to the sea, enlarging the floodplain to its present width. US 12 crosses the Black River a short distance upstream from its confluence with the Chehalis, about halfway between mileposts 38 and 37. Looking north, you can see the Black Hills, which rise steeply to elevations above 1,000 feet (300 m). They are made almost entirely of Crescent basalts.

Several good exposures of the Crescent basalts show up along the highway west of Oakville. Look for a quarry and a small double roadcut at milepost 34. The best exposure of the basalt, however, lies across from a pull-out a quarter mile (400 m) east of milepost 29. This exposure consists mostly of brecciated basalt, but if you look closely you can find some weakly developed columns. Although the rocks appear to be inclined eastward, this orientation is an artifact of fracturing.

Just west of milepost 28, you encounter the first in a series of northward-dipping exposures of the Lincoln Creek Formation, deposited in a shallow sea and along its coast in the later part of Eocene and early Oligocene time. These roadcuts, called the Porter Bluffs, extend off and on to milepost 26 and are unquestionably the best exposures of this rock unit along any of Washington's highways. They consist mostly of tuff-rich siltstone, some of which contains fossils of Eocene age. Netting covers many of the exposures to help mitigate

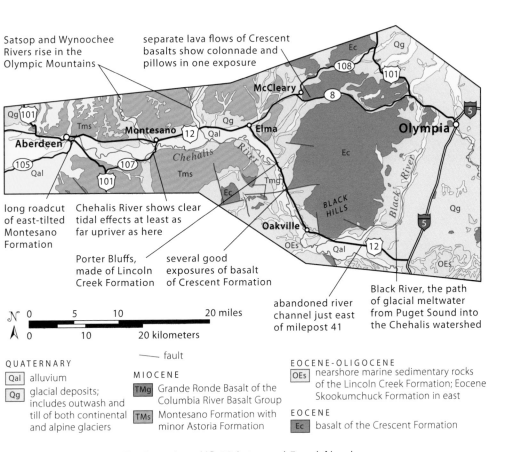

Satsop and Wynoochee
Rivers rise in the
Olympic Mountains

separate lava flows of Crescent
basalts show colonnade and
pillows in one exposure

McCleary

Montesano Elma

Olympia

Aberdeen

Chehalis River

long roadcut
of east-tilted
Montesano
Formation

Chehalis River shows clear
tidal effects at least as
far upriver as here

Porter Bluffs,
made of Lincoln
Creek Formation

several good
exposures of basalt
of Crescent Formation

Oakville

BLACK
HILLS

Black River

abandoned river
channel just east
of milepost 41

Black River, the path
of glacial meltwater
from Puget Sound into
the Chehalis watershed

N   0   5   10   20 miles

0   10   20 kilometers

———— fault

QUATERNARY

Qal alluvium

Qg glacial deposits;
includes outwash and
till of both continental
and alpine glaciers

MIOCENE

TMg Grande Ronde Basalt of the
Columbia River Basalt Group

TMs Montesano Formation with
minor Astoria Formation

EOCENE

Ec basalt of the Crescent Formation

EOCENE-OLIGOCENE

OEs nearshore marine sedimentary rocks
of the Lincoln Creek Formation; Eocene
Skookumchuck Formation in east

Geology along US 12 between I-5 and Aberdeen.

the ever-present hazards of rockfall and landslides. About a half mile (0.8 km) west of milepost 26, you pass a slide scar and gain a good view of the Chehalis River floodplain.

The floodplain widens considerably at milepost 20, by the community of Elma, with the convergence of several tributaries. West of Elma, US 12 crosses over glacial outwash and passes wetlands and abandoned river channels, some of which can be accessed by side roads. The Satsop River, a quarter mile (400 m) west of milepost 16, is fed by tributaries that drain an area of nearly 300 square miles (780 km²), including parts of the Olympic Mountains. The Wynoochee River, which the road crosses on the west side of Montesano, also rises in the Olympics. At Montesano, the Chehalis River has a tidal range that can exceed 9 feet (2.7 m); you can drive 1 mile (1.6 km) south on South Main Street (WA 107) to a bridge across the Chehalis. You can see a small exposure of outwash gravels by a gated road on the south side of the highway at milepost 7.

Roadcut of east-tilted Montesano Formation. Inset shows fossil bed and its approximate location.

Just east of Aberdeen, a dramatic roadcut of east-tilted sandstone of the Montesano Formation frames the north side of the highway. These rocks were deposited in a shallow sea between about 10 to 5 million years ago during late Miocene time. A large pull-out is available for eastbound travelers only. Directly across from a point about 100 feet (30 m) west of milepost 1, you can see a bed full of clam fragments and pebbles near the bottom of the roadcut. Elsewhere, look for the many rounded concretions that formed within individual beds and stand out in relief from the more easily erodible parts of the rock. Numerous drains stick out from the rock to help mitigate the landslide hazard.

US 101
OLYMPIA—PORT ANGELES
120 miles (193 km)

US 101 between Olympia and Port Angeles provides an outstanding perspective of the effects of the huge ice lobes that advanced and then retreated from the Puget Sound area. The most recent of these glacial advances, called the Vashon, was at its maximum about 16,500 years ago. You'll see numerous sedimentary deposits related to the ice, as well as evidence for the glacial rebound that took place after the ice retreated.

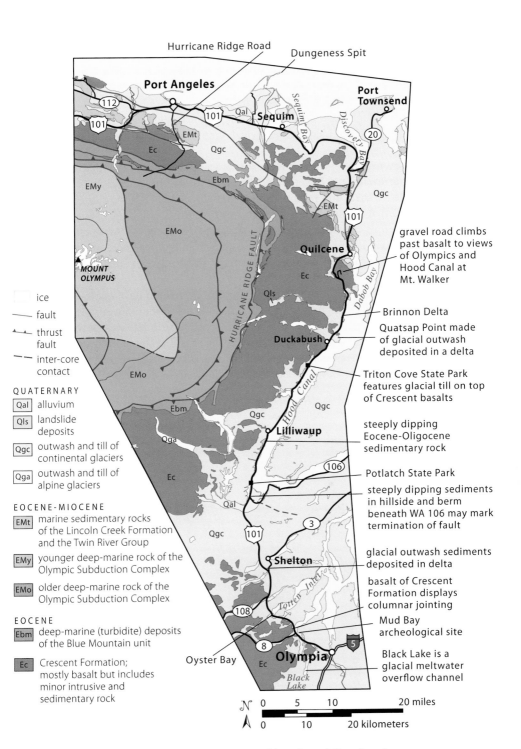

Geology along US 101 between Olympia and Port Angeles.

Just 2 miles (3.2 km) west of Olympia, you pass the turnoff to Black Lake, which fills a valley on top of the low divide between Puget Sound and the Black River. The Black River empties into the Chehalis River, which then flows westward to Aberdeen and the Pacific Ocean. As the Puget ice lobe retreated from the southern Puget Lowland, its voluminous meltwater filled the area with water to form Glacial Lake Russell, which emptied southward into the Black River.

Mud Bay, near milepost 363, is the shallow southern edge of Eld Inlet, which empties northward into Puget Sound. The inlet formed by water erosion at the base of the ice sheet and resembles the other sinuous inlets around the Puget Sound area that formed by the same process. An important archeological site was revealed just north of the highway at Mud Bay. Occupied by ancestors of today's Squaxin Island Tribe, the site was preserved by subsidence and subsequent burial of the bay, possibly about three hundred years ago during the magnitude 9 subduction zone earthquake.

Bedrock exposures of basalts of the Crescent Formation lie along the south and west sides of the highway a quarter mile (400 m) west of milepost 362 and just south of milepost 360. The one near milepost 360 exhibits well-developed columnar jointing on its north end, which suggests that the lava flowed over land instead of underwater. More basalt is exposed in the quarry 0.75 mile (1.2 km) north of Oyster Bay, and less than a mile north of that, another roadcut

Southern end of Mud Bay. Forested hills in the background are made of basalt of the Crescent Formation of the Siletz terrane.

lies on the west side of the highway. The southeastern edge of the Olympic Mountains, occasionally visible to the north, consists of the same basalt. Totten Inlet and Oyster Bay, near milepost 356, also formed by erosion beneath the ice sheet, similar to Mud Bay.

You can see a thick accumulation of glacial sediments halfway between mileposts 347 and 346, near Shelton. These sediments accumulated in a delta as streams flowing south from the retreating ice in Hood Canal entered Glacial Lake Russell. The road passes through older sediments near the bottom of Purdy Canyon before crossing the Skokomish River valley. Similar deposits, but inclined to the southwest at angles of about 45 degrees, are exposed in places within the vegetated hillside on the northwest side of the valley at the intersection of US 101 and WA 106. These rocks lie on the south side of a small anticline that may approximate the termination of a fault zone found on the northwest side of the valley. The fault zone may also explain the uplift of the prominent berm that extends to the southeast across the flood-prone valley and divides the extensive wetlands in two. WA 106 follows the top of this berm for just over 1 mile (1.6 km).

Potlatch State Park at milepost 335 gives a beautiful view northwestward up Hood Canal. Several prominent benches of glacial outwash lie on the west side. The ice lobe lay immediately to the east as it moved southward into the sound. While the benches all formed at similar elevations, they now increase in elevation toward the north. This southward tilt resulted from glacial advance and retreat: the weight of the ice initially depressed the crust, but as the ice melted, the crust rebounded. Because the ice thickened northward, the rebound also increased northward, causing the tilt. In 1989, Robert Thorson at the University of Connecticut first used delta deposits, including the one at Shelton, to estimate that the crust rose approximately 4.5 feet per mile (1.4 m/km) toward the north in response to the melting ice.

Halfway between mileposts 335 and 334, you pass a power plant operated by Tacoma Power. Penstocks carry water to the power plant from Lake Cushman, some 4 miles (6.4 km) to the northwest and more than 200 feet (60 m) higher in elevation. A headwall to a large slump behind the penstocks exposes gravels that were derived from a pre-Vashon glaciation.

For about the next 20 miles (32 km) you see a wide range of features. South of Lilliwaup, you pass several exposures of steeply east-tilted sedimentary rocks of Eocene to Oligocene age. These rocks are mapped as part of the marine-deposited Lincoln Creek Formation, which is better known in the southwestern corner of the state and has in some places yielded fossil crabs. At Lilliwaup Bay, look for a small modern delta and mudflats along Lilliwaup Creek at low tide. Immediately north of the bay, the road passes by several well-stratified deposits of glacial outwash; some pre-Vashon gravels are well exposed less than a quarter mile (400 m) north of milepost 324 and near milepost 325. Three miles (4.8 km) farther north, look for a small abandoned quarry exposure of Crescent basalts on the west side of the road. Glacial striations decorate the rock on some of its upper surfaces. More gravels, deposited as part of a glacial delta, are exposed on the west side of the highway just south of the Hamma

Hamma River at milepost 320. For comparison, look eastward downstream to the river's modern delta. For the next several miles north, the road passes several outcrops of Crescent basalts. You can see some rough pillow shapes between mileposts 319 and 318.

Triton Cove State Park, about halfway between mileposts 315 and 314, features glacial till deposited on top the Crescent basalts, some of which display distinct pillow shapes. Look for these exposures on the coast immediately south of the boat ramp. Some smaller exposures of the till and basalt line parts of the driveway into the park. More good exposures of the basalt lie along US 101 for the next several miles north to Duckabush, where the Duckabush River forms another delta as it enters Hood Canal. You can look up the Duckabush River to a spectacular view of basaltic mountains of the eastern outer edge of the Olympics. The prominent cliff at Quatsap Point consists of more glacial sediments deposited in an ancestral Duckabush River delta.

The most dramatic glacial delta deposit along this route lies directly across from milepost 306. Named the Brinnon Delta, it displays beautiful

Bedrock exposure at Triton Cove State Park. A distinct contact between glacial till and the underlying basalt of the Crescent Formation runs across the photo about 1 foot (0.3 m) above the geologist.

Glacial outwash deposits of the Brinnon Delta.

northeast-inclined beds from deposition at the steep edge of the delta. Above these beds lie flat-lying beds, deposited after the delta front had migrated. Geologists at the Washington Division of Natural Resources suggest that the unconsolidated nature of this deposit indicates it accumulated during the retreat of the most recent Vashon glaciation.

The road turns inland between mileposts 304 and 303 and climbs more than 600 feet (180 m) to Walker Pass, which separates Mt. Walker to the northeast from Buck Mountain to the southwest, both of which are made of Crescent basalts. Several large roadcuts expose the basalt on the descent into Quilcene. From the pass, you can take a well-graded gravel road 4.2 miles (6.8 km) to the top of Mt. Walker, which rises nearly 2,000 feet (610 m) above the pass. Its north overlook offers views of peaks of the outer edge of the Olympics, while its south overlook offers views over Hood Canal to Mt. Rainier. Along the road are numerous outcrops of Crescent basalts.

North of Quilcene, US 101 rises into a scraggly forest growing on glacial till, which overlies Eocene marine sedimentary rocks. Some good but not easily accessible exposures of the Eocene rock form a series of ledges just north of milepost 285 and at milepost 284. A more accessible but much smaller exposure lies off the highway near Quilcene. To get there, drive a half mile (0.8 km) down Center Road, which intersects US 101 in the center of Quilcene. There, the rocks consist of siltstone and micaceous sandstone that have in places yielded pieces of wood.

US 101 encounters Discovery Bay just north of milepost 283 and then Sequim (pronounced "Squim") Bay just south of milepost 271, both of which are primarily developed in glacial sediments. Both inlets open northward to the Strait of Juan de Fuca and have steep sides that drop rapidly to their bottoms. Discovery Bay is the deepest, reaching depths greater than 200 feet (60 m) near its entrance. Most likely these inlets originated in a way similar to those in Puget Sound, eroded by meltwater flowing beneath the lobe of glacial ice.

The town of Sequim sits on gravels left by an early channel of the Dungeness River, which now lies about 2 miles (3.2 km) farther west; US 101 crosses the modern channel about halfway between mileposts 263 and 262. Some of these older gravels are exposed along the entrance ramp from River Road at milepost 263. The Dungeness River built a combination alluvial fan and delta complex that visibly bulges northward into the Strait of Juan de Fuca. At its north end, Dungeness Spit has grown northeastward along the edge of the strait.

Dungeness Spit

Extending like a long wishbone about 5.5 miles (8.9 km) out from the coast, Dungeness Spit consists of sand and gravel eroded from the high coastal bluffs at its west end. It is the largest of a complex of three spits that protect the waters behind them and create habitat for a wide variety of wildlife, from migratory birds to salmon and steelhead. The two other spits, Graveyard Spit and the small Cline Spit, divide Dungeness Bay into two shallow lagoons but are closed to the public.

Spits form because prevailing winds drive waves that transport sand along the coastline. In the case of Dungeness Spit, as well as Ediz Hook (the spit immediately north of Port Angeles), the winds come primarily from the west and so drive sand toward the east. Dungeness Spit is a little more complicated because the Dungeness River delta provides additional sediment from the south, and occasional storms from the northeast temporarily reverse the general direction of sediment migration. The overall effect of these complications is the growth of Graveyard Spit, which extends southward from the middle of Dungeness Spit almost to the coastline. Most spits, including Dungeness Spit, display a smooth front side that faces the ocean and a somewhat irregular back side that faces the shallow lagoon; the front side is continually smoothed by the waves, whereas the back side is shaped partly by sediment that washes over the spit during storms.

If you look at the beach cobbles, you'll notice that many are flattened as a result of being repeatedly washed back and forth across the front of the beach. In general, the larger cobbles make up the slightly elevated backbone of the

Aerial photo (looking southwest) of Dungeness Spit and the Olympic Mountains. Note the cloud of sediment behind the spit that comes from the Dungeness River.

spit that is reached only by storms, whereas sand dominates the lower reaches that receive daily wave action. You might also notice their variability, with numerous granitic rocks, gneisses, and metamorphosed volcanic rocks thrown into the mix. The same rock types make up much of the gravel in the bluffs at the west end of the spit.

The bluffs consist of glacial outwash material, which explains its diversity of rock types. Granites, gneisses, and metamorphosed volcanic rocks are not found in the Olympics but are very common in the North Cascades and southern Canada. As glaciers carried material southward, streams flowing off the ice carried the sediment to where it is now. A closer look at the bluffs reveals the presence of crossbedding and shallow channels, telltale signs of a river environment.

Comparisons of the spit to early photographs and maps show how the spit has grown through time. Researchers at Western Washington University determined that it grew an average of 14.4 feet per year (4.4 m/yr) between 1855 and 1985. Narrow beach ridges near the tip of the spit, formed by the addition of new material, indicate the spit is still growing.

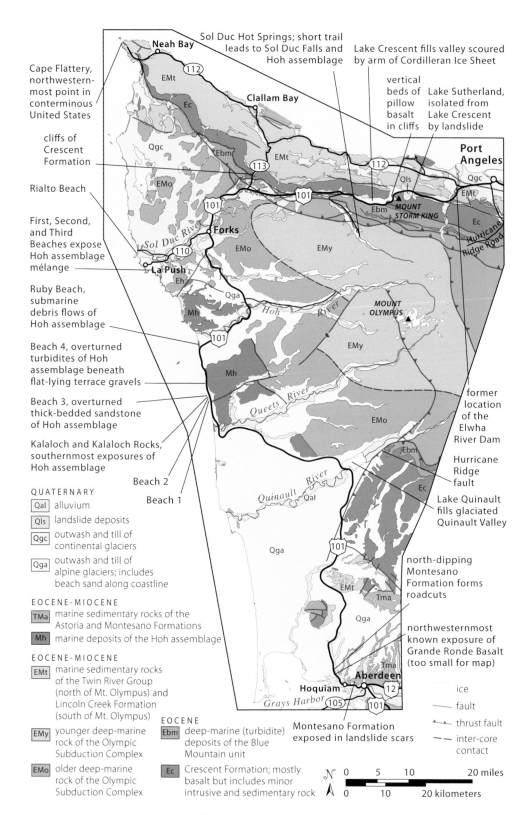

Cape Flattery, northwestern-most point in conterminous United States

cliffs of Crescent Formation

Rialto Beach

First, Second, and Third Beaches expose Hoh assemblage mélange

Ruby Beach, submarine debris flows of Hoh assemblage

Beach 4, overturned turbidites of Hoh assemblage beneath flat-lying terrace gravels

Beach 3, overturned thick-bedded sandstone of Hoh assemblage

Kalaloch and Kalaloch Rocks, southernmost exposures of Hoh assemblage

Beach 2

Beach 1

Sol Duc Hot Springs; short trail leads to Sol Duc Falls and Hoh assemblage

Lake Crescent fills valley scoured by arm of Cordilleran Ice Sheet

vertical beds of pillow basalt in cliffs

Lake Sutherland, isolated from Lake Crescent by landslide

Port Angeles

former location of the Elwha River Dam

Hurricane Ridge fault

Lake Quinault fills glaciated Quinault Valley

north-dipping Montesano Formation forms roadcuts

northwesternmost known exposure of Grande Ronde Basalt (too small for map)

Montesano Formation exposed in landslide scars

Neah Bay

Clallam Bay

Forks

La Push

Aberdeen

Hoquiam

MOUNT STORM KING

MOUNT OLYMPUS

QUATERNARY

Qal   alluvium

Qls   landslide deposits

Qgc   outwash and till of continental glaciers

Qga   outwash and till of alpine glaciers; includes beach sand along coastline

EOCENE–MIOCENE

TMa   marine sedimentary rocks of the Astoria and Montesano Formations

Mh   marine deposits of the Hoh assemblage

EOCENE–MIOCENE

EMt   marine sedimentary rocks of the Twin River Group (north of Mt. Olympus) and Lincoln Creek Formation (south of Mt. Olympus)

EMy   younger deep-marine rock of the Olympic Subduction Complex

EMo   older deep-marine rock of the Olympic Subduction Complex

EOCENE

Ebm   deep-marine (turbidite) deposits of the Blue Mountain unit

Ec   Crescent Formation; mostly basalt but includes minor intrusive and sedimentary rock

ice

—— fault

▲▲ thrust fault

– – inter-core contact

N   0   5   10   20 miles
    0   10   20 kilometers

Geology along US 101 between Port Angeles and Aberdeen.

US 101
PORT ANGELES—ABERDEEN
164 miles (264 km)

Between Port Angeles and Aberdeen, US 101 traverses the north and west sides of the Olympic Mountains, providing access to some of Washington's most active geology. Ongoing subduction and accretion of material to Washington's coast drives recent uplift and its attendant erosion. US 101 offers outstanding bedrock exposures along Lake Crescent and at Ruby Beach, Beach 4, and Beach 3, all of which are north of Kalaloch.

Hurricane Ridge Road in Olympic National Park

Beginning at the park visitor center at an elevation of 250 feet (76 m), the Hurricane Ridge Road rises through the Peripheral rocks to end within the Olympic core at an elevation above 5,000 feet (1,500 m), all in fewer than 18 miles (29 km).

For the first 4 miles (6.4 km) the road climbs steadily southward up a slope of glacial till, most of which is forested. With the exception of two comparatively poor outcrops near milepost 4, you start seeing exposures of Crescent basalt 0.4 mile (640 m) past milepost 7. Above there, the outcrops and views get progressively better. Lookout Rock, at milepost 9, provides a wonderful view into the deep valley of Morse Creek. Rowland Tabor, who mapped the Olympic Mountains in the 1970s and 1980s for the US Geological Survey, reported granitic boulders in Morse Creek and its tributaries up to elevations of 3,500 feet (1,070 m). As these granites have no local sources, they must have been brought here by ice. However, Morse Creek and its tributaries show distinctively V-shaped profiles typical of river valleys, rather than the U-shaped profiles characteristic of glacial valleys. Tabor suggested that the ice sheet in the Strait of Juan de Fuca temporarily dammed Morse Creek, creating a deep lake, and icebergs calving off the ice floated the boulders up the valleys.

Three short tunnels cut through the basalt just south of Lookout Rock at milepost 9. Much of the basalt exhibits distinctive pillow shapes from erupting underwater. The basalt also contains zones of highly brecciated pillows because underwater eruptions are fraught with explosions. The most spectacular outcrops occur between mileposts 10 and 11. About a quarter mile (400 m) above milepost 10 and just below milepost 11, look for some red sedimentary rocks among the pillows. These rocks are red limestones, colored from oxidized iron and manganese derived from the adjacent basalt. They formed from lime-rich mud during times of relatively quiet volcanism and contain fossils of tiny single-celled marine creatures called foraminifera. The fossils help date this

Near its top, Hurricane Ridge Road winds past exposures of Crescent basalts and Blue Mountain unit into the core of the Olympic Peninsula, the Olympic Subduction Complex. Mt. Olympus in the background.

Pillow basalt of the Crescent Formation along Hurricane Ridge Road.

part of the Crescent Formation as middle Eocene and show that it was deposited in a deep ocean.

Steeply inclined turbidite deposits form roadcuts beginning just above milepost 14. Turbidites are deposited by turbid flows of sediment at the bottom of an ocean. They typically consist of alternating beds of sandstone and shale, each combination of which represents a single depositional event. As the coarser material settles out of suspension first, it is overlain by progressively finer material. If you look closely at the rocks, you can see this gradual decrease in grain size upward, which corresponds to the direction in which the rocks get younger. Along this part of the Hurricane Ridge Road, the turbidites get younger toward the northeast, down the road. They are part of the Blue Mountain unit of the Crescent Formation, which is mostly older than the basalts.

The abrupt bases of the sandstone beds and the gradual grading into shales indicate these turbidites of the Blue Mountain unit become younger to the right in the photo.

Milepost 16 marks the trace of the Hurricane Ridge fault, which carried the Peripheral rocks over the Olympic core. For such a big and important structure, however, the fault is very difficult to discern. Turbidites on both sides of the fault look nearly identical. However, a close inspection reveals that the rocks above the upper end of the pull-out tend to be more deformed and are slightly metamorphosed, as indicated by a subtle sheen and the numerous visible flakes of white mica. These rocks are part of the Olympic core. The rocks below the milepost show less deformation and fewer visible flakes of mica and are part of the Peripheral rocks. Between them lies a zone of faulting that contains slivers of both the core and Peripheral rocks.

The last stretch of road to the parking lot on Hurricane Ridge passes by numerous outcrops of the core rocks, which display a more intensive sheen, and hence degree of metamorphism, than those adjacent to the fault. Many of these rocks tend to break into pencil shapes, the result of two pervasive planes of weakness in the rock, called *cleavage*, that intersect at high angles. Such cleavage planes are common features during metamorphism. From the visitor center, you gain outstanding views to the southwest over the deep, glaciated Elwha River toward the Bailey Range and Mt. Olympus. All the rock you can see in that direction belongs to the Olympic core.

For the first few miles out of Port Angeles, you drive over glacial till from the most recent ice advance, which reached a peak about 16,500 years ago. Just east of milepost 241, look for north-tilted sedimentary rocks of the Aldwell Formation, deposited on top of the Crescent basalts in a deep ocean during the middle part of the Eocene. About 40 miles (64 km) to the southeast on Hood Canal, similar sedimentary rocks are interbedded with the Crescent basalt to indicate at least parts of the two rock units formed at the same time. A quarter mile (400 m) west of milepost 240, you can see Crescent basalts on the east side of the highway.

Halfway between mileposts 240 and 239, you cross the bridge over the Elwha River, the site of the largest dam removal project in US history as of this book's writing. Two dams, the 105-foot-high (32 m) Elwha Dam, built in 1913, and the 210-foot-high (64 m) Glines Canyon Dam, built in 1927, were removed in stages between 2011 and 2014. The Elwha Dam backed up Lake Aldwell almost to US 101, whereas the Glines Canyon Dam lay about 5 miles (8 km) farther upstream and backed up Lake Mills. Both dams were built for hydro-electric power, and neither offered any type of structure to allow the passage of fish upstream.

Besides blocking fish migration, the reservoirs behind the dams trapped sediment. At the time of dam removal, the Elwha River had built deltas into the two reservoirs that totaled about 27.5 million cubic yards (21 million m³) of sediment. Lake Mills, being upstream, accumulated most of the sediment. The calm reservoir waters absorbed sun energy to make them unnaturally warm, decreasing their oxygen-carrying capacity and affecting plant and animal life downstream.

Dam removal had to be gradual to prevent the entire ecosystem from being flooded with sediment. Even so, by 2013, after partial dam removal, the Elwha River had built a new delta into the Strait of Juan de Fuca that was more than 800 feet (240 m) long. Today, the ecosystem is recovering, and the salmon and steelhead are beginning to find their way back up the river.

Looking down the Elwha River from the highway bridge. The large volume of sediment in the channel, now being transported to the sea, originally accumulated behind the Glines Canyon Dam, 5 miles (8 km) upriver.

More substantial roadcuts of the Crescent basalts lie on the north side of the highway beginning at about milepost 234 and continuing westward for more than a half mile (0.8 km) along the shore of Lake Sutherland. Lake Sutherland and the much larger Lake Crescent, immediately to the west, fill a valley eroded by an arm of the Cordilleran Ice Sheet. They used to be one continuous lake that drained eastward down Indian Creek into the Elwha River, but they were separated by a series of landslides sometime after retreat of the ice. The first landslides came from high on Mt. Storm King to the south and consisted primarily of blocks of Crescent basalts; later landslides came from the north and consisted of sedimentary rock. Today, Lake Sutherland still drains down Indian Creek, but Lake Crescent, now 80 feet (24 m) higher, drains westward and then northward into the Lyre River.

You can see some of the large boulders that came down in the landslides along East Beach Road, which joins US 101 at milepost 232. Between there and milepost 231, some landslide blocks are so big that they look like regular roadcuts. If you follow East Beach Road 4 miles (6.4 km) to its end, you can access some trailside and lakeshore exposures of the Aldwell Formation by hiking twenty minutes along the Spruce Railroad Trail. From that side of the lake you also gain a nice view of the landslide.

At the national park boundary, near milepost 231, you first see Lake Crescent as well as cliffs along its shoreline. The easternmost exposures consist of Aldwell Formation, but these give way within a quarter mile (400 m) to resistant Crescent basalts that continue well past milepost 229. One of the safest and perhaps most interesting places to see the rocks is at the pull-out at Sledgehammer Point, about a half mile (0.8 km) south of milepost 230. There, you can see vertical beds of pillow basalt cut by a series of white veins. These veins consist mostly of the mineral laumontite, a type of zeolite. Just south of the point, more high cliffs consist of basalt.

Lake Crescent Lodge and the national park ranger station rest on a flat bench of alluvium deposited by Barnes Creek, which drains the valley on the south side of Mt. Storm King. A short distance west, you pass into older marine sedimentary rocks, which crop out in places along the road all the way to the west end of the lake. Across the lake to the north, basalt of the Crescent Formation forms prominent cliffs. Directly back to the east, Mt. Storm King rises to an elevation of 4,500 feet (1,370 m), nearly 4,000 (1,200 m) feet above the lake.

The highway crests a low pass about 0.75 mile (1.2 km) west of milepost 220 at the intersection with the road to Sol Duc Hot Springs. This developed resort area and important trailhead lie along the Sol Duc River. A half mile (0.8 km) hike leads to Sol Duc Falls, where the river plunges some 50 feet (15 m) over steeply dipping rocks of the Olympic core. It is one of two major hot springs in the national park, the other being Olympic Hot Springs, about 10 miles (16 km) to the east. Both hot springs issue from rocks of the Olympic core, with water exceeding 115°F (46°C).

Rialto Beach and First, Second, and Third Beaches

From just north of Forks on US 101, you can drive about 15 miles (24 km) southwestward on WA 110 (La Push Road) to Rialto Beach, in Olympic National Park, or La Push, a small community on the coast within the Quileute Indian Reservation. Just south of La Push, you can take short hikes through old-growth forest to First, Second, and Third Beaches. The road follows a low ridge of mostly glacial outwash deposits between the Sol Duc and Bogachiel Rivers, which join to form the Quillayute River. The only roadside bedrock exposure consists of sandstone of the Hoh assemblage at the entrance to Rialto Beach. The beaches, however, display a variety of excellent exposures of the Hoh assemblage, Miocene-age rocks of the Olympic Subduction Complex.

At Rialto Beach, you can hike 2 miles (3.2 km) north to Hole in the Wall, a small sea arch of Oligocene-to-Eocene sandstone. Southward toward the mouth of the Quillayute River, you can see several small islands, most of which consist of brecciated Hoh assemblage. The one exception is the northernmost Little James Island, which is made of glacial till.

At First Beach, you can see dramatic sea stacks and cliffs of Hoh assemblage. The cliffs along the southwest end of the beach lie beneath an uplifted and tilted marine terrace. Ocean waves eroded the flat surface of the terrace when it was at sea level, but it has since been uplifted and tilted, probably in the past 100,000 years. To access the rocks, you can take a short trail from an overlook a short distance below the Second Beach parking lot. The sea cliffs consist mostly of tilted beds of conglomerate, and the terrace is developed on beds of sand and gravel.

Sea arch at Second Beach.

Trailheads along WA 110, 1.1 and 2.5 miles (1.8 and 4 km) east of La Push, access Second and Third Beach, respectively. Both require short hikes. At Second Beach, you'll see numerous sea stacks and a sea arch made of Hoh sandstone as well as breccia. A landslide comes right down to the beach on its north side. At Third Beach, you can find abundant mélange, formed when different rock types became mixed together. This mixing can occur by a variety of mechanisms, such as during a submarine landslide or while sediments on the seafloor are carried into a subduction zone. Look for blocks of greenstone, chert, sandstone, siltstone, and limestone all mixed together within a shaly matrix in the landslide-prone slopes southward along the beach. A large landslide carried material right down to the water about a quarter mile (400 m) to the south. Near the north end of Third Beach, you can also find oil seeps and broken-down oil-drilling equipment when the tide is low.

West of the Sol Duc Hot Springs intersection, US 101 follows the Sol Duc River almost all the way to Forks. The river meanders between high ridges of Crescent basalts to the north and mostly core rocks to the south. Look for occasional high-elevation exposures of both. The highway crosses the river several times as it traverses a forested, logged, and replanted landscape. Just west of milepost 211, look for exposures of the Blue Mountain unit, marine sedimentary rocks near the base of the Crescent Formation.

South of Forks, US 101 follows the Bogachiel River over glacial outwash and alluvium for about 7 miles (11 km). The Bogachiel River rises high on Bogachiel Peak, some 25 miles (40 km) to the east. As no part of its watershed is presently glaciated, the river is not subject to the seasonal flooding and high sediment pulses of most other Olympic rivers. Its fishing season, which features steelhead, chinook, and coho salmon, lasts longer than that of rivers fed by glaciers. US 101 crosses a wide divide between the Bogachiel and Hoh watersheds between mileposts 181 and 180.

Between mileposts 179 and 178, Upper Hoh Road leads 18 miles (29 km) up the Hoh River valley to a ranger station and trailhead for the Hoh Rain Forest and Blue Glacier. The Blue Glacier blankets much of Mt. Olympus. About 1 mile (1.6 km) south, the road crosses the Hoh River, which it follows to the coast. Note the turquoise hue of the river, caused by fine glacial sediment, or rock flour, suspended in the water.

Ruby Beach, just west of milepost 165, offers the first of several detailed views of the rocks that make up the Hoh assemblage, some of the rocks that were most recently accreted to the edge of North America. The beach area is divided in two by Cedar Creek, which issues from a steep-walled valley. Just south of the creek, a small sea stack offers a picturesque window eroded through its narrow frame. It consists of highly fractured sandstone. A close look at the rock shows that many of the sand grains themselves consist of tiny rock fragments. The most interesting bedrock, however, is exposed in sea stacks and cliffs just north of the creek. These rocks consist mostly of large, angular, and partially rounded particles deposited in underwater debris flows on an ocean floor. If you look closely at them, you will see fragments of Crescent basalts, Hoh sandstone, and scattered pieces of white and red limestone. Some fragments are more than 1 foot (0.3 m) across.

The name Ruby Beach comes from the occasional patches of reddish garnet-rich sand. Rowland Tabor of the US Geological Survey suggested that many of these garnets were weathered from sandstones of the Olympic core. He found that many of those sandstones contain grains of garnet that were deposited as part of the original sediment. Their ultimate sources were probably garnet-bearing metamorphic rocks of the North Cascades. Some of the garnet of Ruby Beach probably also came directly from the North Cascades as glacial outwash during the Pleistocene.

Between mileposts 164 and 163, a pull-out offers a clear view to Destruction Island, a remnant of sandstone of the Miocene-age Hoh assemblage that sits more than 3 miles (4.8 km) off the coast.

Beach 4, a half mile (0.8 km) south of milepost 161, hosts a dramatic angular unconformity between steeply tilted sandstone and shale of the Miocene Hoh

Debris flow deposits make up the Hoh assemblage at Ruby Beach. The large white rock embedded in the dark sea stack on the left side of the photo is a block of limestone.

assemblage and flat-lying gravels deposited during the Pleistocene Epoch. You can see graded beds in the sandstone, where sediment settled to the seafloor out of suspension; coarser particles settled first, followed by progressively finer ones. These graded beds indicate the bedrock here is overturned, so that it actually becomes younger toward the west. In other places you can find cross-bedding, as well as good examples of soft-sediment deformation, irregular folding that occurs while the sediment is still soft and malleable.

The overlying gravels are significant because they were deposited at sea level during the later part of Pleistocene time, when sea levels were appreciably lower. Here the gravels are about 20 feet (6 m) above present-day sea level. Their presence indicates uplift of the Olympic coastline as a consequence of ongoing subduction and accretion just 100 miles (160 km) to the west.

A short trail from a pull-out about a quarter mile (400 m) south of milepost 160 leads to Beach 3, where you can see thick-bedded sandstone of the Hoh assemblage. Most of the rock is coarse grained and dips some 45 degrees eastward. Like the rocks at Beach 4, these beds are overturned, becoming younger toward the west. Look for beds containing *rip-up clasts*, fragments of already-deposited finer material that was torn up during storms and redeposited with the sand.

The southernmost exposures of the Hoh assemblage along US 101 show up at Kalaloch as the low sea stacks called Kalaloch Rocks. The coastal bluffs north and

Angular unconformity at Beach 4. Tilted, overturned beds of sandstone of the Miocene-age Hoh assemblage have been planed off and covered with Pleistocene gravels.

south of Kalaloch consist mostly of silt and clay with a great deal of plant material, deposited between about 70,000 to 17,000 years ago in shallow lakes and bogs.

Beaches 1 and 2, immediately south of Kalaloch, consist mostly of sand and beach cobbles, the coarser material sitting higher up on the beach closer to the bluffs. These beaches are devoid of bedrock exposures. The cobbles, most of which consist of sandstone, were derived from the adjacent bluffs. In turn, the bluffs consist of sand and gravel that was eroded from bedrock exposures in the high peaks and transported as glacial outwash.

Just south of the Queets River, near milepost 152, US 101 turns eastward and follows glacial outwash and till for 26 miles (42 km) to Lake Quinault. Some scattered exposures of yellowish soil, derived from the glacial deposits, show up along the side of the road. The Queets River heads on a glacier on the southeast side of Mt. Olympus, more than 35 miles (56 km) to the northeast, and forms a dramatic glacial valley for most of its length. It drains more than 200 square miles (520 km²) of land, almost all of which is in the national park. Beginning near milepost 134 and continuing to Lake Quinault, you catch glimpses through the trees of a high ridge of the Olympic core rocks to the north.

Lake Quinault fills the glaciated Quinault Valley, which formed partly along the Hurricane Ridge fault. Look for a nice view of the lake where you cross the Quinault River at Amanda Park. Until 2011, the Quinault River was fed by the Anderson Glacier, but the glacier has since melted back enough to no longer be a steady source of water. On the northwest side of the lake, you can see a high ridge of core rocks, whereas basalt of the Crescent Formation forms the high ridge southeast of the lake.

From Lake Quinault to Hoquiam and Aberdeen, US 101 continues over a low-relief surface of forest-covered glacial deposits with very little bedrock exposed. For the first 6 miles (10 km) or so, you can see the high Quinault Ridge to the east, made of Crescent basalts, and then south of Humptulips the road skirts hills made of younger Eocene sedimentary rock. One good exposure of north-dipping, Miocene-age sandstone and siltstone of the marine Montesano Formation exists on the west side of the highway less than a quarter mile (400 m) north of milepost 89. These rocks form most of the hills you see as you drive through Hoquiam on to Aberdeen. An exception consists of the Grande Ronde Basalt, the northwesternmost-known outcrop of this lava flow of the Columbia River Basalt Group. It's beautifully exposed on the north side of milepost 89 at the corner of US 101 and WA 109 Spur. Westbound travelers can see numerous additional exposures of the Montesano Formation exposed in landslide scars between mileposts 87 and 86.

Aberdeen and Hoquiam lie on the edge of Grays Harbor, an estuary and tidal flat complex at the mouth of the Chehalis River. The estuary extends over an area of approximately 90 square miles (235 km²), about 60 percent of which is exposed as tidal flats during low tides. Grays Harbor provides a major staging area for migratory shorebirds, with hundreds of thousands of birds stopping in spring to feed and rest before their long journey north. A national wildlife refuge occupies a stretch of the north shore just a couple miles west of Hoquiam.

US 101
ABERDEEN—ASTORIA, OREGON
78 miles (126 km)

While the geology along this southern segment of US 101 is far subtler than the other two segments, it offers some instructive bedrock exposures as well as views of some critical aspects of Washington's geology. For example, the giant estuaries and tidal flats of Willapa Bay offer some of our most compelling evidence for the large-magnitude earthquakes that periodically shake the coast.

As you cross the Chehalis River on the south side of Aberdeen, look upstream to see a large roadcut of east-dipping Montesano Formation along the north side of the river and US 12. More of the Montesano Formation, which was deposited in a shallow sea during Miocene time, crops out on the southwest side of US 101 halfway between mileposts 80 and 79. Look for numerous dark circular features within the outcrop. They likely formed because circulating groundwater precipitated minerals around inclusions, such as fossils or irregular grains, and discolored that part of the rock. Unlike concretions, which are more resistant to weathering and erosion, these features seem to weather and erode just as easily as the surrounding rock.

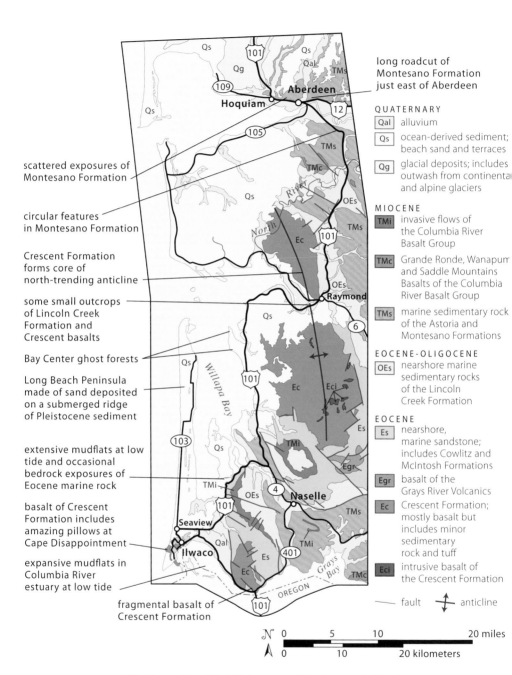

long roadcut of
Montesano Formation
just east of Aberdeen

QUATERNARY

| Qal | alluvium |

| Qs | ocean-derived sediment; beach sand and terraces |

| Qg | glacial deposits; includes outwash from continenta and alpine glaciers |

MIOCENE

| TMi | invasive flows of the Columbia River Basalt Group |

| TMc | Grande Ronde, Wanapum and Saddle Mountains Basalts of the Columbia River Basalt Group |

| TMs | marine sedimentary rock of the Astoria and Montesano Formations |

EOCENE-OLIGOCENE

| OEs | nearshore marine sedimentary rocks of the Lincoln Creek Formation |

EOCENE

| Es | nearshore, marine sandstone; includes Cowlitz and McIntosh Formations |

| Egr | basalt of the Grays River Volcanics |

| Ec | Crescent Formation; mostly basalt but includes minor sedimentary rock and tuff |

| Eci | intrusive basalt of the Crescent Formation |

― fault   ✛ anticline

scattered exposures of
Montesano Formation

circular features
in Montesano Formation

Crescent Formation
forms core of
north-trending anticline

some small outcrops
of Lincoln Creek
Formation and
Crescent basalts

Bay Center ghost forests

Long Beach Peninsula
made of sand deposited
on a submerged ridge
of Pleistocene sediment

extensive mudflats at low
tide and occasional
bedrock exposures of
Eocene marine rock

basalt of Crescent
Formation includes
amazing pillows at
Cape Disappointment

expansive mudflats in
Columbia River
estuary at low tide

fragmental basalt of
Crescent Formation

N

0   5   10   20 miles

0   10   20 kilometers

Geology along US 101 between Aberdeen and Astoria.

The road passes only a few more bedrock exposures between the circular features and Raymond. Small exposures of the Miocene-age Astoria Formation lie on the south side of the road a quarter mile (400 m) north of milepost 77 and immediately south of milepost 75. At milepost 67 you can see Oligocene-age Lincoln Creek Formation exposed by small landslides. These different rock units don't look appreciably different as you drive by at 55 miles per hour (90 km/h) because they're both light-colored, fine-grained sandstones without any noticeable bedding. Both were deposited mostly in shallow marine environments. Up close, however, the sandstone of the Astoria Formation typically contains flecks of white mica, whereas sandstone of the Lincoln Creek Formation typically contains shards of volcanic glass. The rocks get older toward the south because they are on the eastern limb of a north-trending anticline that is cored by Crescent Formation.

Raymond sits at the confluence of the Willapa and South Fork Willapa Rivers, both of which are influenced by tides. You can see wetlands and small tidal sloughs of this large estuary after crossing the Willapa River halfway between mileposts 60 and 59. After passing through Raymond and crossing the Willapa River's south fork, US 101 turns westward and follows the edge of the estuary toward South Bend. At milepost 57, look for a mostly overgrown roadcut of sandy Lincoln Creek Formation; across the slough to the northwest, basalt of the Crescent Formation makes up most of the ridge. You can see a large quarry exposure of the basalt north of the river from near milepost 56; the low hills south of the river are also made of the basalt.

These exposures of basalt mark the large fold that trends roughly north-south through the area. This fold has arched the rocks upward, bringing the older, deeper rocks closer to the surface. This fold also explains the sedimentary deposits you occasionally see along the highway for about the next 10 miles (16 km). These deposits consist of the same material you see in the many sloughs and tidal flats but have been uplifted to form the hills along the coast. Because they are Pleistocene in age, they tell us that at least some of the arching along the fold has taken place since then. The best exposure forms a low cliff on the east side of the highway just south of milepost 48. Another good exposure lies directly across from the heritage marker pull-out a quarter mile (400 m) north of milepost 46.

Many small coastal rivers enter Willapa Bay, including the Niawiakum River near milepost 44 and the Naselle River near milepost 30. These rivers are influenced by the tide for some distance upriver. The bay and its tributaries were nontidal river valleys prior to about 15,000 years ago, during glacial times when sea levels were lower. As the glaciers melted, however, the valleys were flooded by the rising sea levels.

Several features at Willapa Bay indicate that large-magnitude earthquakes affected the area in the recent past. The most visible of these features is the stand of dead cedar trees near Bay Center. Called a ghost forest, these trees formed a healthy stand until the large-magnitude earthquake of 1700 caused the coastal area to subside below sea level. To reach the ghost forest, go west for 1.8 miles (2.9 km) on Bay Center Road, which intersects US 101, 0.4 mile (640 m)

Trees of the ghost forest near Bay Center drowned when this part of the coast subsided below sea level following a large earthquake in 1700.

south of milepost 42. Just before reaching the forest, you pass over a recently uplifted marine terrace.

The other main indicators of large earthquakes are tsunami deposits. The most definitive deposits typically consist of a layer of ocean-derived sand between an underlying layer of peat and an overlying layer of intertidal mud. A tsunami, a series of huge waves generated by the earthquake, carries the sand onshore, depositing it over the peat. Because peat forms in marshes above sea level and intertidal mud forms at or below sea level, the sequence reflects coastal subsidence, similar to what drowns ghost forests. In the Willapa Bay area alone, five known marsh and lake sites preserve tsunami deposits from three hundred years ago, although these are not visible along the highway. More than fifty sites have been described from northern California to northern Vancouver Island.

US 101 intersects WA 4 just south of milepost 29 and heads westward along an arm of the Naselle River, which empties northward into Willapa Bay. The steep hill on the north side of the road displays overgrown exposures of Lincoln Creek Formation; a much better exposure of these rocks lies directly south of the bridge at milepost 26. You enter the Willapa National Wildlife Refuge near milepost 24 and pass spectacular mudflats and sloughs for the next 5 miles (8 km). Some good bedrock exposures of Eocene marine rocks show up on the east side of the road along this stretch.

You get your first glimpse of the Columbia River just south of milepost 18 as you cross a low pass into the Columbia River's watershed. At low tides, the river displays its own expansive mudflats. Just south of milepost 16 the road branches, and you can continue westward on US 101 to Seaview, the Long Beach Peninsula, and Cape Disappointment or head straight south on Alternate US 101 for a more direct path to the river. Either way, you pretty much stay on Columbia River alluvium until just east of milepost 3 near Fort Columbia State Park. There, basalt of the Crescent Formation forms Chinook Point. Good exposures of these same rocks form the base of the cliff at the intersection with WA 401, directly at the north end of the Astoria-Megler Bridge that leads 4.1 miles (6.6 km) south to Oregon. If you look at these rocks closely, you'll see that the basalt is broken into countless tiny fragments, probably from its explosive interactions with seawater.

Naselle River near its mouth at Willapa Bay.

Long Beach Peninsula and Cape Disappointment

Like a long finger parallel to the coast, the Long Beach Peninsula extends from the mouth of the Columbia River northward for more than 26 miles (42 km). Averaging only about 2 miles (3.2 km) in width, the peninsula faces open ocean waters of the Pacific to the west and quiet waters of Willapa Bay to the east. Contrary to its initial appearance, the peninsula is not a spit but rather a ridgeline that was nearly inundated by rising sea levels after the Pleistocene. The peninsula's surface, however, is made mostly of sand derived from the Columbia River to the south, so it does resemble a spit in many ways. Dunes behind the beach form a series of ridges parallel to the coastline. Between the dunes lie peat deposits, formed from compacted vegetation in the peninsula's extensive wetlands. Cores from drilling show that these surface deposits sit above a much thicker sequence of marine-deposited silt, sand, and gravel of probable Pliocene age.

Cliffs of basalt of the Crescent Formation beneath the visitor center and lighthouse at Cape Disappointment. Inset shows pillow basalt.

The only exposed bedrock on the peninsula lies at its south end at Cape Disappointment. There, in the cliffs below the visitor center, you can find beautiful examples of brecciated, massive, and pillow basalt flows of the Eocene-age Crescent Formation. At the north end of Benson Beach, near the campground, you can clamber over phenomenal pillow basalts, possibly the best examples in the state.

The bedrock slopes northward into the subsurface beneath the sediment of the Long Beach Peninsula. A well drilled near the north end of the peninsula reached bedrock at a depth of 1,400 feet (430 m).

WA 4/WA 401
Longview—Naselle—Astoria-Megler Bridge
69 miles (111 km)

Longview occupies the floodplain where the Cowlitz River empties into the Columbia. The floodplain sediments rest on a variety of bedrock, most notably basalt flows of the Grays River Volcanics. These rocks make up the bases of the surrounding hills, such as those directly north of the bridge across the Cowlitz River and Mt. Solo, just southwest of the highway between 38th and 50th Avenues. The Grays River Volcanics resemble the basalt of the Crescent Formation in many ways. Both are basaltic and erupted mostly on the seafloor.

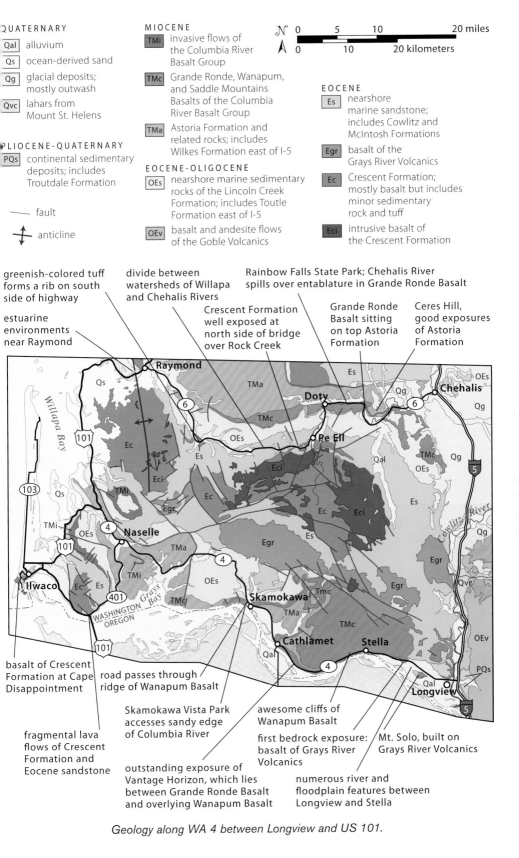

QUATERNARY

Qal — alluvium

Qs — ocean-derived sand

Qg — glacial deposits; mostly outwash

Qvc — lahars from Mount St. Helens

PLIOCENE-QUATERNARY

PQs — continental sedimentary deposits; includes Troutdale Formation

—— fault

✛ anticline

MIOCENE

TMi — invasive flows of the Columbia River Basalt Group

TMc — Grande Ronde, Wanapum, and Saddle Mountains Basalts of the Columbia River Basalt Group

TMa — Astoria Formation and related rocks; includes Wilkes Formation east of I-5

EOCENE-OLIGOCENE

OEs — nearshore marine sedimentary rocks of the Lincoln Creek Formation; includes Toutle Formation east of I-5

OEv — basalt and andesite flows of the Goble Volcanics

EOCENE

Es — nearshore marine sandstone; includes Cowlitz and McIntosh Formations

Egr — basalt of the Grays River Volcanics

Ec — Crescent Formation; mostly basalt but includes minor sedimentary rock and tuff

Eci — intrusive basalt of the Crescent Formation

greenish-colored tuff forms a rib on south side of highway

estuarine environments near Raymond

divide between watersheds of Willapa and Chehalis Rivers

Crescent Formation well exposed at north side of bridge over Rock Creek

Rainbow Falls State Park; Chehalis River spills over entablature in Grande Ronde Basalt

Grande Ronde Basalt sitting on top Astoria Formation

Ceres Hill, good exposures of Astoria Formation

basalt of Crescent Formation at Cape Disappointment

road passes through ridge of Wanapum Basalt

fragmental lava flows of Crescent Formation and Eocene sandstone

Skamokawa Vista Park accesses sandy edge of Columbia River

outstanding exposure of Vantage Horizon, which lies between Grande Ronde Basalt and overlying Wanapum Basalt

awesome cliffs of Wanapum Basalt

first bedrock exposure: basalt of Grays River Volcanics

Mt. Solo, built on Grays River Volcanics

numerous river and floodplain features between Longview and Stella

Geology along WA 4 between Longview and US 101.

However, Grays River Volcanics range in age from about 42 to 37 million years old, which makes them substantially younger than the 56- to 49-million-year-old rock of the Crescent Formation.

Driving through Longview, you pass the north edge of Lake Sacajawea at the corner of Kessler Boulevard and WA 4. Originally called Fowler's Slough, this narrow, curved lake was a former channel of the Cowlitz River. It was dredged in the 1920s and landscaped to form Longview's signature city park. At 38th Avenue, WA 4 parallels Coal Creek Slough (also called Cut-Off Slough), which meanders over the flat floodplain just northeast of the highway. Near and west of milepost 56, you see more wetlands and sloughs along Coal Creek, which eventually joins the Columbia River near Stella, near milepost 51. Even though the Columbia is 60 river miles (100 km) from its mouth, its flow here is influenced by tides that range from about 2 feet (0.6 m) to as much as 4 feet (1.2 m) in height. These tides also affect the low-gradient tributaries such as Coal Creek.

Halfway between mileposts 54 and 53, you'll see basaltic lava flows of the Grays River Volcanics. If you look closely at the rock, you'll spot white crystals composed of radiating fibers. These crystals are zeolites, precipitated in air pockets in the lava after it had solidified. Steel netting, to mitigate the rockfall hazard, covers some of the roadside cliffs.

Between a quarry exposure across from milepost 52 and Stella are cliffs of the Grande Ronde Basalt member of the Columbia River Basalt Group. The rocks dip gently southwestward because they lie on the north limb of a syncline that has arched the rocks downward toward the Columbia River. The overlying

Two flows of Wanapum Basalt are separated by a reddish soil that formed on top of the lower flow before the upper flow erupted.

Wanapum Basalt member of the Columbia River Basalt Group comes down to road level between mileposts 52 and 51. Some of the best places to see this basalt are from the broad shoulder along the highway beginning at Mill Creek Road, 0.3 mile (480 km) west of milepost 48. You can see two flows of the Wanapum Basalt, separated by a layer of soil that formed on the lower (older) flow. The lower flow displays beautiful columnar joints.

The road continues to follow the river and pass cliffs of Wanapum and Grande Ronde Basalt until Cathlamet, near milepost 35. Forested landslide deposits dominate the landscape for 1 to 2 miles (1.6–3.2 km) on either side of milepost 45. At milepost 40, look for outcrops of sandstone and mudstone. These sedimentary rocks, called the Vantage Horizon, mark a period of inactivity between eruptions of the Grande Ronde and Wanapum Basalts. The best exposure lies a quarter mile (400 m) west of milepost 40, directly beneath a flow of the Wanapum Basalt. Because it dips eastward, the rocks get older toward the west, and just a short distance in that direction you enter the Grande Ronde Basalt. At Cathlamet, WA 4 passes over tidal sloughs and wetlands of the Columbian White-Tailed Deer National Wildlife Refuge for about 7 miles (11 km) to Skamokawa, where the road cuts inland through a ridge of Wanapum Basalt.

At Skamokawa Vista Park, you can access the Columbia River and inspect a large river-deposited sandbank along its shore. Much of this material probably originated from the 1980 eruption of Mount St. Helens. For about the next 8 miles (13 km) north, WA 4 passes through heavily vegetated hills of the Oligocene-age Lincoln Creek Formation, mostly sandstone and siltstone deposited in a shallow sea and along its coast. Between mileposts 21 and 20, look for a good exposure of thin-bedded siltstone. The spherical nodules in the rock are concretions. Circulating groundwater precipitated minerals around an object, such as a fossil, making that part of the rock more resistant to weathering and erosion. Individual concretions are easy to find along the base of the exposure.

WA 4 passes through the Miocene-age Astoria Formation as it descends back to the Columbia River. Some exposures of the rock show up on the west side of the highway just west of milepost 17 and about a quarter mile (400 m) west of milepost 15. Miller Point Road leads about 1 mile (1.6 km) to Grays Bay, which forms an inlet on the Columbia River about 12 square miles (30 km²) in area. Approximately half of the bay consists of exposed mudflats during low tides.

From Naselle, northbound travelers follow the Naselle River to the estuary at Willapa Bay. Southbound travelers can take WA 401 southward for a quicker approach to the Astoria-Megler Bridge and Oregon. This route follows the drainage of the north-flowing South Naselle River, another tributary of the Naselle, to a divide less than a quarter mile (400 m) north of the Columbia River. You reach the Columbia near milepost 4, and the highway hugs the north shore all the way to the bridge. At Cliff Point, halfway between mileposts 3 and 2, you pass cliffs of east-dipping Eocene marine siltstone. A wide pull-out allows parking. Even better exposures of this rock form cliffs a quarter mile (400 m) to the west, but there is nowhere safe to park.

Older Eocene-age basalt-rich sandstone holds up the steep ridge near milepost 1. You can park at the Dismal Nitch Rest Area, named for the adjacent cove that sheltered the Lewis and Clark Expedition just before they reached the Pacific. The sandstone shows through the vegetation in only a few places. It was deposited alongside the younger parts of the Crescent Formation, which forms the basement rock of the Coast Range. The Crescent is exposed sporadically for the remaining distance to the bridge. Along here it consists of brecciated and intact lava flows as well as basalt-rich sandstone. A double roadcut and pull-out 0.4 mile (640 m) west of milepost 1 offers a chance to inspect some of the brecciated basalt alongside the river.

Brecciated basalt of the Crescent Formation exposed along WA 401. Inset photo shows close-up of the rock.

WA 6
CHEHALIS—RAYMOND
52 miles (84 km)
See map on page 57.

This route through the Willapa Hills follows two river valleys: the eastern portion parallels the Chehalis River almost to its source, and the western portion parallels the Willapa River almost to its mouth. Being confined to river valleys, however, the road passes only a few bedrock exposures. The Grande Ronde member of the Columbia River Basalt Group forms a prominent ridge for much of the first 15 miles (24 km) or so.

West of I-5, WA 6 crosses the Chehalis River within the first quarter mile (400 m) and continues southwestward through farmland developed on the floodplain, with the tree-covered ridge of Grande Ronde Basalt to the south. You see the first roadcut of basalt where the highway passes through the ridge about a quarter mile (400 m) west of milepost 45. Then for the next several miles, the highway sweeps out a large curve to the south, around the side of Ceres Hill, which is made of Astoria Formation overlain by Grande Ronde Basalt.

You can see several good exposures of the Astoria Formation by going north on Ceres Hill Road for just over 2 miles (3.2 km). Ceres Hill Road intersects WA 6 about 0.3 mile (480 m) east of milepost 42. The road passes a parking lot at a Rails-to-Trails access point at 1.5 miles (2.4 km) and then becomes gravel as it climbs the hill via a series of switchbacks. The best exposure lies just uphill from the second switchback.

Outcrop of Astoria Formation of Miocene age on Ceres Hill Road.

About halfway between mileposts 39 and 38, WA 6 bends sharply westward at some good exposures of the base of the Grande Ronde Basalt. You know these exposures mark its base because sandstone of the Astoria Formation lies directly beneath the basalt at the roadcut's south end. Between them, look for a dark-gray ancient soil deposit, formed on the Astoria Formation prior to eruption of the basalt. Within the basalt itself, you can see interesting weathering rinds, where the outer edges have changed colors because of weathering processes while the inner part of the rock retains its fresh, original color.

West of the base exposure, you can see two exposures of colonnade, an interior part of a lava flow that displays prominent cooling fractures that look like columns. One nice off-highway exposure lies along the Chehalis River about a quarter mile (400 m) down River Road, which intersects WA 6 about halfway between mileposts 38 and 37. A more dramatic but less peaceful site lies just south of milepost 37, where a large quarry exposes several flows of the basalt.

At Rainbow Falls, a quarter mile (400 m) east of milepost 35, the Chehalis River pours over a ledge that displays the entablature part of a flow interior. Entablature differs from colonnade in that it consists of many closely spaced cooling fractures, which give the basalt an irregular, blocky appearance. It can form above or below colonnade. A wide pull-out and good exposures along the river banks allow for easy inspection, but please leave your rock hammers in

Chehalis River flowing over Grande Ronde Basalt at Rainbow Falls State Park.

the car. A bridge across the river about 1 mile (1.6 km) farther west takes you 1.2 miles (1.9 km) to the state park.

At Doty, WA 6 turns sharply southward and enters a short canyon that cuts through the basalt and opens southward onto Pe Ell Prairie, a stretch of floodplain. Look for an outcrop and a small quarry of basalt at the canyon mouth, a quarter mile (400 m) south of milepost 33.

At the small community of Pe Ell, you pass into the Eocene-age McIntosh Formation, which consists mostly of sandstone and shale deposited in a deep ocean. Unfortunately, these rocks are not clearly exposed anywhere along the highway. In several places, however, they are interbedded with volcanic rocks that are exposed, including basalt that is probably part of the Crescent Formation as well as some tuffs. While basalt or tuffs were erupting in one part of the ocean basin, sediments of the McIntosh Formation were accumulating in another. The first of these volcanic rock exposures consists of altered tuff directly opposite milepost 27 on the southwest side of Pe Ell. From there, the highway leaves the Chehalis River and follows Rock Creek westward through the poorly exposed McIntosh Formation. Look for well-exposed Crescent basalts just north of milepost 25 at the bridge over Rock Creek. After crossing a broad, almost imperceptible divide at milepost 22, the highway descends along a tributary of the Willapa River. Less than a quarter mile (400 m) west of milepost 12, look for a good bedrock exposure of tuff. Like the tuff near Pe Ell, this one is interbedded with the McIntosh Formation.

The floodplain of the Willapa River grows wider and flatter over the remaining distance to Raymond. Look for increasing wetlands as well as tidal channels as the river becomes increasingly affected by tides. Raymond sits only about 5 miles (8 km) from the river's mouth at Willapa Bay, Washington's largest estuary at more than 260 square miles (670 km²). About half the bay is exposed as mudflats during low tide.

WA 8
OLYMPIA—ELMA
22 miles (35 km)
See map on page 31.

WA 8 is one of the best highways in Washington for viewing basalt of the Crescent Formation. As the highway crosses the Black Hills, it passes roadcuts that display undersea-erupted pillow basalt, brecciated and massive flows, and land-erupted columnar basalt. This fast, divided highway has ample space along its shoulder for emergency pull-outs.

Southwestward from US 101, the road climbs steadily up a valley filled with glacial deposits and reaches its first good roadcut of massive and brecciated flows at milepost 19. Continue a half mile (0.8 km) farther southwest and you can see well-expressed columnar jointing with some staining parallel to the vertical fractures. More rock exposures are visible in this vicinity for north-bound lanes.

The upper basalt flow of this Crescent Formation exposure shows columns, indicative of flowing over land, whereas the lower flow is brecciated and contains some pillows, indicative of erupting underwater.

More roadcuts begin just south of milepost 13 and continue intermittently until milepost 8. Look for massive and brecciated flows in the large roadcuts near milepost 13. Just south of milepost 9, a long roadcut displays a flow of brecciated pillows beneath a younger flow that contains some small but well-developed columns.

At McCleary, near milepost 7, the road passes into younger sedimentary rocks of the Lincoln Creek and Montesano Formations, although these rocks are not exposed anywhere along the rest of the drive to Elma.

WA 112
US 101—NEAH BAY AND CAPE FLATTERY
65 miles (105 km)

The northernmost highway on the Olympic Coast, WA 112 hugs the coastline in places, offering beautiful views of the Strait of Juan de Fuca and access to some outstanding coastal bedrock exposures. For the most part, the rock consists of Oligocene-to-Eocene sedimentary rocks of the Twin River Group, most of which were deposited in a deepwater environment close to the continental margin after the accretion of the Crescent Formation. These rocks dip northward and so get younger toward the coast.

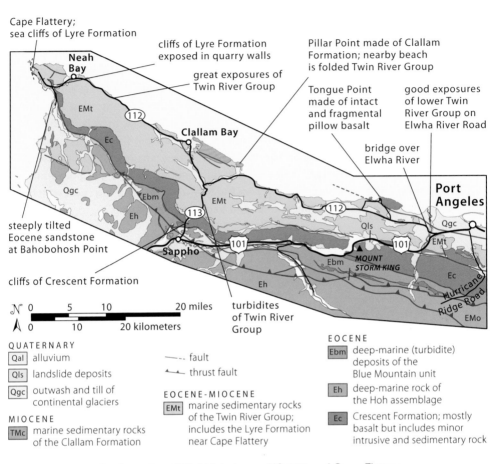

Cape Flattery; sea cliffs of Lyre Formation

cliffs of Lyre Formation exposed in quarry walls

great exposures of Twin River Group

Pillar Point made of Clallam Formation; nearby beach is folded Twin River Group

Tongue Point made of intact and fragmental pillow basalt

good exposures of lower Twin River Group on Elwha River Road

bridge over Elwha River

Neah Bay

EMt

Ec

Qgc

Clallam Bay

Ebm

EMt

Eh

113

112

Qls

EMt

Port Angeles

Qgc

steeply tilted Eocene sandstone at Bahobohosh Point

cliffs of Crescent Formation

Sappho

101

101

Ebm

MOUNT STORM KING

Ec

Eh

turbidites of Twin River Group

EMo

𝒩 0   5   10   20 miles
𝐀 0   10   20 kilometers

QUATERNARY
Qal alluvium
Qls landslide deposits
Qgc outwash and till of continental glaciers

MIOCENE
TMc marine sedimentary rocks of the Clallam Formation

—-- fault
▲—▲ thrust fault

EOCENE-MIOCENE
EMt marine sedimentary rocks of the Twin River Group; includes the Lyre Formation near Cape Flattery

EOCENE
Ebm deep-marine (turbidite) deposits of the Blue Mountain unit
Eh deep-marine rock of the Hoh assemblage
Ec Crescent Formation; mostly basalt but includes minor intrusive and sedimentary rock

Geology along WA 112 between US 101 and Cape Flattery

West of its intersection with US 101, WA 112 crosses a bridge over the Elwha River, with thick-bedded sandstone of the Twin River Group forming cliffs below. Less than a quarter mile (400 m) upstream, the Elwha Dam once formed Lake Aldridge, and about 8 miles (13 km) upstream, the Glines Canyon Dam formed Lake Mills. The dams were constructed in 1913 and 1927, respectively, for hydroelectric power but were removed between 2011–2014 to allow the ecosystem to recover. See the US 101 road guide for a more detailed description of the dam removal and its geologic effects.

Just east of milepost 59, you pass Elwha River Road, which leads north and east in less than 1 mile (1.6 km) to another bridge across the Elwha River and good exposures of the lower Twin River Group. WA 112 heads westward and northward into younger rocks. Conglomerate and sandstone of the middle part of the Twin River Group show up on the south side of the highway between

Crescent Bay

A right turn near milepost 54 leads about 4 miles (6.4 km) northward to Crescent Bay, where the Crescent Formation was first described in detail. From the picnic area at Salt Creek County Park, you can walk along the coast on sedimentary rocks that were deposited as part of the Crescent Formation during periods of relatively inactive volcanism. These rocks consist mostly of sandstone with a high proportion of basalt grains, derived from the nearby basalt. The many irregular features on the bedding surfaces are feeding tracks left by creatures on the seafloor. You can also find beds of pebble conglomerate that contain pieces of chert, eroded from the nearest chert-bearing rock, which crops out only in the San Juan Islands and Vancouver Island.

Feeding tracks in basaltic sandstone of the Crescent Formation.

Tongue Point, reached from the western edge of the campground, forms a promontory of Crescent basalt that frames the eastern side of Crescent Bay. There, you can see intact as well as brecciated pillow basalt, in some places cut by white veins of zeolite minerals. You can also see granitic boulders, left as erratics by the retreating ice sheet, which filled the Strait of Juan de Fuca and flowed westward to the sea.

Crescent basalt also forms the other promontory at the west edge of Crescent Bay. A thrust fault carried the basalt southward over sandstone of the Twin River Group, which occupies the low area immediately to the south. You can see some of the sandstone in the cliffs along the road near the west end of the sandy beach, although stinging nettles prevents close inspection. Because this sandstone lies south of the fault, you can use its location to help see the orientation of the fault, which must continue westward offshore but somewhat parallel to the coast. Geologists at the US Geological Survey map this offshore fault to a spot northwest of Cape Flattery. From here, Crescent Beach Road leads south back to WA 112, rejoining it between mileposts 51 and 50 near the small community of Joyce.

mileposts 58 and 57. From about milepost 55 to 49, you get occasional views southward to a series of forested ridges held up by resistant rock. In general, the lower and closer ridges consist of the lower Twin River Group, whereas the highest and most distant ridge consists of Crescent Formation.

WA 112 follows a partially forested, partially logged landscape developed on glacial till until it descends to the coast near Pillar Point, accessed by a short road just west of milepost 30. Pillar Point consists of cliffs of Miocene sedimentary rock of the Clallam Formation, but the bedrock at the beach access consists of the youngest part of the Twin River Group. If you walk around and inspect the rock you can see that it is folded: at the east side of the beach near the parking lot, the rock dips steeply northwestward, whereas the rock farther west and exposed at low tide dips gently southeastward. Several large blocks of sandstone and conglomerate have fallen onto the beach from the nearby cliffs and allow a convenient inspection of sedimentary features. Granitic boulders also lie on the beach—glacial erratics carried here by the ice in Pleistocene time.

Between Pillar Point and Clallam Bay, WA 112 follows the drainage of the Pysht River until the highway intersects WA 113, a quarter mile (400 m) east of milepost 23. From there the road passes over glacial till until it drops into the valley of the Clallam River. Clallam Bay is lined with sand, most of which makes up a spit that separates the Clallam River from the ocean. You can cross the river on a footbridge at the community beach and access the spit. Slip Point, made of delta deposits of the Clallam Formation, forms the prominent headland on the bay's east side.

Cape Flattery

You enter the Makah Reservation just before milepost 0 and then Neah Bay. The short trail for Cape Flattery, Washington's northwesternmost point, starts at a trailhead about 8 miles (13 km) west of Neah Bay. You need to get a Makah recreational use permit, available from any local business, to park at the trailhead, but it's well worth it. Along the way to the trailhead, you pass some large quarries cut into the Lyre Formation, the same rock unit that makes up Cape Flattery. The Lyre Formation was deposited in Eocene time, after the accretion of the Siletz terrane but before the deposition of the Twin River Group. The rock is significant because its grains are generally coarse and angular, which indicate they eroded from a nearby source. Some conglomerate beds visible from the end of the trail at Cape Flattery contain boulders. Individual grains may consist of a variety of rock types, including chert, granite, and metamorphosed volcanic rocks, none of which could have come from the Olympic Mountains. Researchers generally agree that these rocks were derived from

The Lyre Formation along the coastline at Cape Flattery.

Close-up of Lyre Formation near Cape Flattery. Note the coarse, angular nature of its grains.

Vancouver Island to the north. They were deposited in a deep basin that once occupied the area now filled by the Strait of Juan de Fuca.

While returning from Cape Flattery, you can take Hobuck Road 1.8 miles (2.9 km) to Bahobohosh Point on Makah Bay to see steeply tilted sandstone deposited in deep water in the Eocene Epoch. The rock displays beautiful honeycomb patterns, caused by the repeated wetting and drying of the rock by salt water. As the water dries, tiny salt crystals grow within the rock's pore spaces and loosen the sand grains, eventually forming a small pocket that can collect more water. You can also find pieces of limestone, possibly derived from Vancouver Island, in some of the sandstone.

WA 112 hugs the edge of the bay past Middle Point to gain an unobstructed view of the area. To the east lies Clallam Bay and Slip Point; to the west lies Sekiu Bay and Sekiu Point. Both Middle Point and Sekiu Point consist of bedrock of the Twin River Group, some of which shows up in the vegetated hillside just west of milepost 16. West of the town of Sekiu, the highway turns inland where it follows glacial till for a few miles before heading back to the coast, where you get some beautiful views northward to Vancouver Island.

From the beach at Shipwreck Point just west of milepost 6 to Neah Bay, you see numerous exposures of the middle part of the Twin River Group—in cliffs, landslide scars, and sea stacks. One particularly unusual tooth-shaped exposure about a half mile (0.8 km) west of milepost 6 consists of sandstone with a resistant bedding plane forming a cap. A small pull-out allows inspection of it and the nearby rock, which contains some beds of conglomerate, numerous concretions, and Oligocene-age fossils. Look also for the prominent sea stacks just west of milepost 2, and the large flat-topped one a quarter mile (400 m) east of milepost 0.

Sea stack of the middle part of Twin River Group, with mountains of Vancouver Island in the distance. Notice the eastward dip of the sedimentary beds.

WA 113
US 101—WA 112
10 miles (16 km)
See map on page 65.

WA 113 connects US 101, at the small community of Sappho, to WA 112, which follows the Olympic Peninsula's northern coastline. Because WA 113 runs roughly north-south and the rocks dip generally northward, northbound travelers pass through progressively younger rocks. At Sappho, the road crosses glacial outwash punctuated by hills made of poorly exposed Eocene sedimentary rocks of the Blue Mountain unit of the Crescent Formation. Just south of milepost 2, however, you start seeing excellent exposures of Crescent basalt, which holds up a high east-west ridge. Look for numerous well-defined pillows.

Near milepost 4, Beaver Lake and its wetlands occupy the alluvium-filled valley between Crescent basalt to the southeast and the younger Twin River Group sedimentary rocks to the northwest. You don't see the sedimentary rocks until you encounter a series of good exposures halfway between mileposts 6 and 7. They consist mostly of interbedded sandstones and shales, which give the outcrop a distinctly striped appearance. These rocks, called *turbidites*, generally form when turbid flows of sediment pour into a deep-sea basin and settle out from suspension. The coarser grains settle first, followed by progressively finer ones, hence the alternating sandstone and shale beds. If you look closely at some of the sandstone beds, you can see that they, too, are graded, with the coarsest sand at the bottom.

The remaining 3 miles (4.8 km) of the trip to WA 112 head mostly downhill into the Pysht River valley that is mostly filled with glacial till.

Turbidites of the Twin River Group.

PUGET LOWLAND

Puget Lowland is an apt description for the elongate trough between the Coast Range to the west and the Cascades to the east. Today, the lowland is mostly occupied by Puget Sound, an arm of the sea, but it has only been 12,000 years since a mighty lobe of the Cordilleran Ice Sheet receded from the trough. A thick blanket of glacial deposits covers the bedrock, but a few rock exposures, together with geophysical studies and some deep drilling, tell us that the lowland has been a subsiding basin since the accretion of Siletzia in Eocene time. In plate tectonic terms, the Puget Lowland is a fore-arc basin because it is located between the Cascade volcanic arc and the Cascadia subduction zone. The arc developed in late Eocene or early Oligocene time because of renewed subduction of ocean crust after the accretion of Siletzia, and it stretches southward as far as Lassen Peak in northern California. Other lowlands occupy the fore-arc area to the south as well, most notably the Willamette Valley, which includes the Portland Basin and reaches as far south as Eugene.

The general sequence of bedrock units below the glacial layers in the Puget Lowland has the basalt of Siletzia's Crescent Formation at the base, overlain by sandstone and mudstone of the Oligocene-to-Miocene Blakeley and Blakely Harbor Formations. Farther east are Eocene rocks of the Puget Group, some of which contain coal deposits.

Bedrock is also exposed at the northern end of the lowland in the San Juan Islands. The rock types and structures in the islands are similar to the thrust-bounded terranes on the western side of the North Cascades, so geologists consider the archipelago to be a western extension of those terranes. The bedrock of the islands survived intense sculpting and erosion by the glaciers that repeatedly advanced into and retreated from the lowland.

Glacial Deposits and Landforms of the Puget Lobe

Repeated advances of the Puget Lobe are responsible for the surficial deposits and topography of the Puget Lowland. As the ice melted, it released mud, sand, gravel, and other sediments that now mantle the much harder basement of bedrock beneath. The sediments vary in thickness from zero, where bedrock is exposed, to more than a mile thick (several kilometers). If the mantle of glacier-related materials was stripped off, we would see an irregular topography of mountains and valleys sculpted by ice.

Tidal flats at the mouth of the Nisqually River grade into the waters of Puget Sound.

Careful studies of the Pleistocene-age sediments in the lowland have shown that the Puget Lobe advanced as many as six—perhaps seven—times beginning sometime between 2 and 1.6 million years ago. The sediments in the lowland have been divided into glacial and interglacial units. Glacial units, mainly till and outwash, record the southward advance of the lobe. The lobe retreated northward in interglacial periods, but deposition of sediments in lakes and

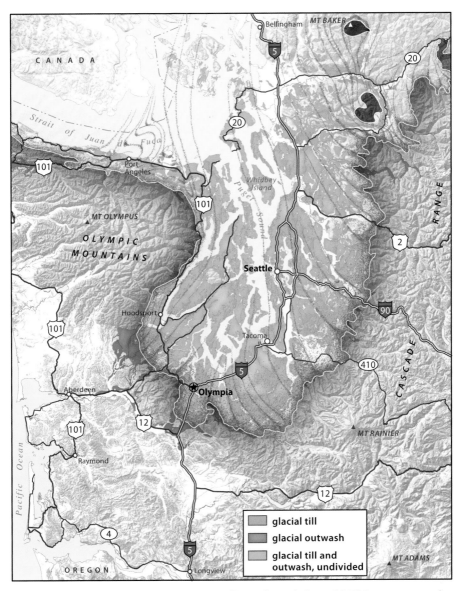

The maximum extent of the Puget Lobe in Puget Sound about 16,500 years ago and the lobe's deposits. –Map courtesy of the Washington Department of Natural Resources, Division of Geology and Earth Resources

rivers continued. Differentiating all the glacial and interglacial units is difficult because each resembles previous and subsequent units. Radiocarbon dating and lab analyses have proven to be the keys to unlocking the age and sequence of the units.

The last advance of ice is called the Vashon Stade. A major glaciation can have multiple advances, each known as a stade. The Vashon glacial advance lasted from about 19,000 to 16,000 years ago, and it probably looked similar to every other maximum extent of ice during the Pleistocene Epoch. Although the Vashon ice left behind a splendid record of sediment and landforms, it largely obliterated the surficial evidence of earlier glaciations. Most of the topography we see today in the Puget Lowland resulted from the advance and retreat of the Vashon ice. Scouting the exposures of Vashon glacial sediments, geologists have constructed the sequence we would likely find on a typical hill in Seattle. At the bottom of the sequence is the Lawton Clay, a layer of dense mud and silt deposited in glacial lakes in front of the advancing Vashon ice. Next is the Esperance Sand, glacial outwash deposited in rivers issuing from the front of the

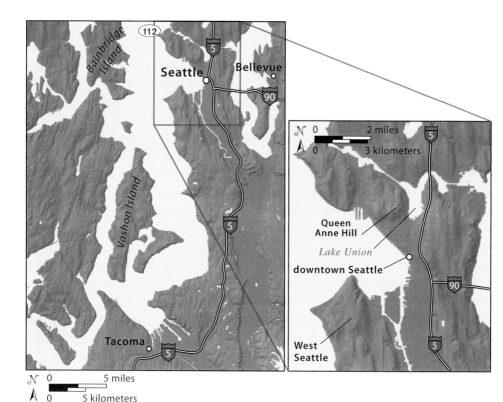

Shaded relief map of north-south-oriented drumlins in and around the Seattle area.
–Map courtesy of Washington Department of Natural Resources, Division of Geology and Earth Resources

advancing ice. The uppermost unit is the Vashon till, deposited directly from the melting ice. The till is dense, mud-rich sediment full of variably sized fragments and boulders of rock.

Because glacial ice is full of embedded rock debris, it is able to erode, abrade, and sculpt the bedrock and other glacial sediments beneath it, including till. Keep in mind that glaciers, or lobes, never move backward. Ice is continually flowing toward the snout of the glacier, no matter if the glacier is advancing or retreating. Advance and retreat depend on the balance between snow added at the head of the glacier and loss of ice due to evaporation and melting at the snout. If the glacier loses ice faster than it gains ice, it retreats, even while it is still moving forward. Therefore, the sculpting of the material beneath the Puget Lobe occurred during both advances and retreats. The most characteristic expressions of glacial erosion are the elongate ridges, called *drumlins*, that were sculpted in glacial sediments below the Vashon ice. The orientation of drumlins is a valuable indication of the direction the ice was moving. Rivers of meltwater flowing beneath the ice eroded some of the valleys in the landscape.

Geologic Hazards

When admiring the inviting landscape in the Puget Lowland—waterways, lakes, and forested hills—residents are well aware of geologic hazards that do or could have profound social and economic costs. Landslides on bluffs and hillslopes are a constant concern, especially after heavy rains that saturate the soil and underlying glacial sediments. The Lawton Clay is a well-known culprit. The clay is an impermeable barrier to water percolating downward, and materials above the clay can slide along it. Another ever-present potential hazard are volcanic mudflows, called *lahars*, emanating from volcanoes in the Cascades. Mt. Rainier, in particular, has produced lahars that reached the Puget Lowland in the not-so distant past, but so too have Mount St. Helens, Glacier Peak, and Mt. Baker. About 5,000 and 500 years ago, respectively, the Osceola and Electron Mudflows originated high on Mt. Rainier and flowed down river valleys that now host attractive suburban towns such as Enumclaw and Buckley.

The Puget Lowland is also at high risk for earthquakes from several sources. The Seattle fault zone is a system of east-west-striking, south-dipping reverse faults. Studies have documented slip on the system large enough to have generated a major earthquake in the lowland about 1,100 years ago, but there are not enough data to establish how frequently earthquakes from this system might recur. By far the most significant seismic hazard would emanate from the Cascadia subduction zone, often described as a *megathrust* because of its immense size. Slip on faults in the Juan de Fuca Plate deep beneath the lowland gave rise to the Nisqually earthquake, centered near Olympia in 2001, and earlier events in 1949 and 1965. The Puget Lowland, like all of western Washington, can also expect severe shaking from a large earthquake generated by slip on the Cascadia megathrust, even though the epicenter of the Big One would lie offshore somewhere between northern California and northern Vancouver Island.

GUIDES TO THE PUGET LOWLAND

INTERSTATE 5

SEATTLE—CANADIAN BORDER

110 miles (177 km)

For about the first 50 miles (80 km) north of Seattle, I-5 affords instructive views of nearby glacial features as well as more distant views of the Cascade Range. Just north of downtown, look west to Queen Anne, a Seattle neighborhood that occupies a large drumlin. At an elevation of 450 feet (137 m), it's one of the city's highest spots. Lake Union, which occupies the low area east of Queen Anne, fills an area eroded into glacial till. During the peak of the most recent glacial advance, about 16,500 years ago, the ice in the Seattle area reached 3,000 feet (900 m) in thickness, about five times the height of the Space Needle. Numerous other lakes, including Green and Haller Lakes to the north, also occupy low areas in the till. As I-5 continues toward Everett, it passes between and over numerous other drumlins, streamlined by the ice in a north-south direction parallel to the glacial flow.

Near milepost 193, the landscape opens up with views eastward over the Snohomish River to the Cascade Range. The Snohomish River, which drains the Cascade Range, occupies a broad floodplain cut through the till and empties into Possession Sound, where it forms a delta. The highway crosses the main river channel near milepost 195, where the floodplain has become a delta complex with a variety of distributary channels, waterways that flow away from the main channel of water. I-5 crosses the delta and its various channels for the next 4 miles (6.4 km). Joanne Bourgeois, of the University of Washington, and her colleague Samuel Johnson, of the US Geological Survey, found evidence in the delta sediments for at least three earthquake events. The oldest of these earthquakes likely occurred on the Seattle fault zone between AD 900 and AD 930 and was accompanied by a tsunami. The huge waves deposited a thin layer of sand directly on top of a vegetation-rich marsh deposit. At Marysville (milepost 199) the interstate climbs off the delta and back into glacial deposits.

The Stillaguamish River flows beneath the highway just shy of 2.5 miles (4 km) north of the Smokey Point Rest Area. This river forms a few miles to the east at Arlington by the confluence of its north and south forks; downstream it divides into distributaries and forms a small delta where it enters Puget Sound. At milepost 210, I-5 crosses a small tributary called Pilchuck Creek (mistakenly identified as the North Fork by the highway sign) and ascends back into glacial till. At milepost 211 northbound travelers can look west to see an exposure of the till on a steep hillside just west of the highway. To visit the exposure, take exit 210 and follow the frontage road north on the west side of the highway for just under 1 mile (1.6 km). The exposure consists mostly of pebbles and cobbles with a few small boulders suspended in unstratified, fine-grained sediment.

Just north of milepost 219, the interstate drops onto the wide delta of the Skagit River. Formed from sediment deposited by the river where it empties

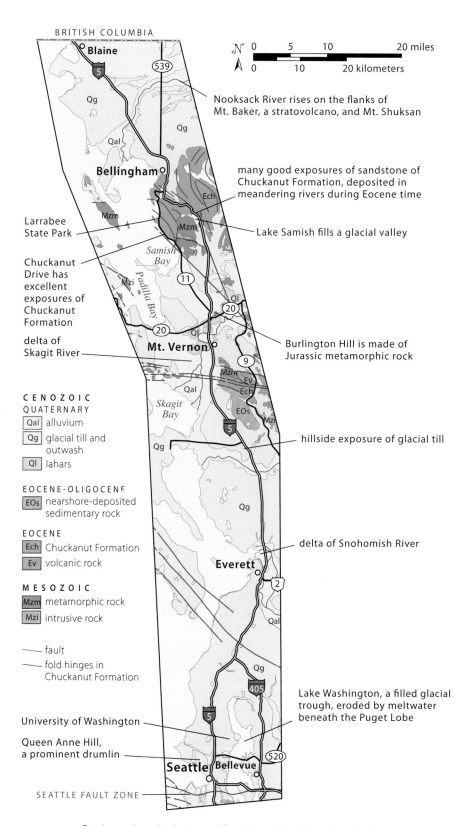

BRITISH COLUMBIA

Blaine

5   539

Qg

Qg

Qal

Bellingham

Ech

many good exposures of sandstone of
Chuckanut Formation, deposited in
meandering rivers during Eocene time

Nooksack River rises on the flanks of
Mt. Baker, a stratovolcano, and Mt. Shuksan

Larrabee
State Park

Mzm

Mzm

Lake Samish fills a glacial valley

Chuckanut
Drive has
excellent
exposures of
Chuckanut
Formation

Samish
Bay

Mzi

Padilla Bay

11

Ql

20

delta of
Skagit River

20   Ql

Mt. Vernon

9

Burlington Hill is made of
Jurassic metamorphic rock

Mzm   Ev
Ech

C E N O Z O I C
QUATERNARY
Qal alluvium
Qg glacial till and
      outwash
Ql lahars

Qal

Skagit
Bay

EOs

Mzi

5

hillside exposure of glacial till

Qg

EOCENE-OLIGOCENE
EOs nearshore-deposited
       sedimentary rock

Qg

EOCENE
Ech Chuckanut Formation
Ev volcanic rock

delta of Snohomish River

MESOZOIC
Mzm metamorphic rock
Mzi intrusive rock

Everett

2

—— fault
—— fold hinges in
       Chuckanut Formation

Qal

University of Washington

Queen Anne Hill,
a prominent drumlin

Qg

405

5

Lake Washington, a filled glacial
trough, eroded by meltwater
beneath the Puget Lobe

Seattle   Bellevue

520

SEATTLE FAULT ZONE

Geology along I-5 between Seattle and the Canadian border.

Glacial till exposed in the hillside across from milepost 211.

into Puget Sound, this delta provides some of the richest farmland in the state. In map view, the delta forms a triangle, with its apex pointed up the Skagit River floodplain and its outer side forming an irregular arc, punctuated by bedrock islands and intervening bays. Skagit Bay in the south, Padilla Bay in the middle, and Samish Bay in the north offer extensive mudflats and valuable wetlands habitat. To give a sense of scale, the highway traverses the eastern edge of the delta plain in a north-south direction for nearly 15 miles (24 km).

The Skagit River, which I-5 crosses just north of Mt. Vernon, rises in the Cascades of southern British Columbia and provides more than 30 percent of the freshwater that enters Puget Sound. The river and its delta lost some 70 to 80 percent of its estuarine habitat since the 1850s because of extensive land reclamation required by agriculture. Beginning in 2007, restoration began here and on other deltas in Puget Sound.

Burlington Hill, directly east of exit 231 (Chuckanut Drive), rises abruptly about 400 feet (120 m) above the delta. It consists of highly faulted, Jurassic-age rock that includes pillow basalt. Just over 1 mile (1.6 km) to the northeast lies Sterling Hill, made of similar bedrock. Some researchers argue that these rocks are part of a mélange that marks the boundary between two crustal blocks of the North Cascades (see page 152 of the North Cascades). North of the Samish River, I-5 climbs through hills of glacial deposits that sit on top of a completely different rock, the Jurassic-age Darrington Phyllite. This low-temperature metamorphic rock formed from shale. A good exposure of the phyllite appears on the east side of the interstate about a quarter mile (400 m) north of milepost 240.

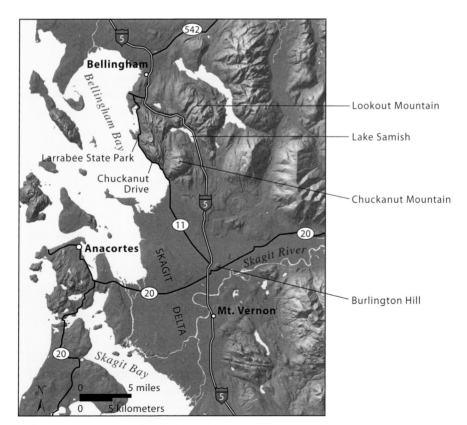

Shaded relief map of Skagit River delta and surrounding area. –Shaded relief courtesy of Washington State Department of Natural Resources, Division of Geology and Earth Resources

About a half mile (0.8 km) north of milepost 244, the interstate passes its first exposure of Eocene-age sandstone of the Chuckanut Formation, and shortly thereafter it begins to pass numerous good exposures of the river-deposited sandstone all the way to Bellingham. The rocks mostly dip steeply away from the highway, so the bedding is somewhat difficult to see. Many more outstanding and accessible exposures of the Chuckanut Formation line Chuckanut Drive and are present in Larrabee State Park along the coast.

Lake Samish fills a glacial valley between Chuckanut Mountain on the west and Lookout Mountain on the east, both of which are made of Chuckanut Formation. These mountains are part of a series of northwest-trending folds that created long northwest-trending ridges of the more-resistant rock and intervening valleys of the less-resistant rock. Folded Mesozoic metamorphic rocks lie immediately below the Chuckanut Formation and occupy the low-relief area on the west side of Lake Samish.

Between Bellingham and the Canadian border, I-5 angles northwestward across a low-relief surface covered by glacial till. Just north of milepost 263, it crosses the Nooksack River, which rises near Mt. Baker and Mt. Shuksan of the North Cascades. On clear days, you gain a spectacular view of Mt. Baker, 30 miles (48 km) away.

Chuckanut Mountain and Larrabee State Park

Chuckanut Drive hugs the east sides of Samish, Chuckanut, and Bellingham Bays along the edge of Chuckanut Mountain, connecting the lowlands of the Skagit River delta with Bellingham. The rocks along this road consist mostly of cliffs of the Chuckanut Formation, deposited in a meandering river during the Eocene Epoch, between about 55 to 35 million years ago, long after the accretion of the Cascades terranes. Altogether, the formation reaches more than 26,000 feet (8,000 m) in thickness, which makes it one of the thickest accumulations of nonmarine sedimentary rock in North America. A good look at the rock shows that it consists mostly of feldspar-rich sandstone and siltstone, and some shale, conglomerate, and coal. The Chuckanut Formation is well known for its troves of plant fossils, including large palms, many of which are on display at Western Washington University. Along Chuckanut Drive, the rocks dip steeply because they lie on limbs of two large folds that define the shape of Chuckanut Mountain.

Beneath the Chuckanut Formation you can see exposures of the Darrington Phyllite, deposited as shale in a deep ocean during the Jurassic Period. It was later metamorphosed to phyllite, and both it and the Shuksan Greenschist were accreted to North America as the Easton terrane.

To see these exposures, take exit 231 from I-5 and head northwest on WA 11 (Chuckanut Drive). The first 9 miles (14 km) pass over the low delta plains of the

Cliffs of Chuckanut Formation along Chuckanut Drive.

Characteristic honeycomb weathering of Chuckanut sandstone along the shore at Larrabee State Park.

Skagit River. Just north of milepost 9, where you cross the submerged mouth of Colony Creek, the road begins to hug the side of Chuckanut Mountain. Bedrock for the next 1.5 miles (2.4 km) to Oyster Creek consists of Darrington Phyllite. A thin slice of gabbro of the Shuksan Greenschist has been mapped in the drainage of Oyster Creek, but these rocks are poorly exposed. You cross into Chuckanut Formation just north of Oyster Creek and follow it all the way to Bellingham.

Few pull-outs allow a safe inspection of the rocks along the road, but Larrabee State Park, between mileposts 14 and 15, provides easy access to numerous coastal exposures in a beautiful setting. You can also see nice examples of honeycomb weathering and granitic glacial erratics. The honeycomb weathering forms in permeable sandstone next to the bay as a result of the growth of salt crystals during the rock's countless wetting and drying cycles. As the salt crystals grow, they crack the rock to allow the infiltration of even more water and the growth of more crystals. Eventually a small pocket forms in the rock that can accelerate the process even further.

INTERSTATE 5
SEATTLE—OLYMPIA
61 miles (98 km)

I-5 follows a nearly continuous urban corridor from Washington's largest city to its capital, skirting the eastern and southeastern edges of Puget Sound. The highway crosses glacial till and outwash shaped into drumlins and troughs beneath ice of the Puget Lobe. Most recently, the lobe advanced over the area and reached a climax about 16,500 years ago. Near Seattle, this ice was about 3,000 feet (900 m) thick, and near southeast Tacoma it reached about 1,800 feet (550 m) thick. Some of the drumlins rise more than 400 feet (120 m) above the surrounding area. Between the drumlins lie deep glacial troughs, the shallower ones now occupied by rivers and the deeper ones occupied by the lakes and inlets of Puget Sound.

Studies of the drumlins and glacial troughs indicate that they formed beneath the glacial ice at approximately the same time, during glacial retreat when the ice front was melting back. The drumlins consist of glacial till deposited directly by the glacier. In some places, the till rests on bedrock, but it mostly lies on a thick blanket of glacial outwash, sand and gravel deposited by rivers flowing off the advancing front of the glacier. The troughs, which can reach depths of more than 800 feet (240 m) below sea level, are also younger than the outwash, because they cut into the outwash. The troughs were eroded by meltwater rivers flowing beneath the glacier rather than by scouring and deepening of already-formed valleys by the glacial ice.

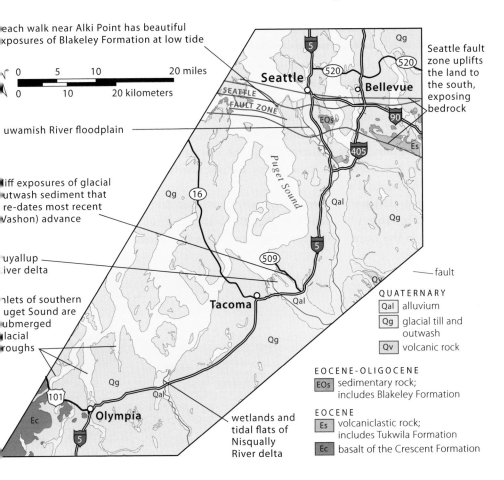

each walk near Alki Point has beautiful xposures of Blakeley Formation at low tide

Seattle fault zone uplifts the land to the south, exposing bedrock

uwamish River floodplain

liff exposures of glacial utwash sediment that re-dates most recent Vashon) advance

uyallup iver delta

nlets of southern uget Sound are ubmerged lacial roughs

0 5 10 20 miles
0 10 20 kilometers

Seattle

SEATTLE FAULT ZONE

Bellevue

EOs

Puget Sound

Qg (16)

Qal

Qg

Tacoma

Qal

Qg

Qg

Olympia

101

Ec

5

Qal

wetlands and tidal flats of Nisqually River delta

— fault

QUATERNARY
Qal alluvium
Qg glacial till and outwash
Qv volcanic rock

EOCENE-OLIGOCENE
EOs sedimentary rock; includes Blakeley Formation

EOCENE
Es volcaniclastic rock; includes Tukwila Formation
Ec basalt of the Crescent Formation

Geology along I-5 between Seattle and Olympia

At the I-90 interchange immediately south of downtown, I-5 crosses the northern edge of the Seattle fault zone, a south-dipping reverse fault that strikes approximately east-west across the Puget Lowland. The zone is about 6 miles wide (10 km), so I-5 doesn't cross the southern edge until between mileposts 158 and 159. Don't expect to see direct evidence of the fault zone because most of what we know comes from field evidence away from the highway and from seismic data. Still, the fault's important because we know one strand of it slipped about 1,100 years ago. The fault separates the Seattle Basin to the north from the Seattle uplift to the south. Nearly all of the Seattle area's bedrock exposures lie within the fault zone, such as near milepost 160 and at Alki Point in West Seattle.

Just south of downtown, I-5 offers a great view westward to the prominent drumlin that defines West Seattle. It reaches an elevation of 520 feet (158 m), the highest spot in Seattle. Its north-south streamlined shape characterizes the

Alki Point and Me-Kwa-Mooks Park

At low tides, you can see bedrock and peat beds along the shoreline of Puget Sound in West Seattle, accessed via Beach Drive. Just south of Alki Point, steeply dipping sandstone and shale of the Blakeley Formation are exposed at the water's edge. These alternating beds of graded sandstone and shale were deposited by turbidity currents on deep parts of a submarine fan complex in Oligocene time. More Blakeley Formation makes up the knob behind the lighthouse at Alki Point, but those rocks are not nearly as well exposed as those along the beach walk.

Immediately behind the beach walk, houses sit on top of a marine terrace, uplifted by the earthquake on the Seattle fault zone 1,100 years ago. If you poke around in the sandy soil on the terrace you can find occasional clamshells and barnacles.

Blakeley Formation exposed beneath beach gravel at Alki Point. The road and public walkway on the right side of the photo follow the uplifted marine terrace.

One mile (1.6 km) south of Alki Point, along the shoreline, Me-Kwa-Mooks Park offers a view of folded peat and silt beds at low tide. These beds were deposited between 28,000 and 23,000 years ago, prior to the last glacial advance in this area. The peat bogs developed in a floodplain during an interglacial time. The fold, a doubly plunging syncline, is attributed by most researchers to movement on the Seattle fault zone.

drumlins throughout the region because they formed parallel to ice flow. The steep hillside on the east side of the highway marks the edge of another drumlin. Notice the occasional glimpses of bedrock near the base of this drumlin near milepost 160. The bedrock is sandstone of the Blakeley Formation, deposited during Oligocene time in a submarine fan complex and uplifted in the Seattle fault zone.

Between these two drumlins is the floodplain of the Duwamish River, occupied by Boeing Field, an airport used primarily for cargo planes. Similar to many of the outlets along the south edge of Puget Sound, the Duwamish lowland formed beneath the ice as a glacial trough but has since filled with enough sediment to lie just above sea level.

About halfway between mileposts 156 and 155, I-5 crosses the Duwamish River. From there it climbs to a saddle between two more drumlins and then drops back to the interchange with I-405. South of I-405, I-5 climbs the side of another large drumlin with views eastward across the Green and Duwamish Rivers to the Cascades. After leveling off, the highway passes over a fairly uniform landscape of till and outwash until it drops onto the floodplain of the Puyallup River, another filled glacial trough. A short detour at exit 137 (Fife) leads to Marine View Drive, which follows the northeast side of Commencement Bay to cliff exposures of glacial outwash deposits. These sediments, which lie below those from the most recent advance, were deposited by an earlier ice advance in the Puget Sound area.

You cross the Puyallup River about a quarter mile (400 m) north of milepost 134, less than 1 mile (1.6 km) south of where the river empties into Puget Sound at Commencement Bay. The river originates as two forks at the toes of the Puyallup and Tahoma Glaciers on Mt. Rainier and drains nearly 950 square miles (2,460 km^2). It's heavily laden with sediment, as can be seen by the distinct plume at its mouth in the bay. Prior to industrialization, Commencement Bay was a thriving coastal area with tidal flats and wetlands that provided abundant shellfish and salmon. Coastal Salish people called it *Shubahlup,* for "sheltered place," because of the natural protection afforded by the land on either side. After the 1880s, however, when the Northern Pacific Railroad sited the area for the western end of the first transcontinental railroad in the northern United States, accelerated development led to large-scale contamination by organics and heavy metals. In 1983, the area was listed as a Superfund site, and the cleanup continues today.

Between the Puyallup River and Olympia, I-5 crosses glacial outwash and till from the most recent glacial advance, interrupted only by the Nisqually River, just south of milepost 115. The mouth of the Nisqually marks another filled glacial trough, like those of the Puyallup and Duwamish Rivers farther north. Like the Puyallup River, the Nisqually rises on Mt. Rainier, although its drainage basin is much smaller, at just over 500 square miles (1,300 km^2). As you cross the river, look toward Puget Sound to see that its mouth is not developed. It is protected by Nisqually National Wildlife Refuge. To gain access to the refuge and its hiking trails, take exit 116 and follow Mounts Road westward to find an expanse of healthy wetlands and tidal flats along the delta of the Nisqually River.

Aerial view of Commencement Bay showing the sediment plume at the mouth of the Puyallup River.

INTERSTATE 5
OLYMPIA—VANCOUVER
106 miles (171 km)

Olympia sits atop glacial deposits of the Puget Lobe at the very south edge of Budd Inlet, which feeds northward into Puget Sound. Like the other inlets at the south end of Puget Sound, Budd Inlet is narrow and mildly sinuous, and it most likely formed by water erosion at the base of the ice sheet. Numerous kettle lakes, formed in depressions left by melting ice blocks, dot the landscape around Olympia and Tumwater but are generally not easily seen from the interstate.

The glacial deposits rest on basalt of the Crescent Formation, the basement rock of the Washington and Oregon Coast Range. This basalt erupted in seawater between about 56 and 49 million years ago, probably as a series of seamounts or an oceanic plateau, and was accreted to North America from about 51 to 49 million years ago as part of the Siletz terrane. You can visit some outstanding exposures of the basalt at Tumwater Falls Park by taking exit 103. The park offers a half-mile (0.8 km) loop trail along both sides of the rocky

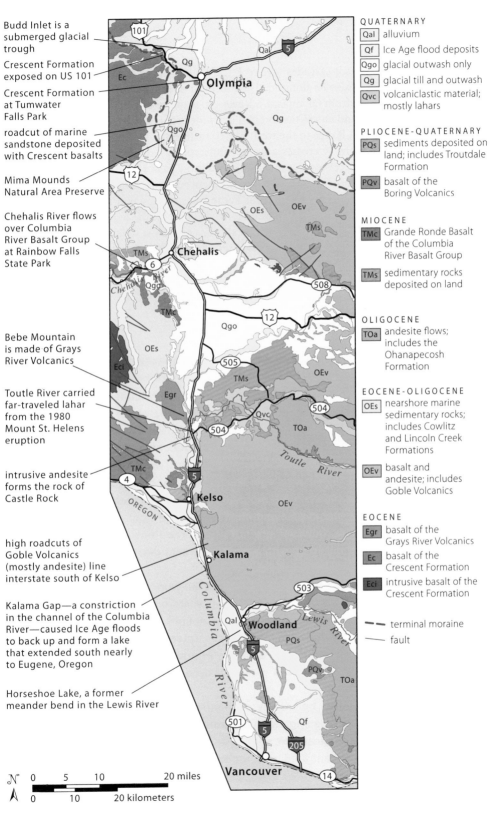

Budd Inlet is a submerged glacial trough

Crescent Formation exposed on US 101

Crescent Formation at Tumwater Falls Park

roadcut of marine sandstone deposited with Crescent basalts

Mima Mounds Natural Area Preserve

Chehalis River flows over Columbia River Basalt Group at Rainbow Falls State Park

Bebe Mountain is made of Grays River Volcanics

Toutle River carried far-traveled lahar from the 1980 Mount St. Helens eruption

intrusive andesite forms the rock of Castle Rock

high roadcuts of Goble Volcanics (mostly andesite) line interstate south of Kelso

Kalama Gap—a constriction in the channel of the Columbia River—caused Ice Age floods to back up and form a lake that extended south nearly to Eugene, Oregon

Horseshoe Lake, a former meander bend in the Lewis River

QUATERNARY
Qal alluvium
Qf Ice Age flood deposits
Qgo glacial outwash only
Qg glacial till and outwash
Qvc volcaniclastic material; mostly lahars

PLIOCENE-QUATERNARY
PQs sediments deposited on land; includes Troutdale Formation
PQv basalt of the Boring Volcanics

MIOCENE
TMc Grande Ronde Basalt of the Columbia River Basalt Group
TMs sedimentary rocks deposited on land

OLIGOCENE
TOa andesite flows; includes the Ohanapecosh Formation

EOCENE-OLIGOCENE
OEs nearshore marine sedimentary rocks; includes Cowlitz and Lincoln Creek Formations
OEv basalt and andesite; includes Goble Volcanics

EOCENE
Egr basalt of the Grays River Volcanics
Ec basalt of the Crescent Formation
Eci intrusive basalt of the Crescent Formation

— — terminal moraine
— fault

0 5 10 20 miles
0 10 20 kilometers

Geology along I-5 between Olympia and Vancouver.

Deschutes River as it cascades toward Puget Sound. Some nondescript exposures of the basalt also lie alongside I-5 just south of Tumwater.

Heading south from Olympia, I-5 leaves the major urban areas of the Puget
Lowland behind and enters the forested corridor between the Cascades to the
east and the Coast Range to the west. For 10 to 15 miles (16–24 km) south of
Tumwater, the highway crosses a gentle landscape covered by glacial till and
outwash. The till becomes dominant as the highway passes through part of the
gigantic terminal moraine of the Puget Lobe between mileposts 97 and 95. The
moraine is noticeably hilly. Numerous kettle lakes occupy depressions in the
moraine, and although they are not directly visible from the highway they are
easily reached by side roads. One good exposure of marine sedimentary rocks,
deposited along with the Crescent basalts, pokes through the glacial deposits
on both sides of the highway about a quarter mile (400 m) north of milepost
96. WA 121, at exit 95, leads a few miles through the moraine to Littlerock and
Mima Prairie, the site of the unusual Mima mounds.

Mima Mounds Natural Area Preserve

Only about fifteen minutes from I-5, the Mima Mounds Natural Area Preserve
presents a somewhat bizarre addition to the landscape of the Puget Lowland.
You'll find the grass-covered Mima Prairie, dotted with hundreds of small,
round hills that range from just 3 feet to about 7 feet high (1–2 m). The prairie
and mounds developed on glacial outwash, which was deposited by meltwater streams about 15,000 years ago. You can find cobbles of a variety of rock
types in the outwash between the mounds and along the prairie edges. Many
of the cobbles are metamorphic and must have come from some distance to
the north. The mounds, however, are not undisturbed outwash. Instead, they
consist of dark sand mixed up with about 50 percent gravel without the sedimentary bedding characteristic of outwash.

Similar-looking mounds appear throughout the world, from near San Diego
to Brazil, and in climates similar to that of western Washington to arid to
subtropical. They occur in the Channeled Scabland of eastern Washington,
where they sit on top of basalt in many places. Internally, other mounds show
significant differences to the ones at Mima Prairie because they rest on and
are made of different materials. A variety of hypotheses have been advanced
to explain them.

The main hypotheses, beautifully summarized in Dave Tucker's *Geology
Underfoot in Western Washington*, range from the purely geological to purely
biological. On the geological side, some researchers argue that existing gravel
and sand can be concentrated into mounds by the repeated wetting and
drying of expandable clays or by seismic shaking. Others have suggested that
the material could accumulate in mounds if deposited by water on top of an

View of the Mima mounds from the overlook.

irregular surface of ice or on the lee side of vegetation that has long since degraded. On the biological side, researchers argue that pocket gophers built the mounds or that they developed as material built up in the spreading roots of vegetation.

An overlook, a 2-mile (3.2 km) trail, and a half-mile (0.8 km) ADA-accessible trail give plenty of opportunity to consider the mounds' origins. You can also inspect an exposed side of a mound at the south edge of the preserve. A nearby shooting range, however, occasionally eliminates any sense of tranquility.

A great thing about this drive is that clear days offer wonderful glimpses and occasional full-on views eastward to four Cascade stratovolcanoes. From north to south, these mountains include Mt. Rainier, Mount St. Helens, Mt. Adams, and Mt. Hood. Notice the smooth summit cone of Rainier, as compared to its comparatively ragged sides. The summit cone formed during a series of eruptions about 2,000 years ago, after a much larger eruption about 6,000 years ago blasted away its top. Mount St. Helens displays a truncated top, its upper 2,000 feet (610 m) having been blown off during its eruption in 1980. Mt. Adams, some 33 miles (53 km) east of Mount St. Helens, looks pretty pristine; its last eruption, about 1,000 years ago, didn't affect the overall appearance of the volcano. Oregon's Mt. Hood, visible as one crosses the Columbia River into Oregon, displays a steeper, less rounded profile, a consequence of large-scale landslides on parts of the upper reaches of the volcano.

Between exit 88 for US 12 West and Chehalis, I-5 follows the Chehalis River upstream for about 12 miles (19 km). The path the Chehalis River takes to the ocean resembles an irregular horseshoe, originating in the Coast Range and flowing northeastward to the edge of the Cascades, then making a broad turn to the north and northwest to flow back through the Coast Range to the Pacific at Aberdeen. Exit 77 for WA 6 gives access to Rainbow Falls State Park, just over 16 miles (26 km) away by road, where the Chehalis River flows over exposures of Grande Ronde Basalt of Miocene age. Just south of milepost 70, I-5 crosses a divide and enters the watershed of the south-flowing Cowlitz River, which flows into the Columbia River at Longview and Kelso.

The Grays River Volcanics, a collection of basalt flows and minor volcanic-rich sedimentary rocks, erupted between 42 and 37 million years ago during late Eocene time. They cover much of the southern Willapa Hills and are considered to be similar in age, composition, and origin to the Tillamook Volcanics in northern Oregon. Along I-5, they form the slopes and crest of Bebe Mountain, immediately west of the Cowlitz River from about milepost 55 to the mouth of the Toutle River, a quarter mile (400 m) south of milepost 52. A good exposure lies along the west side of the interstate, about a quarter mile (400 m) north of milepost 53.

The Toutle River, which flows westward from Mount St. Helens, joins the Cowlitz about a quarter mile (400 m) south of milepost 52. The Toutle River channeled a lahar from the 1980 eruption that, on entering the Cowlitz River, flowed 2.5 miles (4 km) upriver as well as downriver, to where it eventually reached the Columbia River. According to John Cummans of the US Geological Survey, the stream gauge in the nearby town of Castle Rock rose from a height of about 10 feet (3 m) the day before the eruption to 29.26 feet (8.92 m) as the lahar passed through.

Castle Rock occupies a meander bend in the Cowlitz River near milepost 48. It is named for a prominent outcrop of intrusive andesite at the river's edge at the south end of town. The rock is presently undated but is generally considered Oligocene or early Miocene in age. To visit the rock, take exit 48 and drive about 1 mile (1.6 km) north to the edge of town. A similar but much smaller intrusion holds up a narrow ridge about 3 miles (4.8 km) south of town by exit 46; it is easily accessed from the north-directed entrance ramp. Part of the dike is beautifully exposed there, although the rock it intrudes is not.

Just south of milepost 42 at Rocky Point, a great exposure of the Grays River Volcanics frames the west side of the interstate. These rocks are interbedded in places with the Cowlitz Formation that is best exposed in the Willapa Hills. The Cowlitz contains a rich assemblage of fossils, most notably gastropods and bivalves, which enabled Elizabeth Nesbitt of the University of Washington to reliably determine that the formation was deposited in a delta that became deeper westward and northward.

From Kelso, near milepost 36, almost to the mouth of the Lewis River near milepost 21, the highway follows a narrow strip of land between the Columbia River and cliffs of Goble Volcanics. This series of dark-colored andesite flows erupted between about 38 and 32 million years ago. The Goble Volcanics seem

Exposure of basaltic rock of the Grays River Volcanics at Rocky Point.

to be related to the early Cascade volcanic arc and cover much of the landscape between I-5 and Mount St. Helens. Some of the most accessible and interesting exposures of the Goble Volcanics along I-5 show up north and south of Kalama. An unusually dramatic exposure, covered by netting to reduce the rockfall hazard, lies along the east side of the highway near milepost 34. A more accessible exposure for southbound travelers sits next to the off-ramp at exit 32. You can see numerous individual flows, some of which contain the green mineral celadonite filling void spaces. Northbound, you can inspect south-dipping lava flows and interbedded volcanic-rich sedimentary rocks along the off-ramp at exit 30.

If you look westward across the Columbia River, you see part of Oregon's northern Coast Range. The town of Goble, Oregon, for which the Goble Volcanics were named, lies just over 1 mile (1.6 km) away. Kalama marks a noticeable constriction in the Columbia River channel called Kalama Gap. Between about 18,000 and 15,000 years ago, the Ice Age floods raced down the Columbia River's channel; many of these floods originated in Glacial Lake Missoula in Montana. The floods were so enormous that constrictions in the channel functioned as hydraulic dams, which caused them to back up and create temporary lakes. Kalama Gap was one such hydraulic dam and created a lake that filled the Willamette Valley in Oregon almost all the way to Eugene. Glacial melting at the end of the Pleistocene Epoch caused sea level to rise and the river to deposit massive amounts of sediment, the origin of the many islands in today's river channel.

The stretch of highway between Kalama and Woodland seems particularly prone to landslides, as can be seen by several scoop-shaped scars in and around some of the cliffs. Near milepost 24, the road passes the scar, now mostly overgrown, where a slide in 1996 crossed all lanes of traffic on the interstate. A smaller but fresher scar near milepost 21 marks where a landslide closed northbound lanes in December 2015.

Woodland lies on the Lewis River, near milepost 21. Horseshoe Lake, immediately south of the town center, was a meander bend in the Lewis River that was cut off from the river during highway construction in 1940. From the interstate, you can get a quick glance at each arm of this lake just south of milepost 21. The hills to the east consist mostly of Oligocene andesite overlain by Miocene-Pliocene Troutdale Formation, deposits of the early Columbia River. A quarry in the hills to the east exposes basalt of the Columbia River Basalt Group and the underlying Troutdale Formation.

The interstate crosses the east fork of the Lewis River just north of milepost 18 and then passes over a low-relief surface developed on deposits of the Ice Age floods. At mile 0, you gain a wonderful view of Mt. Hood directly up the Columbia River.

WA 20
BURLINGTON—PORT TOWNSEND—US 101
61 miles (98 km); includes one ferry

From Burlington, WA 20 heads directly west into the Puget Lowland and onto Fidalgo Island, then it winds its way southward on Whidbey Island. After crossing Admiralty Inlet on a short ferry ride to Port Townsend, you can continue on WA 20 to US 101 and the Olympics.

West of Burlington, WA 20 crosses southwestward over the flat, cultivated delta of the Skagit River. The landscape is blanketed in some places by the Kennedy Lahar, a mudflow initiated about 6,000 years ago on Glacier Peak, more than 60 miles (100 km) away. Straight ahead to the southwest, you can see a low, forest-covered ridge, which is part of a glacial moraine that runs along much of the eastern side of Fidalgo Island. At milepost 55, the road curves so that Mt. Erie, the island's high point of resistant igneous rock, is nearly straight ahead. About halfway between mileposts 52 and 51 you cross the Swinomish Channel onto Fidalgo Island; Padilla Bay and the easternmost San Juan Islands lie to the north, and the glacial moraine lies to the south.

With the exception of Rosario Head, Fidalgo Island consists of Jurassic- and Cretaceous-age rock that originated as a slice of oceanic lithosphere called an ophiolite. The Fidalgo ophiolite forms the main component of the larger Decatur terrane, which was accreted to western North America about 100 million years ago during the Cretaceous Period. The Decatur terrane encompasses the

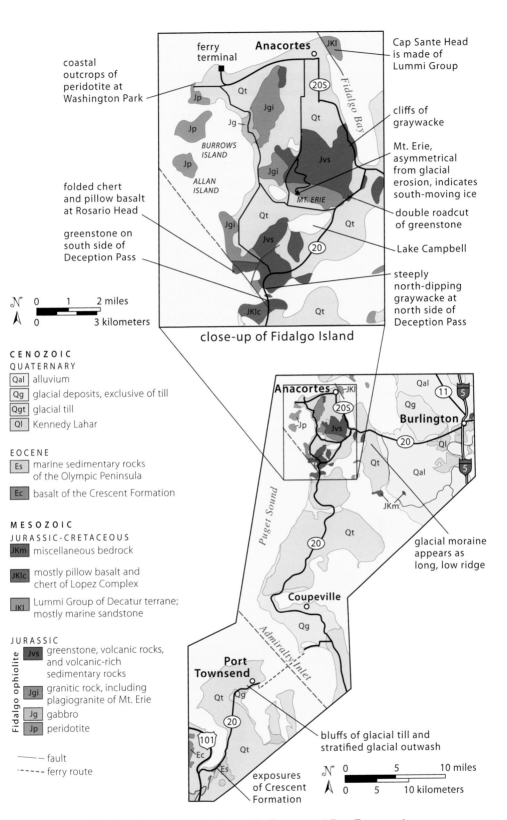

ferry
terminal
Anacortes JKl

Cap Sante Head
is made of
Lummi Group

coastal
outcrops of
peridotite at
Washington Park

Jp Qt
Jgi
Jg
BURROWS
ISLAND
Jp
ALLAN
ISLAND
Jgi

Fidalgo Bay
20S

Qt

cliffs of
graywacke

Jvs

Mt. Erie,
asymmetrical
from glacial
erosion, indicates
south-moving ice

folded chert
and pillow basalt
at Rosario Head

MT. ERIE

double roadcut
of greenstone

greenstone on
south side of
Deception Pass

Jgi
Jvs
Qt

Qt

20

Lake Campbell

steeply
north-dipping
graywacke at
north side of
Deception Pass

JKlc Qt

close-up of Fidalgo Island

N 0   1   2 miles
0   3 kilometers

CENOZOIC
QUATERNARY
Qal alluvium
Qg glacial deposits, exclusive of till
Qgt glacial till
Ql Kennedy Lahar

EOCENE
Es marine sedimentary rocks
of the Olympic Peninsula
Ec basalt of the Crescent Formation

MESOZOIC
JURASSIC-CRETACEOUS
JKm miscellaneous bedrock

JKlc mostly pillow basalt and
chert of Lopez Complex

JKl Lummi Group of Decatur terrane;
mostly marine sandstone

JURASSIC
Jvs greenstone, volcanic rocks,
and volcanic-rich
sedimentary rocks
Jgi granitic rock, including
plagiogranite of Mt. Erie
Jg gabbro
Jp peridotite

Fidalgo ophiolite

—— fault
----- ferry route

Anacortes JKl
20S
Jp
Jvs

Qal
11 5
Qg
Burlington
20
Ql
5

Qt

Qal

JKm

glacial moraine
appears as
long, low ridge

Puget Sound

20

Qt

Coupeville

Qg

Admiralty Inlet

**Port
Townsend**

Jvs

Qt Qg

101

Ec

Qt

20

bluffs of glacial till and
stratified glacial outwash

Es

exposures
of Crescent
Formation

N 0   5   10 miles
0   5   10 kilometers

Geology along WA 20 between Burlington and Port Townsend.

Anacortes and Washington Park

To view Fidalgo Island and its ophiolite, you can go northward on WA 20 Spur to Anacortes and Washington Park. From there, you can take smaller roads southward to Rosario Head and meet up with WA 20 near Deception Pass. North of the WA 20 intersection, near milepost 49S on WA 20S, you pass several outstanding exposures of the graywacke that overlies rocks of the Fidalgo ophiolite. A half mile (0.8 km) north of the milepost, you can pull out at a viewpoint of Fidalgo Bay and Mt. Baker, although trees obscure much of the view during summer months. A slice of the trunk of a 970-year-old Douglas fir displays beautiful tree rings at the overlook. Cap Sante Head, best viewed from Anacortes, lies at the northwest entrance to Fidalgo Bay. It consists of bedrock of the Lummi Group, which was deposited over the top of the graywacke. Between downtown Anacortes and the ferry terminal, you can catch several good views northward to Guemes and Cypress Islands.

Brown-weathering peridotite at Washington Park. View is north-northwestward to Cypress Island on the right, Blakely Island on the left, and Orcas Island in the background behind them.

Just west of the Anacortes ferry terminal, Washington Park hosts beautiful coastal exposures of peridotite, best accessed from Loop Road. Most of the peridotite has been altered to a gray and orangish serpentinite, consisting of serpentine minerals and remnant pyroxene crystals. A basalt dike, which likely fed the overlying lava flows, cuts through the peridotite in a small bay along the park's southern coastline. You can also see several glacial erratics and deposits of till overlying the peridotite along most of the coastline.

To see more of the ophiolite, take Anaco Beach Road south to Marine Drive. You may want to set your odometer to 0 because this stretch of road doesn't have mileposts. You see some beautiful views of Burrows and Allan Islands, which, like Washington Park, are made of peridotite. From mile 1.4 to mile 2.3, you pass exposures of gabbro, including some cliffs between miles 1.9 and 2.1. The safest place to stop is probably at mile 2.3, where a pull-out lies directly across the road from an outcrop.

If you head southwest (right) at the T-junction with Havekost Road, you pass some plagiogranite in just over a quarter mile (400 m). Another half mile (0.8 km) to the south and you can continue southward on Rosario Road, which passes by Rosario Head and rejoins WA 20 near Deception Pass, or you can take Campbell Road toward Mt. Erie.

---

eastern half of the San Juan Islands. The rocks of the Fidalgo ophiolite consist of five main types, each of which is plainly visible along the island's roads. From the bottom of the ocean crust to the top, these rocks consist of peridotite, a rock made of olivine and pyroxene that forms within Earth's mantle; gabbro, an intrusive igneous rock with a composition similar to basalt; plagiogranite, a type of granite that is dominated by plagioclase feldspar; greenstone, which is metamorphosed oceanic basalt; and graywacke and argillite, sedimentary rocks deposited over the top of the ophiolite. Graywacke is a type of sandstone with a high proportion of grains made of rock material as opposed to single minerals, such as grains of quartz sand. Argillite is a slightly metamorphosed mudstone.

WA 20 to Deception Pass

South of Fidalgo Bay on WA 20, you pass over a hilly landscape covered in glacial till, with numerous good bedrock exposures poking through. Within the first quarter mile (400 m), you encounter a small roadcut of dark-gray, slightly metamorphosed graywacke on the east side of the road. Near milepost 47, however, you cross a buried thrust fault, which carried plagiogranite as well as greenstone over the graywacke sandstone. The double roadcut 0.1 mile (160 m) south of milepost 47 consists of the greenstone, but the exposures farther south, to milepost 45, consist of plagiogranite.

Mt. Erie, as viewed to the northwest from Lake Campbell. Note how its northern side (at right) is much less steep than its southern side, an asymmetry caused by glacial erosion.

Near milepost 46 you get a nice view northward across Lake Campbell to the profile of Mt. Erie. The mountain is a classic roche moutonnée, an asymmetrical glacial feature in which ice smooths one side of a bedrock mound and steepens the other. The ice moved up Mt. Erie's gentle northern side and over its top, abrading the rock into a fairly smooth form. As the ice flowed over and down the south side, it plucked fragments of bedrock from the edifice, causing that side to become unusually steep. You can drive to the top of Mt. Erie at a turnoff from the Heart Lake Road, which runs along the west side of Mt. Erie. The road climbs nearly 1,000 feet (300 m) over a distance of 1.7 miles (2.7 km) to awesome views and glacially carved plagiogranite. Look for the glacial striations, or scratches, in the bedrock at the summit parking lot.

South of Lake Campbell, WA 20 passes scattered exposures of greenstone until about milepost 43, where you pass back into graywacke.

Deception Pass is a deep, narrow channel between Whidbey and Fidalgo Islands. Steeply north-tilted graywacke on its north side forms beautiful sea cliffs. Halfway across the passage, however, you cross an island made of more greenstone, which also forms the bedrock across from the parking lot on the Whidbey Island side.

For the 27 miles (43 km) between Deception Pass and the Coupeville ferry terminal, you cross a gently rolling landscape of grass-and-tree-covered glacial deposits. The ferry ride, about 4 miles (6.4 km) across Admiralty Inlet, to Port Townsend, allows views back to bluffs that expose the deposits. Admiralty Inlet reaches a depth of about 300 feet (90 m) about halfway between the two ferry terminals. Bathymetric maps of the area show several deep valleys beneath the water that may mark northwest-trending fault zones.

From the ferry, you can't miss the long, high bluffs behind the ferry terminal at Port Townsend. For the most part, the deposits consist of glacial till, but well-stratified gravel near the west side suggests that at least some of the material was redeposited by water. You get a close look at the deposits as you drive along their base near the ferry terminal.

Rosario Head

From WA 20 at the south end of Fidalgo Island, Rosario Road leads 0.8 mile (1.3 km) to Rosario Head. In contrast to the rest of Fidalgo Island, the area around Rosario Head consists of folded chert and basalt of the Lopez Complex of Jurassic to Cretaceous age. They are faulted against the Decatur terrane as part of the series of thrust faults that shuffled the different terranes of the San Juan Islands during their accretion to North America. Right before the parking lot gates, look for a cliff made of greenstone and adorned with glacial grooves. A low cliff of glacial till frames the south end of Rosario Beach. Walking around Rosario Head, you can see basalt on its south side; well-defined pillows are exposed at low tides. On the north side of the head, you can see outstanding exposures of chert. This rock, made of the silica-rich skeletons of single-celled, floating organisms called radiolaria, is both thinly bedded and beautifully folded. The basalt and chert are faulted against the Decatur terrane greenstone near the parking lot.

Folded ribbon chert at Rosario Head.

For most of the 12.5 miles (20 km) between Port Townsend and US 101, WA 20 crosses unexposed deposits of glacial till. Some good bedrock exposures of basalt of the Crescent Formation frame the east side of the highway about halfway between mileposts 3 and 2. More exposures of the basalt line the beginning of Eagle Mount Road near milepost 2. The basalt formed on the Siletz terrane, which was accreted to North America about 50 million years ago, long after the terranes of the San Juan Islands had been accreted. WA 20 crosses over the boundary fault between Siletzia and the terranes of the San Juan Islands somewhere between Deception Pass and these exposures of Crescent basalt, most likely at Admiralty Inlet.

FERRY GUIDE TO THE SAN JUAN ISLANDS

At least four hundred islands make up the San Juan Islands, which are scattered throughout much of northern Puget Sound between Washington and Vancouver Island. They range in size from Orcas Island, with an area of 57 square miles (148 km^2), to unnamed islands smaller than an acre. The archipelago extends northward into Canada as the Gulf Islands. Ferries from the Anacortes terminal on Fidalgo Island access several of the larger islands, while

Port Townsend rests on bluffs of glacial deposits in the rain shadow of the Olympic Mountains.

residents and visitors generally rely on personal watercraft to access the others. Similar to a road guide, this ferry guide takes you on a route from Anacortes to Friday Harbor on San Juan Island. From there, the route returns to Anacortes via stops at Shaw, Orcas, and Lopez Islands. The order of the stops, which depends on ferry schedule and season, may vary from this guide.

Geologically, the San Juan Islands consist of a wide variety of bedrock that pokes through an extensive cover of glacial deposits, especially along the island shorelines. The bedrock tells a story of terrane formation and accretion to the North American continent. There are five distinct terranes in the San Juan Islands, each of which is bound by faults and has its own geologic history distinct from the others. They consist of Paleozoic and Mesozoic igneous, metamorphic, and sedimentary rock. The San Juan terranes are small, several of which are less than one-hundreth the size of Wrangellia, the much larger terrane to the west that makes up most of Vancouver Island.

Sedimentary rocks of Cretaceous age rest on top of the individual terranes. Based on studies of these younger rocks, most researchers agree that the San Juan terranes were partially subducted, faulted, and added to the edge of North America during the late Cretaceous Period, between 100 to 84 million years ago. This event was likely driven by the collision of Wrangellia with North America. Caught between North America to the east and Wrangellia to the west, the San

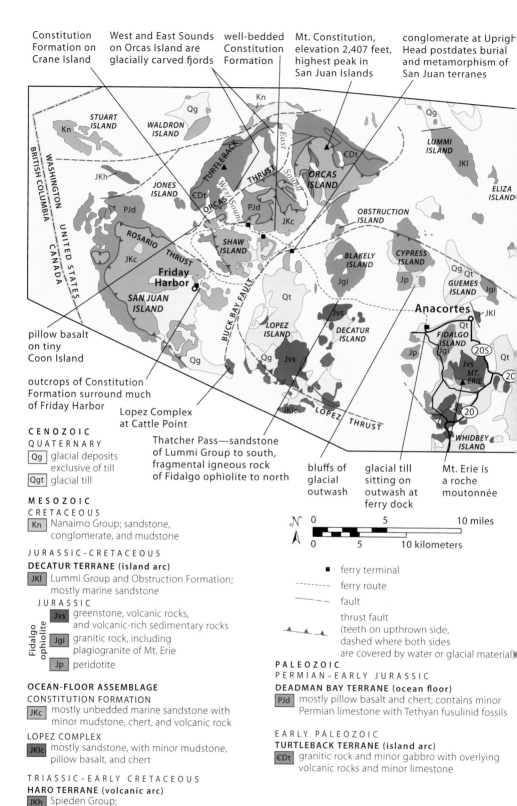

Geology of the San Juan Islands.

Juan terranes piled on top of each other as sheets along a series of thrust faults. As you head westward from Anacortes toward Vancouver Island, you cross into sheets that are progressively lower in the stack and generally made of older rock.

Anacortes to Friday Harbor

While waiting to board the ferry at Anacortes, you can carefully scramble down to the beach to see some glacial deposits directly below the parking lot. The upper part of the deposit consists of till, with scattered boulders and cobbles randomly set in a nonlayered, sandy matrix. Many of the boulders are granitic, and many have weathered out and fallen onto the beach. The lower part of the deposit, however, shows clear bedding that indicates it was deposited by water running off the front of the ice. This outwash was deposited before a glacial advance deposited the till.

From the outside deck of the ferry, you can look back at the glacial deposits. If you look in the opposite direction across the water, you can see cliffs of similar deposits on the southwest edge of Guemes Island.

From the Anacortes terminal, the ferry initially heads straight north toward Cypress Island but turns westward within the first few minutes of the trip. The cliffs at the southern end of Cypress Island consist of well-layered glacial outwash deposits. Most of the island's bedrock is peridotite and dunite of the Decatur terrane, similar to what lines the shores of Washington Park at the northwest tip of Fidalgo Island.

Glacial deposits at Anacortes ferry terminal. Stratified outwash beds lie beneath unstratified till.

As the ferry heads straight west, you can look back at Fidalgo Island and the prominent, glacially carved edifice of Mt. Erie. Looking north up Rosario Strait, you can see Mt. Constitution on Orcas Island, the highest point in the San Juans at an elevation of 2,407 feet (734 m). Mt. Constitution consists of sedimentary rocks of the Constitution Formation, which forms much of San Juan, Shaw, and Orcas Islands. It consists predominantly of marine-deposited sandstone and mudstone, and in some places pillow basalt and chert. Radiolarians, single-celled floating organisms with silica skeletons, are found in the chert and give an age of late Jurassic to early Cretaceous.

The ferry traverses Thatcher Pass between Decatur and Blakely Islands about twenty minutes into the trip. The closeness of the shoreline on either side of the ferry provides probably the best glimpse of the main rock units of the Decatur terrane, which makes up the eastern half of the San Juan Islands. Looking south to Decatur Island, you can see exposures of the late-Jurassic-to-early-Cretaceous Lummi Group, mostly volcanic-rich sandstone deposited on the ocean floor. These rocks also contain younger minerals indicative of metamorphism under the deep but relatively cool conditions found in subduction zones. Throughout the eastern San Juan Islands, the Lummi Group directly overlies the Fidalgo ophiolite, the accreted basement of the Decatur terrane. As you approach the narrow part of Thatcher Pass, the ferry travels near the south edge of Blakely Island. There, you can see cliffs of brecciated igneous rocks of the ophiolite; light-colored exposures of glacial outwash deposits sit on top the ophiolite at the southern tip of the island.

Turning northwestward, the ferry passes either between Willow Island and Blakely Island or between Willow and Frost Islands, depending on the specific route. Like Blakely Island, Willow and Frost Islands consist of the ophiolite. Immediately west looms the much larger Lopez Island. From here you gain some outstanding views northward up the East Sound of Orcas Island to Mt. Constitution and eastward back to Mt. Baker.

About thirty-five to forty minutes into the trip, the ferry rounds the northern tip of Lopez Island, called Upright Head, and begins turning to the southwest. Here, you gain views of two very different late Mesozoic sedimentary rocks. Directly west, Shaw Island consists mostly of Constitution Formation, which, like the Lummi Group of the Decatur terrane, contains minerals that recrystallized in the rock to indicate metamorphism in subduction zones. Directly south, the Lopez Island ferry terminal sits on a younger Cretaceous conglomerate that contains no metamorphic minerals; perhaps it is part of the Nanaimo Group.

Together, these rocks explain why researchers think the terranes were buried, faulted, and uplifted between about 100 and 84 million years ago. Because the Constitution Formation and Lummi Group contain high-pressure/low-temperature metamorphic minerals, they must have been buried in a subduction zone along with the rest of their respective terranes. Moreover, the terranes contain rocks as young as 100 million years, so the burial must have occurred after that point in time. The lack of metamorphic minerals in the younger rocks of Upright Head, however, indicates that they were deposited after the burial and metamorphism. The conglomerate at

Looking north to Blakely Island at Thatcher Pass. Glacial outwash forms the light-colored deposits on the left, whereas igneous bedrock of the Fidalgo ophiolite forms the outcrops on the right.

Cretaceous conglomerate at Lopez Island ferry terminal on Upright Head.

Outcrops of Constitution Formation at Friday Harbor with Turtleback Mountain of Orcas Island forming the skyline on the right. Shaw Island is the low island in front of Turtleback Mountain.

Upright Head is no younger than 84 million years old, but it may be as old as 94 million years. Elsewhere in the San Juan and Gulf Islands, sandstones and conglomerates of the Obstruction Formation also don't contain minerals that grew during metamorphism.

San Juan Island comes into view to the southwest soon after the ferry passes Upright Head. You can see Cattle Point at the southeastern tip of San Juan Island, where the Lopez Complex is thrust over the Constitution Formation. The low cliffs along the shoreline of Lopez Island consist mostly of glacial till.

During the remaining twenty minutes of the trip to Friday Harbor, the ferry passes bedrock of Constitution Formation and some light-colored exposures of glacial deposits on San Juan Island. As the ferry pulls into Friday Harbor, you can see many excellent outcrops of both. University of Washington labs occupy a prominent spot on the north side of the harbor.

Friday Harbor to Anacortes via Shaw, Orcas, and Lopez Islands

The trip back to Anacortes passes deeper into the San Juan thrust sheets on the way to Shaw and Orcas Islands. Exiting Friday Harbor, you see numerous outcrops of the Constitution Formation along the shoreline and good views of Turtleback Mountain on Orcas Island—the views keep getting better as the trip goes on. Less than 20 miles (32 km) to the northwest are some of the Gulf Islands in Canada, made of the sedimentary Nanaimo Group, a Cretaceous submarine fan deposit that formed over the top of the accreted terranes. The Nanaimo Group also shows up on the northern San Juan Islands, including at North Beach on Orcas Island.

About twenty minutes into the trip, the ferry passes through the Wasp Islands, a cluster of about a dozen small islands and rocks that extend off the western peninsula of Shaw Island. They consist mostly of Triassic and Jurassic chert and minor pillow basalt of the Deadman Bay terrane. Look northward for exposed pillow basalt along the south coast of Coon Island. A fault runs between Coon Island and Crane Island, so a different rock—good exposures of Constitution Formation—occur on the south side of Crane Island.

The ferry continues eastward along the north coast of Shaw Island for another fifteen minutes to the Shaw Island ferry terminal. You can look north up West Sound of Orcas Island, an impressive fjord that formed as a glacially carved valley was flooded by rising sea levels at the end of the glacial period. Turtleback Mountain, at the north end of the sound, rises more than 1,500 feet (460 m) above sea level. It consists of early Paleozoic igneous rocks of the Turtleback terrane, the oldest terrane of the San Juan Islands and lowest-known thrust sheet. In contrast, the outcrops along the north shore of Shaw Island all consist of the Constitution Formation. These rocks also frame the east side of the dock at the Shaw Island ferry terminal.

The short ride from Shaw Island to the ferry terminal at Orcas Island stays entirely within the Constitution Formation. A granitic boulder, resting on the sandstone at the east entrance to the harbor, attests to the glaciers that covered the area some 15,000 years ago. As no similar granitic rocks exist in the San Juan Islands, it must have been transported by ice. While at the dock, you can see glacial till along the shoreline to the east and bedrock of the Constitution Formation to the west.

Between Orcas and Lopez Islands, you get some great views of well-bedded Constitution Formation along the south coast of Orcas Island, an unusual sight given the bedding of the rock is typically imperceptible. Directly to the east, Obstruction Island, the namesake for the Obstruction Formation, forms the low island between the east arm of Orcas Island and the north shore of Blakely Island. As you pull into the Lopez ferry terminal, you can look northward up part of the channel of East Sound, an even larger fjord than West Sound. Cliffs frame the west edge of the terminal's bay. With binoculars, you can tell these rocks are conglomerates, probably the Nanaimo Group.

WASHINGTON STATE FERRY
SEATTLE—BREMERTON
1-hour crossing time

Ferry rides provide wonderful unobstructed views of the surrounding landscape, and the short trip between Seattle and Bremerton is easy and instructional. The first part of the trip also can be used for the Seattle-Winslow ferry route.

You initially gain some great views of the Olympic Mountains as you head west past Duwamish Head, which forms the north edge of the large drumlin that makes up West Seattle. The view of the drumlin improves as you head farther out into Puget Sound, although Bainbridge Island soon blocks the view of the Olympics. Alki Point, protruding westward into the sound, hosts a lighthouse and bedrock of the Blakeley Formation.

About twenty minutes into the trip the ferry crosses a deep, narrow channel beneath the water that runs north-south up this part of the sound. It reaches a maximum depth of nearly 800 feet (240 m). The underwater topography is an expression of both subglacial erosion and deposition from when this part of the Puget Lowland lay beneath some 3,000 feet (900 m) of glacial ice.

Restoration Point, which juts eastward into the sound from the southern end of Bainbridge Island, becomes easily visible about twenty-five minutes into the trip. The low-lying forested area with houses is a marine terrace, uplifted about 23 feet (7 m) during a major earthquake on the Seattle fault zone about 1,100 years ago. The fault zone runs east-west, so that uplift on the south side of the fault zone resulted in exposed bedrock at Alki Point and on the south side of Bainbridge Island. The northernmost strands of the fault zone run east-west between Winslow and just north of Alki Point. On the north side of Restoration Point is Blakely Harbor, namesake of the Blakely Harbor Formation of Miocene to Pliocene time. The conglomerate crops out along the harbor, as well as at Blakely Rock, which pokes out of the sound at low tide north of Restoration Point. Try not to confuse the Blakely Harbor Formation with the older Blakeley Formation at Alki Point, even though both are sedimentary rocks deposited in the Cenozoic Era and both are mostly buried beneath the Puget Lowland. Blake Island to the south consists of glacial till, and Mt. Rainier dominates the southeastern horizon on clear days.

The ferry passes between the south end of Bainbridge Island and Point Glover of the Kitsap Peninsula. Look for scattered bedrock exposures on each, although these are mostly obscured by glacial till. These rocks also belong to the Blakeley Formation.

Heading southwestward into Port Orchard Sound for the last fifteen minutes of the ride, you get good views northward up the channel on the west side of Bainbridge Island as well as to the Olympic Mountains. Bremerton, a town of about forty thousand, is built on glacial till. The elongate ridges that define much of the landscape are drumlins, formed by glacial ice. Basalt of the Crescent Formation crops out along WA 3 just southwest of town.

View north to Restoration Point from the ferry deck. The flat surface on which the houses are built is a marine terrace, uplifted 23 feet (7 m) during a major earthquake 1,100 years ago. The snow-capped mountain in the distance is Mt. Baker.

Glaciated peaks in the crystalline core of the North Cascades.
The high mountain in the background is Sinister Peak.

NORTH CASCADES

Chances are that when people hear about the Cascade Range of Washington, they immediately picture the chain of isolated, glacier-clad volcanic cones extending north from the Oregon border to Canada. However, these young volcanoes have been built on older bedrock—the basement of the Cascades—and it is this basement that dominates the North Cascades. South of I-90, most of the exposed bedrock consists of Cenozoic-age volcanic rocks and flows that have emanated from the Quaternary cones of Mt. Adams, Mount St. Helens, and Mt. Rainier. North of I-90, however, volcanic rocks are nearly absent. The volcanic rocks originally covered the bedrock of the North Cascades, too, but erosion stripped them away as the entire range was tilted up to the north. What remains is a diverse assortment of glacially sculpted sedimentary, metamorphic, and plutonic rocks. This bedrock forms the American Alps of North Cascades National Park and the many nearby wilderness areas.

The basement bedrock of the North Cascades is extremely complex with a diverse assortment of rock units: geologic maps resemble the proverbial patchwork quilt. We follow Rowland Tabor and Ralph Haugerud of the US Geological Survey, who provided a useful way forward by dividing the range into three domains juxtaposed by major faults. The Western domain is separated from the Crystalline Core domain to the east by the Straight Creek fault. The core is in turn juxtaposed with the Methow domain by the Ross Lake fault. Within each domain, geologists have grouped rock units into terranes. Each fault-bounded terrane is composed of rock units that share a common history that differs from the histories of other terranes in the domain.

The roadside geologist may feel bewildered by the myriad rock units and terranes and wonder where they all originated. Rest assured, seasoned scholars of the North Cascades ask the same question. Most geologists would probably agree on the following. First, the sedimentary and volcanic rocks in the Western, Crystalline Core, and Methow domains were deposited or erupted somewhere along the western margin of the North American continent. All agree that pieces of oceanic crust and mantle have been mixed with these rocks, but the tectonic settings of the mixing, be they subduction zones or zones of strike-slip faulting, are not yet understood. Second, all agree that the Cretaceous and early Cenozoic plutons represent the roots of magmatic arcs formed by subduction—further evidence that they intruded along the margin of a continent.

But beyond these points of agreement, one outstanding, and perhaps alarming, question remains: Were any of the domains displaced northward along the North American continental margin in late Cretaceous time? There are two schools of thought, the fixists and the mobilists. The fixists hold that the domains in the North Cascades have occupied roughly their present positions relative to continental North America since Cretaceous time, although they agree that the domains were shuffled by tens of miles of right-lateral slip on the major faults. The mobilists respect paleomagnetic data—measurements of the Earth's magnetic field recorded in rocks—indicating that the Mt. Stuart Batholith intruded the Crystalline Core rocks 93 million years ago while they were situated about 2,000 miles (3,000 km) south of their present latitude relative

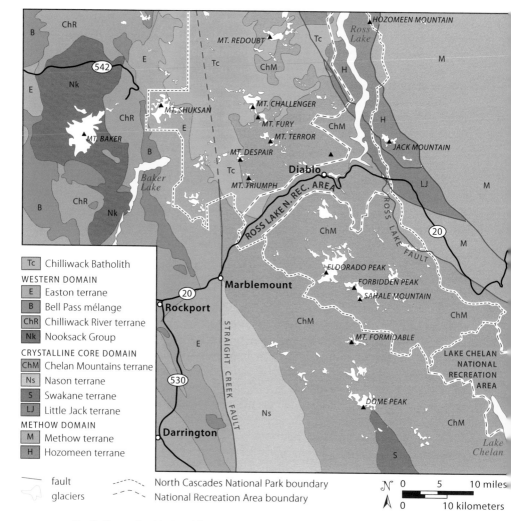

Tc   Chilliwack Batholith

WESTERN DOMAIN
E   Easton terrane
B   Bell Pass mélange
ChR   Chilliwack River terrane
Nk   Nooksack Group

CRYSTALLINE CORE DOMAIN
ChM   Chelan Mountains terrane
Ns   Nason terrane
S   Swakane terrane
LJ   Little Jack terrane

METHOW DOMAIN
M   Methow terrane
H   Hozomeen terrane

— fault
⌒ glaciers
⌐⌐⌐ North Cascades National Park boundary
⌐ ⌐ National Recreation Area boundary

0   5   10 miles
0   10 kilometers

North Cascades National Park spans the three domains of the North Cascades, each consisting of multiple terranes. –Map modified from Tabor and Haugerud, 1999

to North America. This profound controversy involving reconstructing the margin of the continent will continue to motivate research and thought for a long time.

The Western Domain

The Western domain is a folded stack of diverse terranes separated by thrust faults. The domain has two blocks: to the north is the Northwest Cascades thrust system, and to the south are the Eastern and Western mélange belts. The terranes in the San Juan Islands (discussed in the Puget Lowland chapter) may be a western extension of the Western domain. The lowest unit in the North Cascades, the Nooksack Group, consists dominantly of late Jurassic and early Cretaceous turbidites that were deposited on an ocean floor composed of arc-related volcanic rocks of middle Jurassic age. Resting above is the Chilliwack River terrane, sedimentary and volcanic units ranging in age from Devonian to late Triassic. Although metamorphosed to marble, limestone of this terrane has yielded a rich trove of late Paleozoic fossils. The Chilliwack River terrane is older than the Nooksack Group below—a defining characteristic of rock units separated by thrust faults. Some parts of the terrane are upside down, having been folded before they were thrust onto the Nooksack Group.

Above the Chilliwack River terrane is the Bell Pass mélange—a unit that epitomizes the literal meaning of *mélange*, as it is a mixture of a wide variety

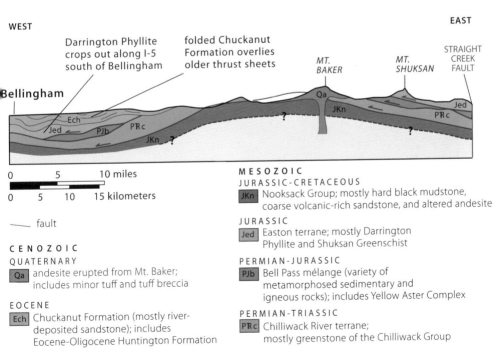

WEST

EAST

Darrington Phyllite crops out along I-5 south of Bellingham

folded Chuckanut Formation overlies older thrust sheets

MT. BAKER

MT. SHUKSAN

STRAIGHT CREEK FAULT

Bellingham

| 0 | 5 | | 10 miles |
| 0 | 5 | 10 | 15 kilometers |

—— fault

C E N O Z O I C
QUATERNARY
Qa andesite erupted from Mt. Baker; includes minor tuff and tuff breccia

EOCENE
Ech Chuckanut Formation (mostly river-deposited sandstone); includes Eocene-Oligocene Huntington Formation

M E S O Z O I C
JURASSIC-CRETACEOUS
JKn Nooksack Group; mostly hard black mudstone, coarse volcanic-rich sandstone, and altered andesite

JURASSIC
Jed Easton terrane; mostly Darrington Phyllite and Shuksan Greenschist

PERMIAN-JURASSIC
PJb Bell Pass mélange (variety of metamorphosed sedimentary and igneous rocks); includes Yellow Aster Complex

PERMIAN-TRIASSIC
PRc Chilliwack River terrane; mostly greenstone of the Chilliwack Group

West-to-east cross section of the Western domain from Bellingham to the Straight Creek fault.

of rock types. Although dominantly composed of sedimentary rocks—sandstone, hard shale called argillite, and minor chert—exposures in the high country are notable for their lack of sedimentary layering. Instead, fragments and blocks of original layers are encased in flaky argillite. Two aspects of the mélange indicate that the blocks and slabs were not part of a single rock unit. First, ages of the rock fragments vary widely. The sedimentary components range from Triassic to early Cretaceous in age. The mélange also contains fragments of the oldest rocks in western Washington: Yellow Aster Complex gneisses, which are metamorphosed Precambrian sediments. Second, uncommon ultramafic rocks, which are pieces of Earth's mantle, have somehow been mixed with rock of the shallow crust. The largest and most prominent of these slices of mantle is the Twin Sisters, easily visible from I-5 as the 9-mile-long (15 km), craggy, red-brown weathering massif on the southwest flank of perpetually snow-and-ice-covered Mt. Baker.

Situated above all these terranes and separated from them by the Shuksan thrust is the Easton terrane, constituting the bulk of Mt. Shuksan, one of the

Deformed, light-colored chert and dark-colored argillite of the Bell Pass mélange in foreground. American Border Peak (on the Canadian border) is made of Chilliwack River terrane and forms the prominent steep-sided peak just left of the photo's center.

most-photographed massifs in the range. The Easton terrane is composed of late Jurassic marine sediment deposited on basaltic oceanic crust, but you can hardly recognize these original rock types. The basalts were metamorphosed in a subduction zone to greenschists and blueschists—the latter especially valued by petrologists because of their colorful blue amphiboles. The marine sediments were metamorphosed into the contorted, dark-gray Darrington Phyllite. Petrologists who have studied the units in the Western domain note how their metamorphism differs from what is recorded in the Crystalline Core domain: the latter was metamorphosed at higher temperatures, deep within the crust, whereas most of the rock units in the Western domain record cooler temperatures indicative of a subduction zone.

A major, northwest-striking fault zone, the Darrington–Devils Mountain fault, separates the Northwest Cascades thrust system from the Western and Eastern mélange belts, exposed in some places along US 2 and I-90. These belts are not to be confused with the Bell Pass mélange of the Western domain. Permian limestone in the mélange belts, and also in the San Juan Islands, contains rare genera of foraminifera called fusulinids that are unlike the Permian fusulinids characteristic of continental North America and even those in the nearby Chilliwack River terrane. They could be evidence that some terranes are indeed exotic to North America and were accreted to the continent during late Mesozoic subduction.

Studies of various parts of the Western domain indicate that the thrusting and metamorphism in the northwest Cascades and the San Juan Islands, to the west, were largely completed by late Cretaceous time, about 100 million years ago. Certainly by about 55 to 50 million years ago crustal extension had superseded the crustal shortening recorded in the thrust system. Up to 26,000 feet (8,000 m) of the Paleocene-Eocene Chuckanut Formation, which unconformably rests on the terranes of the Western domain, was deposited by southwest-flowing streams in an extensional basin. The Chuckanut is renowned for its rich assemblage of early Cenozoic plant fossils that record a climate much warmer than today's.

The Crystalline Core

No matter whether you are driving east or west across the North Cascades, the visual character changes dramatically upon entering the Crystalline Core. The extreme topography and relief—the distance from the highest peaks to the deepest valleys—result directly from the hard nature of the rock. The core consists of metamorphic rocks pierced by granitic plutons. The western boundary of the core is the 155-mile-long (250 km) Straight Creek fault, which extends northward into southern British Columbia. The trace of the fault on geologic maps is indeed straight—a characteristic of strike-slip faults. By matching geologic units in British Columbia with those in the Crystalline Core, geologists have determined that these rocks were offset, or separated, by 56 miles (90 km) of right-lateral slip between about 50 and 35 million years ago. The contrast between the history of the rocks in the Crystalline Core to the east and the Western domain to the west indicates that substantial dip-slip has also occurred on the fault: the east side has evidently been elevated by tens of kilometers relative to the rocks to the west.

The Crystalline Core formed as a magmatic arc above a subduction zone along the western margin of North America in Cretaceous and early Tertiary time. Granitic plutons of the arc intruded four terranes, each of which were assembled before the inception of magmatism about 95 million years ago. They form the bedrock framework on which the arc was constructed.

The Chelan Mountains terrane consists of the Napeequa Schist, metamorphosed ocean-floor basalt and sediments; the Cascade River Schist, metamorphosed arc-related volcanics and sediments; and the Marblemount pluton. The pluton and Cascade River Schist probably formed in a magmatic arc in late Triassic time. Geologists have named large tracts of the terrane the Skagit Gneiss Complex to honor the fact that it largely crops out in the watershed of the Skagit River. The complex consists of two types of gneiss: extensive bodies of deformed granitic rocks, called *orthogneiss*, that intruded the Chelan Mountains terrane, and *banded gneiss*, alternating layers of light granitic rock and dark schist. Much of the banded gneiss qualifies as migmatite, which contains a high proportion of igneous material. Migmatites form under very high temperatures, so some of the igneous material typically comes from partial melting of the rock itself.

Migmatite of the Skagit Gneiss Complex. Note the abundance of igneous (light-colored) material in this otherwise metamorphic rock. Photo is about 10 feet (3 m) across.

Other terranes in the Crystalline Core include the Nason terrane, consisting dominantly of the Chiwaukum Schist derived from early Cretaceous sands and muds. Parts of the Chiwaukum Schist that were more intensely metamorphosed are banded migmatitic gneiss. The Swakane terrane consists of biotite gneiss that formed from the metamorphism of Cretaceous sediments. The Ingalls terrane consists of a Jurassic ophiolite and overlying sedimentary rocks.

Since Peter Misch and his graduate students at the University of Washington began the pioneering studies of the Crystalline Core in the 1950s, the domain has been investigated by a host of geologists, all attracted by its amazing landscapes and unusual history. As you drive through the core, imagine that for most of late Cretaceous time, from about 95 to 60 million years ago, the terranes and invasive plutons you see resided about 25 miles (40 km) below the Earth's surface. They were obviously exhumed and brought to the surface by the removal of the rocks originally above them. The uplift and exhumation of the core was remarkably rapid, occurring between about 50 and 45 million years ago: Rocks in the Skagit Gneiss Complex that were being deformed and metamorphosed deep in the crust about 50 million years ago were also intruded in the upper crust by undeformed plutons, such as the Golden Horn and Cooper Mountain Batholiths, just a couple million years later. Faulting and erosion probably removed the overlying rock.

The eastern boundary of the Crystalline Core domain is the Ross Lake fault zone. The craggy peaks and deep, narrow valleys of the core, with their fir and hemlock, give way to ponderosa and sage in the wide vistas of the Methow Valley. Here also, the metamorphic rocks of the core are replaced by the well-bedded, unmetamorphosed strata of the Methow domain.

The Methow Domain

The oldest rocks of the Methow domain are late Paleozoic and early Mesozoic ocean-floor basalt and chert of the Hozomeen terrane. An ocean of unknown width was present in the domain from late Jurassic to early Cretaceous time. Jurassic volcanic and sedimentary rocks in the Ladner Group and Newby Group are evidence that a volcanic arc developed on or near the Hozomeen ocean floor.

Beginning about 110 million years ago, the ocean basin began to fill with sediments. Detailed research on the source of the sediments revealed an interesting pattern. The oldest strata of the Harts Pass Formation are turbidites consisting of sand derived from granitic plutonic rocks to the *east*—perhaps but not indisputably from the Okanogan Batholith. These strata are overlain by the Virginian Ridge Formation, which contains some clasts, dominantly chert, clearly derived from a terrane—perhaps but not indisputably the Hozomeen terrane—to the *west*. The overlying strata of the Winthrop Formation were again derived from eroded plutonic rocks of the Okanogan Batholith to the *east*. Collectively, the Cretaceous strata of the Methow domain are an instructive example of what geologists term a *linking assemblage*: they are the rock evidence of two disparate terranes bounding an ocean basin.

Sandstone of the Virginian Ridge Formation formed from sand eroded from terranes to the west in Cretaceous time. The cross-laminations in the center of the photo formed by running water.

As the Virginian Ridge and younger sediments filled the ocean basin, it became shallower, and by about 100 million years ago the basin had filled. The youngest sediments were deposited in rivers and overlain by andesitic volcanic rocks. A few 90-million-year-old granitic plutons, which intruded the Methow strata, may have been the magma chambers that fed the volcanic rocks.

Although the Methow strata are unmetamorphosed, roadside observers will see tilted and perhaps even folded beds. In mid-Cretaceous time, east-west shortening, or compression, folded the rocks and formed thrust faults in the sedimentary fill of the Methow Basin. Whether this deformation is related to events in the domains to the west is one of the many controversies concerning the geologic history of the North Cascades. Two prominent, steep fault zones bound the Methow domain: the Pasayten fault on the east and the Ross Lake fault zone on the west. The amounts and the nature of slip on these faults are similarly controversial.

Basins in Eocene Time

The rapid exhumation of the deeply buried Skagit Gneiss Complex of the Crystalline Core domain occurred between about 50 and 45 million years ago. At about the same time, a series of elongate, northwest-trending basins began forming on the southern flank of the Crystalline Core domain. What caused these seemingly unrelated events? Quite literally a commotion in the ocean. Hundreds of miles to the west, along the coast, the accretion of the Siletz terrane occurred about 50 million years ago, as a subduction zone along the coast consumed ocean crust. Oceanic spreading ridges also descended into the subduction zone. This special situation—the consuming of an ocean ridge—can invoke all types of deformation in the crust above, including stretching of the crust and the formation of basins.

The earliest basin, which formed 54 to 48 million years ago, was filled with at least 16,000 feet (5,000 m) of the nonmarine Swauk Formation, consisting of sandstone, conglomerate, and mudstone deposited by southwest- and northeast-flowing streams. Flows of the 51-million-year-old Silver Peak volcanic member are interbedded with Swauk sediments in the western part of the basin. The Swauk Formation was folded, perhaps due to the accretion of Siletzia to the west, and overlain unconformably by the predominantly basaltic Teanaway Formation. The dikes that fed the Teanaway lava flows are a characteristic feature of many outcrops and roadcuts of the Swauk Formation. The Teanaway

Chumstick Formation exposed along US 2 near Wenatchee. The steeply dipping sandstone erodes into ridges made of more resistant beds.

Formation is overlain in turn by the nonmarine sediments of the Roslyn Formation, which is less than 45 million years in age.

From about 49 to 45 million years ago, nonmarine sediments of the Chumstick Formation accumulated in one or more basins within the Chiwaukum graben northwest of Wenatchee. Although some workers hypothesize that the Chumstick Formation is a locally preserved section of a formerly more extensive Eocene unit, we favor the alternative: as much as 40,000 feet (12,000 m) of sediments were deposited by streams and in lakes in fault-bounded basins that subsided due to crustal extension. The Leavenworth and Entiat faults at the west and east margins, respectively, of the Chiwaukum graben were likely active right-lateral faults during deposition.

The Cascade Magmatic Arc

A subduction zone has existed along the west coast of Washington for most of the past 100 million years. A subduction-related magmatic arc generated the late Cretaceous and early Tertiary plutons in the core of the North Cascades. A lapse of about 15 million years occurred in the magmatic arc after the accretion of Siletzia, but the subduction-related magmatic arc reestablished itself in Washington in Oligocene time. The large granitic batholiths—the roots of the arc—are exposed in a roughly north-south line from the Canadian border to I-90. Farthest north, the Chilliwack Batholith is composed of many smaller plutons ranging in age from 35 to 2.5 million years old. The Index and Grotto Batholiths, visible along US 2, are Oligocene in age, and the Snoqualmie Batholith, exposed on I-90, is mostly Oligocene and Miocene but includes smaller, younger bodies. The prominent stratovolcanoes in Washington's Cascade Range are the present-day expression of the magmatic arc.

The North Cascades also preserve some of the lavas and ash flows that erupted on Earth's surface, although unlike the extensively preserved volcanic rocks in the South Cascades, most have been eroded. The present-day Mt. Baker volcano is probably no older than about 40,000 years, but nearby volcanic rocks record earlier arc-related volcanic activity. The 3.7-million-year-old Hannegan caldera and the younger Kulshan caldera, about 1.1 million years old, are classic expressions of shallow magma chambers that explosively vented; the resulting depressions were filled by erupted volcanic materials that are now preserved and exposed.

The ice-clad volcanic cones of Mt. Baker and Glacier Peak remind us that glacial ice has been an important sculptor of the North Cascades that we see today. At its maximum extent from about 20,000 to 15,000 years ago, the Cordilleran Ice Sheet probably covered the core of the North Cascades as far south as Glacier Peak. As the sheet receded, it left extensive deposits of till and outwash. The highest peaks and ranges today preserve small remnants of the originally more extensive but now receding alpine glaciers.

Guides to the North Cascades

SEATTLE—ELLENSBURG
107 miles (172 km)

I-90 between Seattle and Ellensburg is one of the most exciting stretches of interstate anywhere, crossing the fantastically rugged landscape of the Washington Cascades after traversing the east half of the glaciated Puget Lowland. Just east of Seattle, the Lacey V. Murrow Memorial Bridge crosses Lake Washington, a deep trough eroded by meltwater beneath the Puget Lobe. Mercer Island, on the lake's east side, is covered in glacial till that was shaped by the ice into north-south-trending drumlins. The lake's bathymetry shows the island's steep shorelines descending some 50 feet (15 m) or more under water before leveling off into the irregular, till-covered bottom of Lake Washington, which averages just over 100 feet (30 m) deep. The same maps show that East Channel, the short stretch of water between Mercer Island and Lake Washington's eastern shoreline, was a subglacial drainage divide, which separated north-flowing meltwater from south-flowing meltwater. About a half mile (0.8 km) east of the eastern shore, look north to see Mercer Slough and its wetlands. A shallow arm of Lake Washington filled this area before 1917, when the canal to Lake Union was completed and lowered the level of Lake Washington about 9 feet (2.7 m).

The bridge nearly coincides with the northernmost strand of the Seattle fault zone, which crosses the Puget Sound area in an east-west direction and caused a major earthquake, on the order of magnitude 7.0, between about AD 900 and AD 930. The terrace at Alki Point in West Seattle rose 23 feet (7 m) during the quake, and sediments at the mouth of the Snohomish River near Everett show evidence of liquefaction. A landslide occurred during this quake near the southern tip of Mercer Island. Drowned trees that are still vertical adorn a hummocky, underwater landscape.

The Seattle fault zone brings its south side up over its north side as a high-angle reverse fault. The only bedrock in the Seattle area lies within or south of the fault zone because to the north the bedrock has been dropped along the fault and covered by more recent deposits. For example, the steep, forested hill south of the freeway between mileposts 14 and 15 contains Blakeley Formation, but the hills to the north consist entirely of glacial deposits. Lake Sammamish, another filled glacial trough, extends northward between those hills. Similar to Lake Washington, its bottom lies mostly below sea level. Today, the lake spills north into the Sammamish River, which flows into the northern end of Lake Washington. During the Pleistocene, however, this exit was blocked by ice and the lake was much larger.

At Issaquah, you can get a sense of the enormity of Glacial Lake Sammamish. The quarry immediately northeast of the town mines gravel deposited by west-flowing rivers in a delta that emptied into the lake toward the end of glacial

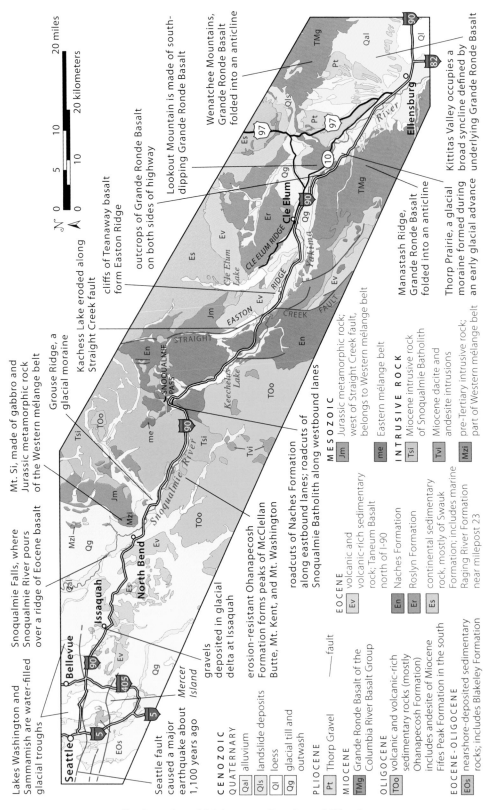

Geology along I-90 between Seattle and Ellensburg.

Lakes Washington and Sammamish are water-filled glacial troughs

Mt. Si, made of gabbro and Jurassic metamorphic rock of the Western mélange belt

Snoqualmie Falls, where Snoqualmie River pours over a ridge of Eocene basalt

Wenatchee Mountains, Grande Ronde Basalt folded into an anticline

Lookout Mountain is made of south-dipping Grande Ronde Basalt

outcrops of Grande Ronde Basalt on both sides of highway

cliffs of Teanaway basalt form Easton Ridge

Kachess Lake eroded along Straight Creek fault

Grouse Ridge, a glacial moraine

Kittitas Valley occupies a broad syncline defined by underlying Grande Ronde Basalt

Manastash Ridge, Grande Ronde Basalt folded into an anticline

Thorp Prairie, a glacial moraine formed during an early glacial advance

Seattle fault caused a major earthquake about 1,100 years ago

gravels deposited in glacial delta at Issaquah

erosion-resistant Ohanapecosh Formation forms peaks of McClellan Butte, Mt. Kent, and Mt. Washington

roadcuts of Naches Formation along eastbound lanes; roadcuts of Snoqualmie Batholith along westbound lanes

20 miles

20 kilometers

N

MESOZOIC

Jm — Jurassic metamorphic rock; west of Straight Creek fault, belongs to Western mélange belt

me — Eastern mélange belt

INTRUSIVE ROCK

Tsi — Miocene intrusive rock of Snoqualmie Batholith

Tvi — Miocene dacite and andesite intrusions

Mzi — pre-Tertiary intrusive rock; part of Western mélange belt

CENOZOIC

QUATERNARY

Qal — alluvium

Qls — landslide deposits

Ql — loess

Qg — glacial till and outwash

PLIOCENE

Pt — Thorp Gravel

MIOCENE

TMg — Grande Ronde Basalt of the Columbia River Basalt Group

OLIGOCENE

TOo — volcanic and volcanic-rich sedimentary rocks (mostly Ohanapecosh Formation) includes andesite of Miocene Fifes Peak Formation in the south

EOCENE

Ev — volcanic and volcanic-rich sedimentary rock; Taneum Basalt north of I-90

En — Naches Formation

Er — Roslyn Formation

Es — continental sedimentary rock, mostly of Swauk Formation; includes marine Raging River Formation near milepost 23

EOCENE–OLIGOCENE

EOs — nearshore-deposited sedimentary rocks; includes Blakeley Formation

— fault

Quarry in delta gravels at Issaquah. Note how the bedding is inclined, indicating the original slope of the delta front where it emptied into a glacial lake.

times, after 16,500 years ago. The main source of the gravel was the East Fork of Issaquah Creek, which I-90 follows east from Issaquah to Preston (exit 23). The gravel blankets the tops of the hills at an elevation greater than 400 feet (120 m), indicating the glacial lake's water was at least that deep. Issaquah's townsite was deeply submerged, as was the whole area. Take exit 17 to gain a closer look at the gravel.

The valley narrows considerably near milepost 18, but bedrock remains largely covered by glacial and river deposits. Halfway between mileposts 26 and 27, an outstanding exposure of east-dipping, volcanic-rich sedimentary rocks of the Eocene-age Tukwila Formation forms cliffs north of the westbound lanes.

Between mileposts 27 and 28, I-90 begins a long descent from the north side of Rattlesnake Mountain into Snoqualmie Valley. Rattlesnake Mountain consists of 38-million-year-old rock of the Mt. Persis Volcanics, which formed during the earliest stages of the modern Cascadia subduction zone. They are exposed on both sides of the highway at milepost 30. Snoqualmie Valley marks the site of another gigantic ice age lake, Glacial Lake Snoqualmie. Its flat lakebed is now the wide, flat, fertile floodplain of the Snoqualmie River. From the town of North Bend (exit 32), the river flows about 4 miles (6.4 km) to Snoqualmie Falls, where it spills over a ledge of Mt. Persis Volcanics, and then about another 25 miles (40 km) northward to its confluence with the

Skykomish River. Together, these rivers become the Snohomish River, which empties into Puget Sound at Everett.

Soaring more than 3,500 feet (1,070 m) above North Bend to the east is Mt. Si. It consists of part of the Western mélange belt, a mixed zone of rock that came together in a subduction zone and was accreted to the western edge of North America in the Cretaceous Period. The bedrock of Mt. Si illustrates the concept of a mélange beautifully, as it consists of a variety of rock types that appear jumbled together. Its steep western front consists of a large block of Jurassic-age gabbro, an intrusive rock, surrounded by a complex of low-grade Jurassic and Cretaceous metasedimentary rock; Little Si, the small peak low on Mt. Si's western front, consists of Jurassic metavolcanic rock.

For the next few miles east of North Bend, I-90 stays on the flat, former bottom of Glacial Lake Snoqualmie, but near milepost 35 it bends southward and begins climbing Grouse Ridge, a glacial moraine built of till deposited

The Snoqualmie River drops nearly 270 feet (82 m) at Snoqualmie Falls into a deep canyon of Mt. Persis Volcanics. The lip of the falls is a basaltic andesite lava flow. In the canyon bottom, you can see lahar deposits.

by ice of the Puget Lobe. Look for gravel in the hillside along the highway. Twin Falls State Park lies in the canyon below, where the South Fork of the Snoqualmie River descends steeply through the moraine. Mt. Washington, made of volcanic-rich sedimentary rock of the Ohanapecosh Formation, is the high, rounded mountain directly to the south. About a half mile (0.8 km) east of milepost 36, bedrock emerges from beneath the till. Its west side consists of accreted Cretaceous-Jurassic metasedimentary rock of the Western mélange belt, whereas its east side consists of the younger Ohanapecosh Formation.

For the next 50 miles (80 km), keep on the lookout for the Ohanapecosh and Naches Formations, both of which are intruded by granodiorite of the Snoqualmie Batholith. The Ohanapecosh, deposited between about 32 and 26 million years ago, marks early stages of the Cascade volcanoes. It's made of volcanic-rich breccia, sandstone, and mudstone, as well as basalt and andesite flows, most of which were altered by hydrothermal activity. It covers more than 155 square miles (400 km²) of Washington's South Cascades and shows only a mild degree of deformation.

The older Naches Formation, dating from about 44 to 40 million years ago, is highly deformed due to folding and faulting and pre-dates the inception of the Cascade volcanoes. The Naches Formation formed during strike-slip faulting that affected the Cascades during the Eocene Epoch. It contains a wide variety of lavas and interbedded, fine- to coarse-grained sedimentary rocks, some of which contain leaf fossils and even small amounts of coal.

From milepost 38 to Snoqualmie Pass, the bedrock mostly consists of poorly exposed granodiorite of the Snoqualmie Batholith, which extends from a few miles south of here northward almost to US 2. It intruded as a series of separate magma bodies between 25 and 17 million years ago. These intrusions likely fed some early eruptions of the Cascade volcanoes.

The granodiorite, which is normally very resistant to erosion, tends to be less resistant than the other rock types in this area, as demonstrated by the overall lack of exposures of granodiorite at low elevations. This phenomenon seems at least partly the result of the heating and consequent strengthening of the older Ohanapecosh and Naches Formations during intrusion of the magma. At exit 42 for Mason Creek Road, you can look south up Alice Creek to see the spike-shaped McClellan Butte on the west side of the drainage and the steep north face of Mt. Kent at the head of the drainage, both of which consist of Ohanapecosh Formation; the granodiorite occupies the eroded valley bottom of Alice Creek. Some good roadcuts of the granitic rock do exist. Along the eastbound lanes, look for exposures about 0.75 mile (1.2 km) east of milepost 48; along the westbound lanes, there is a long stretch of exposed granodiorite between the bridge over Denny Creek, just south of milepost 51, almost to milepost 49.

Beginning at milepost 49 and continuing almost to Snoqualmie Pass, you see Naches Formation lining the eastbound lanes. At the pass, a prominent unnamed peak immediately to the northwest consists of rock of the Snoqualmie Batholith. Northwest of it are higher peaks of the Ohanapecosh Formation. South of the pass, the ski areas and surrounding peaks consist of sedimentary and volcanic rock of the Eocene Naches Formation.

View northward from Snoqualmie Pass, looking up the South Fork of the Snoqualmie River. Cliffs on the left are granite of the Snoqualmie Batholith. Guye Peak, on the right, and most of the ridge behind it are Naches Formation.

East from Snoqualmie Pass, I-90 follows the upper reaches of a glacial valley that pours into Keechelus Lake, at milepost 55 just east of Hyak. With the exception of a large quarry in the Ohanapecosh Formation near milepost 59, most of the exposed bedrock over the next 15 miles (24 km) belongs to the highly variable Naches Formation. At Hyak, for example, you can look northward up Gold Creek Valley to see peaks made of the Naches; those at the head of the valley and on its west side are mostly sedimentary, and those on the right are basaltic. A gigantic pile of boulders by the eastbound entrance ramp from Hyak allows for the inspection of andesitic rocks of the Naches. Sedimentary rocks dominate the Naches bedrock between the south end of Keechelus Lake and Easton, and rhyolitic tuffs and breccias show up as roadcuts just east of milepost 55 and at milepost 68.

Keechelus Lake, whose name derives from a Native American word that means "few fish," forms the headwaters of the Yakima River. In 1917, an earth-fill dam was built near its mouth to increase storage capacity for irrigation. The expanse of tree stumps tell the story of rising lake waters after damming, and then falling lake levels during dry periods. I-90 follows the Yakima River from the lake all the way to Ellensburg.

East of milepost 68, I-90 descends gradually to the town of Easton on the moraine that naturally impounded Kachess Lake, which fills another glaciated valley just to the northeast. The name Kachess derives from a Native American word that means "more fish." The Bureau of Reclamation increased the size of the dam in 1912 to increase its storage capacity for irrigation.

Kachess Lake occupies a glaciated valley eroded along the southern extent of the Straight Creek fault, which extends northward in a nearly straight line into southern British Columbia. With upward of 56 miles (90 km) of right-lateral offset, the Straight Creek fault has widely different rocks on each side. Here, the Naches Formation to the west is faulted against the Eocene Swauk and Teanaway Formations to the east. Andesite flows of the Swauk Formation can be seen as the first cliffs above the east shore of the lake; basalt flows of the Teanaway Formation form the top of the ridge and create spectacular bedded cliffs best viewed across from mileposts 70 to 71. The basalt flows continue to the southeast as Easton Ridge, which frames the northern side of Yakima Valley as far east as exit 80 for Roslyn.

Basalt flows of the Teanaway Formation as seen near milepost 70.

The tree-covered hills that border the Yakima Valley to the south consist of low-grade metamorphic rock of Jurassic age. These rocks include the Shuksan Greenschist and Darrington Phyllite, two rock units that typify much of the Cascade Range farther north near the Canadian border and west of the Straight Creek fault. To visit an exposure of the greenschist, take exit 71 and follow Cabin Creek Road for a half mile (0.8 km) westward.

Just west of milepost 81, the interstate crosses the Cle Elum River, which drains Cle Elum Lake about 5 miles (8 km) to the northwest. Like Kachess and Keechelus Lakes to the west, Cle Elum Lake is a natural lake that fills a glacial valley with a glacial dam that the Bureau of Reclamation enhanced for irrigation.

The town of Cle Elum (exit 84) occupies the northern edge of the Yakima River floodplain, which is covered in glacial outwash and modern alluvium. Just north of town, the forested Cle Elum Ridge consists of Eocene river and delta deposits of the Roslyn Formation. About 3 miles (4.8 km) to the south, Miocene Grande Ronde Basalt, the most voluminous member of the Columbia River Basalt Group, caps South Cle Elum Ridge. The Grande Ronde Basalt also caps Lookout Mountain, which rises over 1,000 feet (300 m) above the valley to the east. Lookout Mountain slopes gently southward parallel to the dip of the basalt.

For the 25 miles (40 km) between Cle Elum and Ellensburg, I-90 generally follows the Yakima River over a surface made of recently deposited river gravels and glacial outwash. Topographically high areas to the south and west consist mostly of the Grande Ronde Basalt. You gain spectacular views northward of the Stuart Range, an uplifted complex of oceanic lithosphere intruded by granitic rocks, between mileposts 88 and 91 and from the Indian John Hill Rest Area. A double roadcut of the Grande Ronde Basalt lies on both sides of the road about a half mile (0.8 km) east of milepost 93. East of here you cross Thorp Prairie, a glacial moraine left by an older, extensive glacial advance about 130,000 years ago. Just south of milepost 95, the highway descends about 200 feet (60 m) through gravels of the moraine just south of milepost 95 and drops onto a terrace made of outwash deposits of this older glaciation.

Ellensburg sits near the middle of the Kittitas Valley, an unusually wide valley along the Yakima River. Its broad floor is covered by river deposits that include modern alluvium, glacial outwash, and a 3.7-million-year-old deposit named the Thorp Gravel. This gravel deposit blankets old river terraces, former positions of the floodplain, now perched 300 to 650 feet (90–200 m) above today's floodplain. Near milepost 100, you can look northeastward to bluffs along the Yakima River capped by the Thorp Gravel. A side trip, described in the US 97 guide, follows WA 10 through the Yakima River Canyon past numerous outcrops.

Kittitas Valley's shape derives from folding of the Columbia River Basalt Group, which lies beneath the river deposits. These basalt flows are folded into large anticlines at Manastash Ridge to the southwest and the Wenatchee Mountains to the northeast; the intervening syncline lies right below the valley. The folds are part of the Yakima fold belt, which extends southward from here to northern Oregon and eastward to the Tri-Cities.

US 2
EVERETT—WENATCHEE
123 miles (198 km)

Between Everett and Wenatchee, US 2 crosses over Stevens Pass through the spectacular landscape of the Cascade Range. For the first 2 miles (3.2 km) east of I-5, US 2 crosses distributary channels of the upper part of the Snohomish River delta, and then at milepost 2 the road turns southward and passes alongside and through north-south-trending drumlins. You can see patches of till showing through some bare spots in the grass near milepost 3.

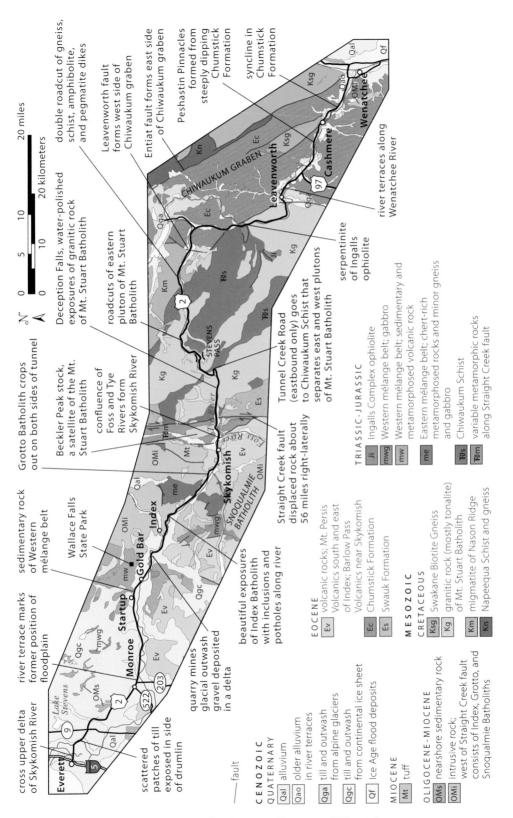

Geology along US 2 between Everett and Wenatchee.

Map labels (clockwise from top):

double roadcut of gneiss, schist, amphibolite, and pegmatite dikes

Leavenworth fault forms west side of Chiwaukum graben

Entiat fault forms east side of Chiwaukum graben

Peshastin Pinnacles formed from steeply dipping Chumstick Formation

syncline in Chumstick Formation

river terraces along Wenatchee River

serpentinite of Ingalls ophiolite

Tunnel Creek Road (eastbound only) goes to Chiwaukum Schist that separates east and west plutons of Mt. Stuart Batholith

roadcuts of eastern pluton of Mt. Stuart Batholith

Deception Falls, water-polished exposures of granitic rock of Mt. Stuart Batholith

confluence of Foss and Tye Rivers form Skykomish River

Beckler Peak stock, a satellite of the Mt. Stuart Batholith

Grotto Batholith crops out on both sides of tunnel

sedimentary rock of Western mélange belt

Wallace Falls State Park

beautiful exposures of Index Batholith with inclusions and potholes along river

Straight Creek fault displaced rock about 56 miles right-laterally

quarry mines glacial outwash gravel deposited in a delta

scattered patches of till exposed in side of drumlin

cross upper delta of Skykomish River

river terrace marks former position of floodplain

Towns: Everett, Monroe, Startup, Gold Bar, Index, Skykomish, Leavenworth, Cashmere, Wenatchee, Stevens Pass, Lake Stevens

Legend:

N
0   5   10   20 miles
0   10   20 kilometers

— fault

CENOZOIC

QUATERNARY

Qal alluvium

Qao older alluvium in river terraces

Qga till and outwash from alpine glaciers

Qgc till and outwash from continental ice sheet

Qf Ice Age flood deposits

MIOCENE

Mt tuff

OLIGOCENE–MIOCENE

OMs nearshore sedimentary rock

OMi intrusive rock; west of Straight Creek fault consists of Index, Grotto, and Snoqualmie Batholiths

EOCENE

Ev volcanic rocks; Mt. Persis Volcanics south and east of Index; Barlow Pass Volcanics near Skykomish

Ec Chumstick Formation

Es Swauk Formation

MESOZOIC

CRETACEOUS

Ksg Swakane Biotite Gneiss

Kg granitic rock (mostly tonalite) of Mt. Stuart Batholith

Km migmatite of Nason Ridge

Kn Napeequa Schist and gneiss

TRIASSIC–JURASSIC

Ji Ingalls Complex ophiolite

mwg Western mélange belt; gabbro

mw Western mélange belt; sedimentary and metamorphosed volcanic rock

me Eastern mélange belt; chert-rich metamorphosed rocks and minor gneiss and gabbro

TRJs Chiwaukum Schist

TRJm variable metamorphic rocks along Straight Creek fault

Perhaps the best reminder of the glaciation, however, is the gigantic glacial erratic near Lake Stevens at the north end of 83rd Ave SE, detailed in Dave Tucker's *Geology Underfoot in Western Washington*. The erratic is about 2.3 miles (3.7 km) from US 2.

From Monroe to Stevens Pass, US 2 follows the Skykomish River and the Tye River, one of the Skykomish's main tributaries. All told, the Skykomish drains a basin of more than 800 square miles (2,100 km²). In the more open landscape at Monroe, you'll notice that the river's floodplain is unusually wide, the consequence of the river being much larger during the Pleistocene. As you cross Woods Creek immediately east of Monroe, look north to see a river terrace perched over an abandoned meander channel of the river. Eastward as the floodplain narrows, look for numerous modern and glacial-age gravel deposits.

In most ways, the small town of Startup at milepost 26 marks the edge of the western Cascades. Not only has the Skykomish River's floodplain narrowed, but you see your first bedrock of accreted terranes. About a quarter mile (400 m) east of town, look for outcrops of the Western mélange belt, accreted during late Cretaceous time, on the north side of the road. Here, these rocks consist mostly of broken sandstone and shale deposited in what was likely a deep ocean environment during Jurassic and Cretaceous time, but as a whole the mélange belt is variable and makes up most of the hills between here and about milepost 31. Haystack Mountain, the prominent forest-covered peak visible about 2 miles (3.2 km) southeast of Gold Bar, consists of gabbro of the mélange. At Wallace Falls, accessed by hiking 1.5 miles (2.4 km) in Wallace Falls State Park, the Wallace River cascades over 265 feet (80.8 m) of the mélange.

I-90 crosses to the south side of the Skykomish River just east of milepost 30, where you can see some beautifully exposed river gravels beneath a river terrace on the north side of the road. Southeast of the bridge, a ridge of gravel that accumulated on the edge of an ice-dammed lake rises several hundred feet above the road. A large quarry is visible through the trees at the chain-up area between mileposts 31 and 32. On clear days, you can look south to Mt. Persis for which the Mt. Persis Volcanics are named. At 38 million years in age, the Mt. Persis Volcanics represent some of Washington's earliest volcanic activity related to the modern subduction zone. To the north, the high cliffs consist of granodiorite of the Index Batholith, which intruded both the Mt. Persis Volcanics and the Eastern mélange belt about 35 million years ago. A small roadcut of the batholith lies on the south side of the road halfway between mileposts 32 and 33.

The Index Batholith was named for the small town of Index, which lies beneath soaring granitic cliffs 1 mile (1.6 km) north of the highway near milepost 36. Just east of the milepost, you pass into numerous outstanding exposures of the granitic rock. A pull-out immediately west of milepost 39 gives easy access to the Skykomish River below a spectacular stretch of rapids. There, you can inspect ledges of the granodiorite, much of which contains inclusions of mafic rock. In some places, rounded potholes attest to

Outcrops of Index granodiorite along the Skykomish River contain scattered dark inclusions.

the erosive power of the river. Exposures of the granodiorite continue until about milepost 42, near the town of Baring, at which point you pass into Triassic-to-Jurassic rocks of the Eastern mélange belt. Look for chert-rich metamorphosed sedimentary rocks behind the railroad tracks a half mile (0.8 km) southeast of milepost 42. These rocks form the hills northeast of the highway for the next couple of miles.

The Grotto Batholith, intruded during Miocene time, forms some beautiful exposures on both sides of the tunnel at milepost 46. In addition to the Eastern mélange belt, the batholith intrudes Eocene volcanic and sedimentary rocks, which show up along the highway as roadcuts between milepost 47 and the town of Skykomish. Some smaller intrusions of the batholith also crop out along the highway, one of which lies directly across from milepost 49.

US 2 crosses the Skykomish River halfway between mileposts 49 and 50, and again at milepost 51, just downriver from the confluence of the Tye and Foss Rivers. Milepost 51 approximately coincides with a strand of the Straight Creek fault, which continues northward into Canada and southward as far as I-90. It separates the Western domain of the Cascades from the Crystalline Core and shows some 56 miles (90 km) of right-lateral offset, which occurred in the last 50 million years. Between mileposts 51 and 53 you pass through a large rock avalanche deposit. Look for numerous large boulders as you drive through this stretch, at least one of which is as large as a house.

If you are headed east, you see the first exposure of the Crystalline Core at milepost 53. This cliff of granodiorite is part of the Beckler Peak stocks, a series of small intrusions related to the Cretaceous-age Mt. Stuart Batholith. A quarter mile (400 m) east of milepost 53, look for highly deformed schist. Sporadic exposures of the schist crop out until you cross the eastern strand of the Straight Creek fault between mileposts 54 and 55.

For almost the entire route between the Straight Creek fault and Leavenworth, 42 miles (68 km) to the east, you'll pass through exposures of the Chiwaukum Schist and the Mt. Stuart Batholith. The Chiwaukum Schist originated mostly as a sequence of sediments deposited in an ocean during Cretaceous time, although it includes a banded gneiss unit that appears to have a more complicated origin. The schist is the primary rock unit of the Nason terrane, which accreted to North America during late Cretaceous time. The Mt. Stuart Batholith is the largest intrusive complex of the Cascades east of the Straight Creek fault. It formed as a series of intrusions between 96.5 and 91 million years ago. The Beckler Peak stocks are the westernmost and smallest of these intrusions. The two main plutons of the batholith, known as the western and eastern plutons, are separated by a narrow zone of the Chiwaukum Schist. For the most part, rocks of the batholith consist of tonalite, which resembles granite except that its feldspar minerals are nearly all plagioclase. The batholith also includes some granodiorite, diorite, and granite.

Deception Falls, 0.6 mile (1 km) east of milepost 56, provides an easy place to escape the highway and take a close look at granitic rock of the western pluton of the Mt. Stuart Batholith. You can see many good, water-polished exposures of the rock, which is cut by thin mafic and felsic dikes. Another off-highway spot with an outcrop is the Iron Goat Interpretive Site, just east of milepost 58, which describes the history of the railway over Stevens Pass. Operated by the Great Northern Railway, the trains initially climbed a series of switchbacks to a tunnel below Stevens Pass. Always subject to dangerous winter avalanches, one of which killed nearly one hundred passengers and crew in 1910, the railway built miles of snowsheds. In 1929, the Great Northern completed work on a lower tunnel and abandoned the high tracks and tunnel, which today form the Iron Goat Trail. As you cross the bridge just west of milepost 59, you can look up the modern railroad tracks to the tunnel entrance.

US 2 crosses Tunnel Creek at the hairpin turn 0.6 mile (1 km) uphill from milepost 60. Several pull-outs sit next to exposures of faulted tonalite of the western pluton of the Mt. Stuart Batholith. Tunnel Creek Road, accessible by eastbound traffic only on the uphill side of the turn, takes you a half mile (0.8 km) to outcrops of Chiwaukum Schist metamorphosed by the heat of the magma of the batholith. These rocks contain a range of metamorphic minerals, including andalusite and garnet. Jerry Magloughlin of Colorado State University found a metamorphic age for the minerals of about 94 million years, approximately the same age as the batholith.

For the next 3.5 miles (5.6 km) to Stevens Pass, US 2 passes numerous excellent exposures of the Mt. Stuart Batholith. A poorly exposed band of Chiwaukum Schist on either side of milepost 62 separates the western pluton

View up railroad track to lower tunnel of the Great Northern Railway. The white stripe high above the tracks is a roadcut of the Mt. Stuart Batholith along US 2.

of the Mt. Stuart Batholith from the eastern pluton. Notice the glacially carved, U-shaped profile of the valley of Tye River, about 400 feet (120 m) below the road, as well as the numerous avalanche chutes.

East of Stevens Pass, US 2 follows the valley of Stevens and then Nason Creek, both mostly filled with glacial deposits. Sporadic bedrock exposures of the eastern pluton start to appear just west of milepost 67, with multiple exposures east of that. A long roadcut less than a quarter mile (400 m) below milepost 69 offers a wealth of features in the tonalite, including narrow bands of schist, several dikes, and scattered inclusions. Some of the inclusions are oval-shaped, possibly as the result of flow of the original magma. Zones of intense fracturing mark the locations of some faults. Near the east end of the roadcut, several faults visibly offset a band of schist. Although parking is possible along the front of the roadcut, a large pull-out just east of its eastern end offers a safer place.

About a quarter mile (400 m) below milepost 72 you pass back into the Nason terrane, represented by exposures of the Chiwaukum Schist. Many of these exposures show that the schist is tightly folded, although at such small scales that you need to look closely at the rock to see the folds. A safe pull-out lies across from a roadcut just east of milepost 75.

Halfway between mileposts 76 and 77, a spectacular double roadcut reveals gneisses that bear an unknown relation to the Chiwaukum Schist. The migmatitic gneiss forms most of Nason Ridge to the north and west. It displays a variety of rocks, including pegmatite and quartz veins that intrude granitic gneiss, schist, and amphibolite. According to Robert Miller at San Jose State University, the intrusive rocks formed during the last stages of crystallization of the Mt. Stuart Batholith. More but smaller outcrops of the gneiss continue southeastward for the next couple miles.

Just east of milepost 80, US 2 crosses the Leavenworth fault, east of which are Eocene rocks of the Chumstick Formation. This feldspar-and-mica-rich sandstone, with minor shale and conglomerate, was probably deposited by rivers that traversed a series of alluvial fans. The sediments accumulated in the northwest-trending Chiwaukum graben, which was dropped down between the Leavenworth fault, on the west, and the Entiat fault, about 8 miles (13 km) to the east. The Swakane terrane, composed largely of the Swakane Biotite Gneiss, lies immediately east of the Entiat fault.

Double roadcut of migmatitic gneiss. You can also find abundant schist and amphibolite here, all cut by a variety of pegmatite dikes and quartz veins. Inset is 10 feet (3 m) wide and shows pegmatite dikes cutting the gneiss.

Between about mileposts 80 and 90, US 2 passes through a more open land-scape eroded into the Chumstick Formation. Unfortunately, you only see one exposure, just north of milepost 88. Near milepost 89, the road approaches the Leavenworth fault, which appears through the trees as the prominent break in slope a short distance to the west. At the entrance to Tumwater Canyon, halfway between mileposts 90 and 91, US 2 crosses back onto rocks uplifted on the west side of the Leavenworth fault. These ultramafic rocks belong to the Jurassic Ingalls Complex, an assemblage of accreted rock of the oceanic lithosphere. A good exposure of these rocks lies on the east side of the highway immediately north of milepost 91, but there isn't a safe spot to stop; a much better place to stop is a half mile (0.8 km) farther south where there is a large pull-out adjacent to the river. The high cliff of serpentinite shows numerous mineral-coated fractures and faults, as well as some blocks of dark-gray biotite schist that are most likely frag-ments of Chiwaukum Schist carried into the Ingalls Complex along faults.

From about milepost 92 to the mouth of Tumwater Canyon at Leavenworth, US 2 is in the Mt. Stuart Batholith. Except for some poor exposures of the Ingalls Complex for a short distance on either side of milepost 93, all roadcuts are tonalite and diorite. Tumwater Dam, just south of milepost 95, backs up Jolanda Lake. Built from 1907 to 1909, the dam generated electricity to power electric locomotives that pulled trains through the old high tunnel beneath Stevens Pass. Initially, the trains were all powered by coal, but the emissions were so awful that the railway turned to electricity.

US 2 crosses to the east side of the Leavenworth fault at the mouth of Tumwater Canyon. Leavenworth, perched on glacial deposits that overlie the Chumstick Formation, offers spectacular views of the mountains west of the fault. Just like Tumwater Canyon, the mountains consist mostly of tonalite of the Mt. Stuart Batholith with fragments of the accreted Nason terrane and Ingalls Complex.

US 2 follows the Wenatchee River from Leavenworth to Wenatchee and passes scattered outcrops of Chumstick Formation almost the entire way. For the first several miles east of Leavenworth, the road passes by orchards growing on river terraces that mark former positions of the floodplain. Keep an eye out for variable angles and directions of tilting in the Chumstick bedding, the result of folding. The slabby outcrops on the west side of the road at milepost 101 and between mileposts 102 and 103, for example, dip steeply to the north-east, whereas the bedding in the cliff along the Wenatchee River just south of milepost 105 is nearly flat. The Peshastin Pinnacles, just north of milepost 109, dip steeply west, providing numerous sandstone walls for rock climbing. (See photo on page 117.)

Near Cashmere, just east of milepost 111, US 2 runs along the base of some steep hills made of an uplifted block of Swakane Biotite Gneiss faulted against conglomerate of the Chumstick Formation. You can see the conglomerate along the mountain front by following Nahahum Canyon Road for 0.3 mile (480 m); at 0.4 mile (640 m), the road turns eastward into the canyon, where you can view some hard-to-reach exposures of the gneiss in less than 1 mile (1.6 km) along the south side.

Aerial view northeastward of Mt. Stuart (S). Granitic rock of the Mt. Stuart Batholith forms the craggy gray mountains, whereas the low-silica rock of the Ingalls Complex ophiolite forms the lower brownish areas. US 2 and the town of Leavenworth (L) are in the deep valley that spans the width of the photo behind (north of) Mt. Stuart.

US 2 passes through a fold in the Chumstick Formation 0.6 mile (1 km) south of milepost 114. This downward-bending type of fold is a syncline, which contains the youngest rock in its core. The north side of the exposure dips southward while the south side dips northward. Just east of milepost 117, you can look up to the northeast and see lake deposits of the Chumstick Formation. These rocks consist of thinly layered shale and thicker beds of sandstone. A quarter mile (400 m) detour up Lower Sunnyslope Road to its intersection with Easy Street takes you to the outcrop.

Wenatchee occupies the floodplain of the Columbia River just downstream of its confluence with the Wenatchee River. The city is built on river deposits, as well as flood deposits from Glacial Lake Columbia. Just a few miles to the north, an abrupt eastward rise in elevation marks the Entiat fault, the eastern boundary of the Chiwaukum graben. The hard gneiss of the Swakane terrane forms hills east of the fault. Just across the Columbia River, Badger Mountain, made of Grande Ronde Basalt, rises to an elevation of 4,254 feet (1,297 m), more than 3,500 feet (1,070 m) above the Columbia River. To the south, Jumpoff Ridge, also made of Grande Ronde Basalt, rises above 5,000 feet (1,500 m) in elevation.

US 97
Ellensburg—US 2 near Leavenworth
52 miles (84 km)

Ellensburg sits on the Yakima River in the Kittitas Valley, formed in a broad downwarp, or syncline, in the bedrock. The first few miles north of Ellensburg on US 97 offer expansive views northeast to the Wenatchee Mountains and southeast to Manastash Ridge, both of which are formed by anticlines on each side of the syncline. Beginning near milepost 142, as US 97 gradually climbs north out of the valley, you start seeing scattered exposures of the Thorp Gravel, which was deposited by an early Yakima River and its tributaries between 4.4

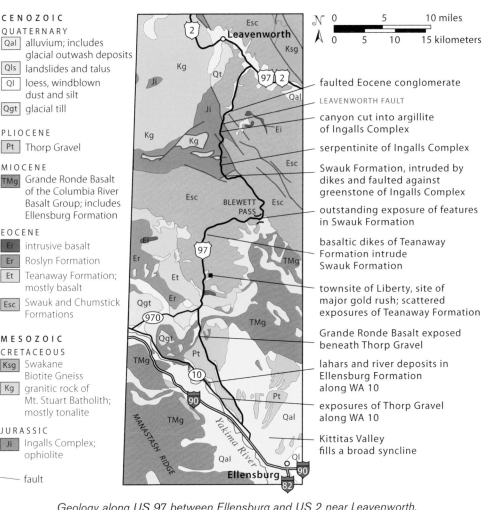

CENOZOIC
QUATERNARY
Qal alluvium; includes glacial outwash deposits
Qls landslides and talus
Ql loess, windblown dust and silt
Qgt glacial till

PLIOCENE
Pt Thorp Gravel

MIOCENE
TMg Grande Ronde Basalt of the Columbia River Basalt Group; includes Ellensburg Formation

EOCENE
Ei intrusive basalt
Er Roslyn Formation
Et Teanaway Formation; mostly basalt
Esc Swauk and Chumstick Formations

MESOZOIC
CRETACEOUS
Ksg Swakane Biotite Gneiss
Kg granitic rock of Mt. Stuart Batholith; mostly tonalite

JURASSIC
Ji Ingalls Complex; ophiolite

— fault

faulted Eocene conglomerate
LEAVENWORTH FAULT
canyon cut into argillite of Ingalls Complex
serpentinite of Ingalls Complex
Swauk Formation, intruded by dikes and faulted against greenstone of Ingalls Complex
outstanding exposure of features in Swauk Formation
basaltic dikes of Teanaway Formation intrude Swauk Formation
townsite of Liberty, site of major gold rush; scattered exposures of Teanaway Formation
Grande Ronde Basalt exposed beneath Thorp Gravel
lahars and river deposits in Ellensburg Formation along WA 10
exposures of Thorp Gravel along WA 10
Kittitas Valley fills a broad syncline

Geology along US 97 between Ellensburg and US 2 near Leavenworth.

WA 10 Side Trip

WA 10 closely follows the Yakima River between Ellensburg and Cle Elum. Since the beginning of the Pleistocene, the river has eroded a canyon through bluffs of the Thorp Gravel, Ellensburg Formation, and Grande Ronde Basalt. Spectacular cliffs of the Thorp Gravel show up along WA 10 between mileposts 102 and 100, about 2 (3.2 km) miles northwest of the US 97 intersection. The Thorp Gravel was deposited by an early incarnation of the Yakima River and its tributaries between about 4.4 and 3.8 million years ago. Look for a prominent ash layer high on the cliff just north of milepost 101.

Bluffs of the Thorp Gravel above WA 10 in the Yakima River valley.

Continuing to the northwest at river level, WA 10 passes a half-mile-long (0.8 km) exposure of Ellensburg Formation at milepost 98 on the north side of the road. These white bluffs consist of several thick, fine-grained beds separated by layers of gravel and sandstone. The fine-grained beds, which contain numerous large fragments of volcanic rock suspended in a muddy matrix, are lahars generated from early volcanoes in the Cascades. The gravel and sandstone beds, which exhibit crossbedding and channels filled with basaltic cobbles, were deposited in rivers during the times between lahars.

WA 10 rises briefly up to a river terrace, then descends back into a narrow stretch of the canyon bordered by cliffs of the Grande Ronde Basalt. Halfway between mileposts 96 and 95 you pass a double roadcut of pillow basalt and orange-brown palagonite. The pillows formed when the basaltic lava flowed into a lake, and the palagonite formed by chemical alteration of volcanic glass, much of which was produced by the sudden chilling of the lava. WA 10 climbs back up through river gravels to an older terrace, then descends back down past some good exposures of Grande Ronde Basalt northwest of milepost 94. Look for nicely developed colonnade and entablature.

Between mileposts 91 and 90, you pass a series of roadcuts of Ellensburg Formation, here consisting mostly of river-deposited gravel and sandstone as well as some ash-rich deposits. Notice the numerous small channels preserved at the base of the gravels; the individual cobbles in the gravels consist mostly of basalt, derived from the underlying Grande Ronde Basalt. Several photogenic normal faults also cut the rocks. At Teanaway, you can return to US 97 via WA 970.

Faulting in a coarse-grained, river-deposited part of the Ellensburg Formation.

and 3.8 million years ago. The gravel forms a broad upland surface that slopes toward the Yakima River at about the same gradient as the nearby streams. Where streams have cut through the gravel, they expose the underlying Ellensburg Formation and Grande Ronde Basalt, which formed at roughly the same time about 15 million years ago. A prominent ledge of volcanic-rich sandstone of the Ellensburg Formation crops out directly across from milepost 146, and the Grande Ronde Basalt is exposed at milepost 147, below the Thorp Gravel. Scattered outcrops of the Grande Ronde Basalt persist almost to Swauk Creek, near milepost 149, just south of the intersection with WA 970.

Between the intersection of WA 970 and Blewitt Pass, US 97 follows Swauk Creek, which hosts the Liberty mining district, one of Washington State's primary gold rush areas. Discovered in 1873, the district hosted gold deposited by hydrothermal fluids in the local bedrock as well as placer gold, found as nuggets and flakes in river sediments. The district includes Williams Creek and the townsite of Liberty, accessed by the Liberty Road between mileposts 152 and 153. Miners initially worked the area by panning the stream gravels, but new prospectors staked out lode claims in the bedrock while others turned to hydraulic mining to more efficiently process the gravel. Some mining continues today, although at a much smaller scale. The area is known for its wire-shaped crystalline gold.

Halfway between mileposts 150 and 151, look for large mounds in the channel of Swauk Creek, a telltale sign of hydraulic mining. Miners broke up alluvial material with high-pressure jets of water and then directed the material through their sluice boxes to retrieve the gold. Although it was far more efficient than panning, hydraulic mining caused a great deal of environmental damage throughout the West and is now carefully regulated.

Several good outcrops of the basalt of the Teanaway Formation line the road from just south of milepost 151 to milepost 152. In addition to basalt lava flows, the formation includes basaltic through rhyolitic pyroclastic material and even some sedimentary rock. The best estimates for its age are about 47 million years. In some places, it contains pocket fillings of chalcedony and is thought to be the source for the valuable Ellensburg Blue Agate.

Look for north-dipping sedimentary rocks of the Swauk Formation, which underlies the Teanaway, 0.6 mile (1 km) north of milepost 152. US 97 continues through the Swauk for almost the next 20 miles (32 km). The Swauk was deposited in meandering rivers between about 54 and 48 million years ago. In some places, however, the Swauk consists mostly of shale, which was probably deposited in lakes, while in others it contains conglomerate that was shed from adjacent highlands into alluvial fans. In general, the Swauk is much more deformed than the Teanaway, as shown by its generally steeper and more variable dip angles at different places along the highway.

In many of the roadcuts, basalt of the Teanaway Formation intrudes the Swauk Formation, mostly as sills parallel to layering, but also as some dikes that cut across layering. Some especially good exposures lie along the road from just south of milepost 158 to milepost 160.

The roadcut 0.4 mile (640 m) north of milepost 163 offers an abundance of sedimentary features in the Swauk Formation as well as a large pull-out on the opposite side of the road. Be especially careful here because of the fast-moving vehicles. You can see abundant crossbeds, coaly plant fragments in gray shale, and several small faults. With a hand lens, you can tell that the sandstone contains abundant feldspar in addition to quartz. At the south end of the exposure, you can see a number of concretions, spherical zones in the rock that are more resistant to weathering and erosion than the surrounding rock. In most cases their resistance comes from a higher degree of cementation.

At Blewett Pass by milepost 164, you cross into the north-flowing watershed of Tronsen Creek, framed by the cliffs of Tronsen Ridge to the east. Tronsen Ridge is made mostly of shales of the Swauk Formation, deposited in lakes. Several of the roadcuts just north of the pass contain abundant shale as well, although a thick-bedded sandstone near milepost 165 suggests this part of the Swauk was deposited in a river.

Roadcut of the Swauk Formation just north of milepost 163. Inset photo shows crossbedded sandstone deposited by running water. Hand lens for scale.

Between mileposts 170 and 172, several outstanding roadcuts of Swauk Formation, cut by basaltic sills, line the highway. The northernmost one, which lies on the east side of the highway, contains both sandstone and conglomerate, as well as several sills, and is faulted against greenstone of the Ingalls Complex of Jurassic age. Highly faulted greenstone is well exposed just north of the parking area on the west side of the road.

For the next 5 miles (8 km), US 97 passes through faulted greenstone, serpentinite, and argillite of the Ingalls Complex, which originated at least partly as an ophiolite, a piece of ocean crust, although it may include parts of an island arc as well. Greenstone typically forms by low-grade metamorphism of basalt; serpentinite forms by alteration of ultramafic rock, which may have originated in Earth's mantle; and argillite forms by low-grade metamorphism of shale or mudstone. Argillite tends to be distinctly undistinctive, as it is hard, fine grained, and dark colored. Good exposures of serpentinite exist on the east side of the road between mileposts 173 and 174, while argillite forms most of the high cliffs that continue to the mouth of the canyon near milepost 177.

The Ingalls argillite ends abruptly at the northwest-trending Leavenworth fault, which uplifted rocks along its southwest side, including the Ingalls Complex, relative to its northeast side. Some 15 miles (25 km) to the northeast, the parallel Entiat fault drops its southwest side down. The down-dropped block of crust between the two faults is called the Chiwaukum graben.

The Swauk Formation (right) is faulted against greenstone of the Ingalls Complex (eroded orangish slope at far left), just south of milepost 172. The greenstone also forms the peak at the left (north) edge of the photo. Several sills (S) intrude the Swauk.

Cliffs of argillite of the Ingalls Complex.

Eocene conglomerate forms the first good rock exposure along the highway north of the Leavenworth fault, at the intersection of Old Blewett Road and US 97, about a quarter mile (400 m) south of milepost 179. Nearly all the cobbles consist of tonalite, a granitic rock in which the feldspar is mostly plagioclase instead of orthoclase. The tonalite was derived from the Mt. Stuart Batholith, which makes up the mountains a few miles to the west. In addition, the outcrop displays many slickensided fault surfaces, polished surfaces with parallel grooves likely formed by slip on the nearby Leavenworth fault.

Most researchers think the conglomerate near milepost 179 marks the edge of the Chiwaukum graben, and that it is coarse-grained material eroded from the rising block and deposited next to the fault. The conglomerate forms a narrow strip along the fault for about 30 miles (50 km). Northeastward, away from the fault, the conglomerate grades into the sandstone of the Chumstick Formation that comes into view near milepost 180. For the most part, the Chumstick consists of light-colored, feldspar-rich sandstone that was probably deposited on alluvial fans.

An alternate view, advocated by Eric Cheney of the University of Washington, argues that the depositional area of the Chumstick Formation was originally far more extensive than it appears now; it only appears localized

Cobbles in this faulted conglomerate of Eocene age consist mostly of tonalite. The red lines are parallel to striations on some of the many small slickensided fault surfaces.

between the Leavenworth and Entiat faults because erosion stripped away the material from the uplifted outer sides of the graben. In fact, Cheney argues that the Roslyn Formation is a finer-grained part of the Chumstick. He argues that the conglomerate near milepost 179 is actually a part of the Swauk Formation, and its position along the Leavenworth fault is fortuitous.

US 97
WENATCHEE—PATEROS
54 miles (87 km)

North of Wenatchee, US 97 follows the Columbia River upstream to the mouth of the Methow River at Pateros. Nearly all the bedrock exposed along the road belongs to terranes of the Crystalline Core of the North Cascades. The route lies so close to the boundary with the Columbia Basin, however, that in many places you can see basalt flows of the plateau perched above the river canyon at high elevations. This stretch of the Columbia River also displays incredible river terraces made of glacial outwash sediments as well as deposits of boulder-rich gravels from the early and late-stage floods from Glacial Lake Missoula.

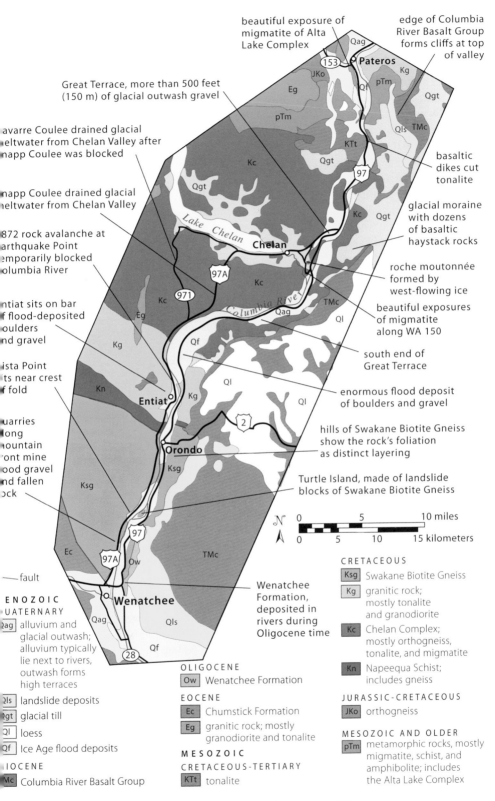

beautiful exposure of migmatite of Alta Lake Complex

edge of Columbia River Basalt Group forms cliffs at top of valley

Great Terrace, more than 500 feet (150 m) of glacial outwash gravel

Navarre Coulee drained glacial meltwater from Chelan Valley after Knapp Coulee was blocked

Knapp Coulee drained glacial meltwater from Chelan Valley

1872 rock avalanche at Earthquake Point temporarily blocked Columbia River

Entiat sits on bar of flood-deposited boulders and gravel

Vista Point sits near crest of fold

quarries along Mountain Front mine good gravel and fallen rock

basaltic dikes cut tonalite

glacial moraine with dozens of basalt haystack rocks

roche moutonnée formed by west-flowing ice

beautiful exposures of migmatite along WA 150

south end of Great Terrace

enormous flood deposit of boulders and gravel

hills of Swakane Biotite Gneiss show the rock's foliation as distinct layering

Turtle Island, made of landslide blocks of Swakane Biotite Gneiss

Wenatchee Formation, deposited in rivers during Oligocene time

N   0   5   10 miles

0   5   10   15 kilometers

— fault

CENOZOIC
QUATERNARY
Qag alluvium and glacial outwash; alluvium typically lie next to rivers, outwash forms high terraces

Qls landslide deposits

Qgt glacial till

Ql loess

Qf Ice Age flood deposits

MIOCENE
Mc Columbia River Basalt Group

OLIGOCENE
Ow Wenatchee Formation

EOCENE
Ec Chumstick Formation

Eg granitic rock; mostly granodiorite and tonalite

MESOZOIC
CRETACEOUS–TERTIARY
KTt tonalite

CRETACEOUS
Ksg Swakane Biotite Gneiss

Kg granitic rock; mostly tonalite and granodiorite

Kc Chelan Complex; mostly orthogneiss, tonalite, and migmatite

Kn Napeequa Schist; includes gneiss

JURASSIC–CRETACEOUS
JKo orthogneiss

MESOZOIC AND OLDER
pTm metamorphic rocks, mostly migmatite, schist, and amphibolite; includes the Alta Lake Complex

Geology along US 97 between Wenatchee and Pateros.

At Wenatchee, US 97 crosses the Columbia and turns north, first passing exposures of the Wenatchee Formation, deposited in rivers during Oligocene time. Within 1 mile (1.6 km), however, you pass into metamorphic rock of the Swakane terrane, which lies directly beneath the Wenatchee Formation. The Swakane terrane consists of biotite gneiss and forms the banded cliffs on both sides of the river for about the next 12 miles (19 km). The gneiss originated as a sedimentary rock about 72 million years ago and was rapidly buried and metamorphosed within the next 4 million years.

Between mileposts 134 and 135 the road wraps around Turtle Island, a hummocky landslide deposit of Swakane Biotite Gneiss. Its rocky profile resembles a turtle when viewed from the south. The landslide most likely occurred when Glacial Lake Columbia broke through its ice dam as one of the late-stage Ice Age floods, sometime before 15,000 years ago. A gravel bar, deposited from this flood, forms the low rib below the rocks, apparently deposited on the protected downcurrent side of the rocks.

North of the turnoff for US 2, the mileposts revert back to those for US 97, beginning with milepost 214. At about milepost 215, US 97 crosses a fault zone, north of which is the Entiat pluton, which forms the cliffs along the highway for the next several miles. The rock consists of tonalite, which resembles granite except that its feldspar content is mostly plagioclase instead of orthoclase. The Entiat pluton intruded 73 to 72 million ago at a depth of about 12 miles (19 km). You pass numerous good outcrops of the tonalite for the next several miles until just north of milepost 219, where they become buried by flood gravels from one of the early Ice Age floods. These gravels form an enormous deposit on the large inside bend of the river, in many places decorated with large granitic boulders. Its flatness makes it resemble a river terrace, but it is altogether different. River terraces consist of river deposits on former floodplains, left stranded as the river cut deeper into its channel. These deposits formed from immense floods depositing material high on the sides of the valley.

View looking upstream at Turtle Island in the Columbia River channel. The island is a landslide deposit of jumbled blocks of Swakane Biotite Gneiss. The low-lying area on its down-current side is a gravel bar deposited by an Ice Age flood.

A haystack rock, made of basalt of the Columbia River Basalt Group, a few miles up McNeil Canyon Road. Other haystack rocks of the Okanogan moraine lie in the background.

At milepost 226, US 97 descends back to the river, passing the first exposures of the Chelan Mountains terrane, which makes up all the bedrock exposures between here and about 10 miles (16 km) north of the town of Chelan. The terrane consists of metamorphosed oceanic and intrusive rock. It includes a great deal of migmatite, a metamorphic rock with a large proportion of igneous material, either from the injection of new magma into the hot, deforming rock or partial melting of the original rock. Across the river, on the inside bend, lies another flat, terrace-like surface made of more flood deposits.

The Great Terrace, consisting mostly of glacial outwash, extends intermittently from about milepost 226 upriver to Pateros and beyond. The Great Terrace is a true river terrace because it marks the former position of the river. Since the gravel was deposited, the Columbia River has cut down through the deposits, leaving them stranded at higher elevations. Looking upward to the east, you can see that the top of the valley is rimmed with Grande Ronde and Wanapum Basalts of the Columbia River Basalt Group. You can see a huge block of basalt that fell from the rim just south of milepost 234, and another at the intersection with McNeil Canyon Road, just east of the US 97 bridge over the Columbia.

A 7-mile (11 km) round-trip drive up McNeil Canyon Road takes you into the terminal moraine of the Okanogan Lobe. It's a wonderfully strange landscape of rolling hills made of till punctuated by dozens of gigantic basalt boulders. The boulders, informally called haystack rocks, are part of the till, but they stand out in relief because of their size.

US 97 crosses the Columbia River halfway between mileposts 234 and 235. A short side trip to Chelan on WA 150 takes you up a steep grade past an astounding roadcut of migmatite that contains parts that are clearly metamorphic, other parts that are clearly igneous, and other parts that are difficult to distinguish. The cliff is not safe to inspect up close because of falling rock, but near the top of the grade you can turn right into a parking area that offers a safe place to see migmatite and tonalite in the same outcrop. In addition, the rock is covered in glacial striations and has been abraded by glacial ice into the classic asymmetric shape of a roche moutonnée. Its steeper western side resulted from movement of ice of the Okanogan Lobe from the Columbia Valley westward into the Chelan Valley.

North of the US 97 bridge, the highway passes through cliffs of migmatite of the Chelan Complex for the next couple of miles at river level. From milepost 238 to 239, the road follows the river directly below the Great Terrace, which rises some 500 feet (150 m) above. It passes light-colored gneiss of the Chelan Mountains terrane north of its intersection with US 97A. An excellent exposure of the gneiss, cut by numerous basaltic dikes, is in the abandoned quarry at milepost 242.

Migmatite of the Chelan Complex near the intersection of US 97 and WA 150.

View looking upriver over US 97 to the Great Terrace, which forms the high flat area near the center of the photo.

The Chelan Complex is intruded by tonalite, which forms hills west of the highway between mileposts 245 and 248. You can see numerous basaltic dikes cutting the light-colored tonalite in the cliffs immediately north of milepost 247. North of about milepost 248, however, US 97 crosses into another set of metamorphic rocks intruded by the tonalite. Called the Alta Lake Complex, these rocks were also metamorphosed during late Cretaceous time and show up intermittently the rest of the way to Pateros. An outstanding exposure of these rocks lies on the south side of the road a quarter mile (400 m) up WA 153. East of the river, the Columbia River Basalt Group continues to form the rim of the Columbia Valley. The Great Terrace is unusually striking in the vicinity of Pateros, where it forms a flat bench that rises some 600 feet (180 m) above the river.

US 97 ALTERNATE
WENATCHEE—CHELAN
34 miles (55 km)
See map on page 143.

Similar to US 97 along the east side of the Columbia River, US 97A passes the southeast edge of the Crystalline Core domain, overlain in places by deposits of the Ice Age floods. However, this route encounters far less traffic than US 97 and offers more opportunities to pull safely off the road.

North of Wenatchee, the road passes high cliffs of Swakane Biotite Gneiss for the first 13 miles (21 km). Several small quarries that mine material shed from the cliffs as well as flood gravels line the road for the first few miles. The gneiss, which is rich in the black mica biotite and prominently banded in places, is the deepest known part of the Crystalline Core. It originated as sedimentary rock about 72 million years ago and had been buried and metamorphosed by 68 million years ago, a period of only 4 million years! The gneiss forms a broad fold, with noticeably south-dipping foliation for the first few miles and noticeably north-dipping foliation beginning near milepost 206. Pegmatite sills that intruded parallel to the foliation make the fold a little easier to see. Numerous pegmatite dikes also intrude across the foliation.

Vista Point, halfway between mileposts 205 and 206, sits near the crest of the fold. Here you can safely inspect some of the rock, which includes quartz veins and some interesting crosscutting relations between different generations of pegmatite. You can also see examples where pegmatite pinched and swelled because of the high-temperature deformation in the rock. The view across the river includes Turtle Island, a giant landslide deposit of jumbled Swakane Biotite Gneiss. The grassy slope immediately downstream from the rocks is a gravel bar deposited and shaped by at least one of the late-stage Ice Age floods. Because gravel bars typically accumulate on the protected downstream sides of barriers, the landslide must have occurred before the floods that deposited the gravel.

Cliffs north of Vista Point exhibit north-dipping Swakane Biotite Gneiss. A grassy area near milepost 213 hides a fault zone that brings a different metamorphic rock—called the Napeequa Schist—against the gneiss. This unit is poorly exposed, however, as the next set of cliffs to the north consist of granitic rock of the Entiat pluton, which intruded 72 million years ago. Because the feldspar content of this rock consists almost entirely of plagioclase, geologists

The crest of the fold in Swakane Biotite Gneiss as seen from Vista Point. The foliation to the left (south) of the geologist dips southward, whereas the foliation to the north of the geologist dips northward.

refer to it as tonalite to distinguish it from true granite. Between about mileposts 214 and 221, the bedrock consists of gray cliffs of the tonalite, which are mostly homogeneous in appearance but in some places exhibit foliation. The small town of Entiat is built on a gravel bar deposited by an early Ice Age flood. You can see numerous large boulders strewn about the landscape.

Earthquake Point, at the heritage marker a quarter mile (400 m) north of milepost 218, marks the site of a rock avalanche that broke off the cliffs during the estimated magnitude 6.8 earthquake of 1872. Some of the debris, which temporarily blocked the Columbia River, is visible on both sides of the highway. You can also see some basaltic dikes in the cliffs above the pull-out. More dikes are easily visible at milepost 219. The dikes are likely Eocene in age and may be related to similar ones along US 2 in Corbaley (Pine) Canyon across the river and about 6 miles (10 km) to the south. Directly across the river from Earthquake Point is the south end of a huge bar constructed of gravel and boulders deposited by one of the early Ice Age floods.

North of milepost 221 is migmatite of the Chelan Mountains terrane. Migmatite forms at high temperature from mixed metamorphic and igneous origins. Migmatite makes up much of the terrane, which also contains abundant igneous rock and metamorphosed sedimentary rock. A spectacular exposure of the migmatite forms the wall over the highway tunnel a half mile (0.8 km) north of milepost 224.

Lake Chelan

In the Salish language, the name Chelan means "deep water." At a depth of 1,486 feet (453 m), which reaches 388 feet (118 m) below sea level, Lake Chelan is the third-deepest lake in the United States. It's also long and narrow, with a length that exceeds 50 miles (80 km) and opposing shorelines that average just 1 mile (1.6 km) apart. The southeast end of the lake backs up against a gigantic dam of glacial outwash and till. In 1927, the Washington Water Power Company constructed Lake Chelan Dam for hydroelectric power and raised the level of the lake 20 feet (6 m).

Lake Chelan occupies a long, narrow glacial valley, carved by multiple glacial advances into bedrock of the Crystalline Core. While it's easy enough to imagine a large glacier flowing out of the North Cascades to deepen and scour the valley sides, it might come as a surprise that most of the ice came from repeated advances of different lobes of the Cordilleran Ice Sheet. The Skagit Lobe was fed by ice that overtopped the Cascade Range and flowed southeastward down the valley, while the Okanogan Lobe flowed northwestward into the Chelan Valley from the Columbia Valley. Alpine glaciers had an effect, especially in the upper reaches of the valley, but their impact wasn't nearly as profound as that of the continental ice sheet.

The many glacial features near the eastern end of the lake came from the Okanogan Lobe. The glacial till, for example, contains giant boulders of Columbia River Basalt Group, which crops out north and east of Lake Chelan but not to the northwest in the North Cascades. West-flowing ice also shaped roche moutonnées. Knapp and Navarre Coulees, along the southern margin of the lake, were carved by the voluminous meltwater as it escaped from between the mouths of the two ice lobes. Researchers argue that Knapp Coulee formed first, when the Okanogan Lobe first advanced into the Chelan Valley and blocked the valley's natural outlet. Navarre Coulee formed later, after the Okanogan Lobe advanced farther westward to block the path down Knapp Coulee.

Although road access is limited to the lower reaches of the lake, boat tours and a ferry service from Chelan access backcountry trails.

North of the tunnel, the road enters Knapp Coulee, a narrow, steep-sided valley with no stream. It leads to a low pass near milepost 228 and then down to Lake Chelan. Meltwater flowing from two different glaciers in the Chelan Valley eroded the coulee. To the northwest lay a tongue of the Skagit Lobe, and to the southeast lay a tongue of the Okanogan Lobe, both of which came from the Cordilleran Ice Sheet. The Okanogan Lobe blocked the valley's natural outlet.

As you pass through the town of Chelan you can see numerous flat areas perched at various levels around the lake. Called kame terraces, they consist of glacial outwash and some till that was deposited between the edge of glacial ice and the valley walls. Notice also that the bedrock consists mostly of tonalite of the Chelan Mountains terrane instead of migmatite.

North of Chelan, US 97A passes by gneiss of the Chelan Mountains terrane and emerges on the top of the Great Terrace, a former floodplain of the Columbia River made mostly of glacial outwash. US 97A descends the terrace and intersects US 97 about a half mile (0.8 km) north of milepost 239.

WA 20
BURLINGTON—TWISP
140 miles (225 km)

You could spend a lifetime on WA 20, the North Cascades Highway. Indeed, some geologists have spent much of their careers studying the North Cascades, a landscape of wildly spectacular scenery and equally complex geology. The highway traverses all three domains of the North Cascades and their bounding fault zones. The Western domain consists of folded and thrust-faulted accreted terranes with rocks that have mostly been metamorphosed at moderate to low

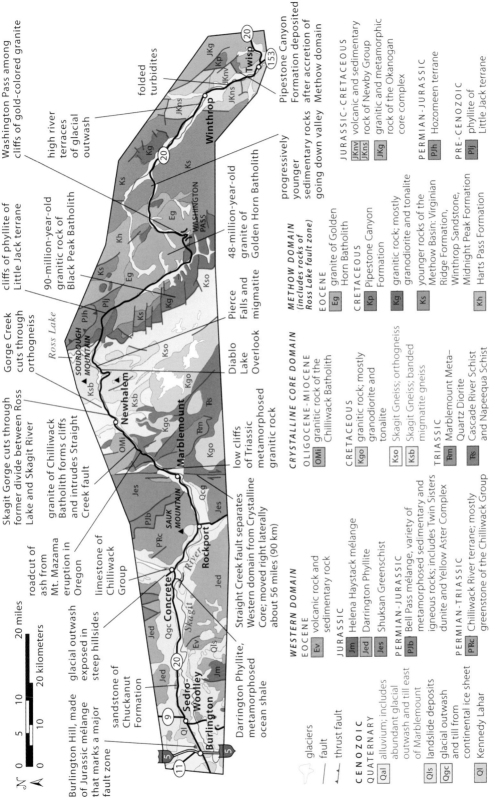

Geology along WA 20 between Burlington and Twisp.

temperatures and intruded in places by Mesozoic and Cenozoic granitic rocks. The Crystalline Core domain consists of high-temperature metamorphic and igneous rocks, which originated as the roots of a magmatic arc. The Methow domain consists of thrust-faulted and folded low-grade metamorphic and sedimentary rocks.

Burlington to Marblemount

East of Burlington, WA 20 follows the meandering Skagit River. Glacial ice of the Puget Lobe flowed eastward some 30 miles (48 km) up the river's floodplain from the Puget Lowland. Sometime after the ice retreated, the Kennedy Lahar, a volcanic mudflow from Glacier Peak some 60 miles (100 km) away, covered the glacial deposits in much of the valley. The lahar is generally hidden by vegetation and urban development, but it covers most of the valley between Burlington and a few miles west of Sedro Woolley.

Just east of I-5, WA 20 passes the south side of Burlington Hill, made of highly deformed Jurassic metasedimentary and metavolcanic rocks of the Helena Haystack mélange. The mélange formed along a large fault zone that separates the two blocks of the Western domain. To the northeast lies the Northwest Cascades thrust system, whereas to the southwest lies the Eastern and Western mélange belts of the Cascades. WA 20 crosses the fault west of Sedro Woolley and heads eastward through the Northwest Cascades thrust system. A quarry on the south side of Stirling Hill, north of the road at milepost 62, exposes some of the metavolcanic rock within the fault zone.

Between mileposts 71 and 72, bedrock exposures consist of the Chuckanut Formation. This rock unit, best viewed near Bellingham, was deposited in rivers during Eocene time, after metamorphism of the Western domain. The floodplain narrows as you head farther east. About a quarter mile (400 m) east of milepost 85, you pass an outcrop of intrusive rocks of the Yellow Aster Complex on the north side of the road. These rocks are the some of the oldest rocks of the North Cascades, pre-dating the Devonian in age. They exist as large, isolated blocks within the Bell Pass mélange but are mostly hidden by the forest.

The high ridges north and south of the road consist of Darrington Phyllite, originally sediments deposited in the deep ocean during the Jurassic Period but later metamorphosed. Although this rock doesn't show up right along the highway, you can visit a large outcrop of it 1 mile (1.6 km) south of the town of Concrete. Take Concrete Sauk Valley Road 0.9 mile (1.4 km) south to where it crosses the Skagit River; the outcrop lies next to the bridge on the south side.

The ocean crust that sediments of the Darrington Phyllite were deposited on was metamorphosed to form the Shuksan Greenschist, which forms some of the mountains along the road farther east. Together, these rocks make up the Easton terrane, which forms the uppermost thrust sheet of the Northwest Cascades thrust system. The presence of the blue amphibole mineral glaucophane in some of the rocks indicates that metamorphism occurred under conditions of high pressures and low temperatures, a combination found only in subduction zones.

Concrete, marked by prominent cement-storage silos on the west edge of town, was incorporated as a town in 1909. It produced cement, created mostly from 330-million-year-old marble mined from outcrops along the southern shore of Lake Shannon, just 1 mile (1.6 km) north of town. The marble is part of the highly deformed Chilliwack Group of the Chilliwack River terrane, which forms the high peaks and ridges on the north and northeast sides of town, including the prominent Sauk Mountain. The terrane consists of metamorphosed volcanic and sedimentary rocks of a volcanic arc that existed in Devonian through Triassic time. A thrust fault separates this terrane from the Easton terrane to the south, which has been thrust over the Chilliwack River terrane. To see the underlying bedrock of the Chilliwack River terrane, head north on South Dillard Avenue on the east side of town for a quarter mile (400 m). Limestone crops out beneath the bridge over the Baker River.

The hills just north and south of Concrete are made of glacial outwash and till left by the ice that flowed up-valley from the Puget Lowland. You can see excellent exposures of the glacial deposits by turning north at the silos and driving 1.5 miles (2.4 km) up Burpee Hill Road.

Glacial outwash sediments near Concrete, deposited by meltwater flowing from ice that moved eastward, up the Skagit River valley.

East of Concrete, greenstone of the Chilliwack River terrane shows up intermittently along the highway between mileposts 92 and 93 and between mileposts 98 and 99. A large pull-out lies opposite an exposure of the greenstone at milepost 98. Although the rock looks pretty nondescript from a distance, a close look with a hand lens shows that it is highly fragmented, probably from when the rock originally formed as a submarine lava. Between milepost 102 and Marblemount, you pass by cliffs of Shuksan Greenschist of the Easton terrane.

Marblemount to Ross Lake

At the east edge of Marblemount, you cross the Straight Creek fault at the Skagit River and pass into the Crystalline Core domain, where the hardest rocks of the Cascades form its most rugged landscapes. Intrusive rocks, mostly granitic in composition, formed in the roots of ancient volcanic arcs, similar to today's Cascade volcanoes, and younger intrusive rocks cut across the older ones. Metamorphism occurred from late Cretaceous to Eocene time.

WA 20 follows the Straight Creek fault northward from Marblemount over forest-covered river deposits. You gain some nice views of peaks in the Western domain made of Shuksan Greenschist. Halfway between mileposts 109 and 110 you encounter outcrops of a Triassic-age, metamorphosed granitic rock called the Marblemount Meta–Quartz Diorite. A good pull-out lies just east of the outcrop at milepost 110. Look for a metamorphic layering (a foliation) defined by white and green micas in a parallel arrangement. In some places you can see the foliation folded into tiny corrugations. The presence of the green mica chlorite indicates that minerals continued to grow in the rock in the presence of water as the rock was cooling down. The highway passes more outcrops of the meta-quartz diorite all the way to the boundary of North Cascades National Park, just west of milepost 112.

Just east of the park boundary, you cross a fault and enter the Cascade River Schist, many good outcrops of which appear on the north side of the highway near milepost 114. The schist originated as shale that was deposited on the quartz diorite in a volcanic arc prior to metamorphism.

Look for a roadcut of volcanic ash on the north side of the road 0.7 mile (1.1 km) east of milepost 114. This ash originated in the cataclysmic eruption of Mt. Mazama that collapsed to form Crater Lake in Oregon 7,700 years ago. It's unusually thick in this location because runoff and tributaries carried the ash into a lake, where the ash settled to the bottom. The lake formed behind a landslide that temporarily dammed the Skagit River about a half mile (0.8 km) downstream.

Between mileposts 115 and 118, WA 20 encounters granitic rock of the Chilliwack Batholith, which intrudes the Cascade River Schist and most of the major fault zones, including the Straight Creek fault. The batholith consists of multiple intrusions that range in age from 35 to 2 million years, most of which are granitic and cover nearly 400 square miles (1,036 km²) of land. It formed the early roots of the magmatic arc along the Cascadia subduction zone. At milepost 118, the high cliffs to the northeast are part of the Skagit Gneiss Complex.

The small town of Newhalem occupies the narrow eastern edge of the Skagit River's floodplain at the mouth of the Skagit River Gorge. Seattle City Light, which operates the Skagit hydroelectric project, owns and operates the town. Electricity comes from three dams in the narrow gorge: Gorge, Diablo, and Ross Dams.

The gorge did not exist before Pleistocene time. The upper Skagit River, beneath what is now Ross Lake, flowed northward into the Fraser River, and the lower Skagit River flowed westward into the Puget Lowland. The mountainous area of today's gorge divided the two watersheds. During the ice ages, each time a glacier advanced southward into the upper Skagit River, it dammed the north-flowing waters, forming a lake that overflowed to the south. The voluminous meltwater of many advances and retreats cut the gorge. Once connected, the upper Skagit reversed its flow to pour southward through the gorge and down the lower Skagit River.

WA 20 follows the gorge as far as the Ross Dam trailhead at milepost 133. Along the way it passes an uncountable number of awesome exposures of the Skagit Gneiss Complex, the main rock unit of the Crystalline Core. The complex consists of two parts: light-colored orthogneiss, formed by metamorphism of intrusive igneous rock, and darker-colored banded gneiss, formed by metamorphism of both intrusive and sedimentary rock. Some of the original sedimentary rock was deposited as far back as the Triassic Period. Donna Whitney, a metamorphic petrologist at the University of Minnesota, determined that the gneiss formed at temperatures greater than 1,300°F (700°C)and at depths of 18 to 19 miles (29–30 km) in the crust.

Between Newhalem and about milepost 125, outcrops consist of orthogneiss. A good place to stop is Gorge Creek, halfway between mileposts 123 and 124, where erosion has cut a slot-like gorge into the orthogneiss. The creek falls steeply from a glacial cirque into the Skagit River Gorge. Its steepness originated in the rapid downcutting of Skagit River Gorge, which left the glacial cirque stranded above the valley.

The pull-out 0.1 mile (160 m) east of milepost 125 offers the first good chance to see the banded gneiss and compare it to the orthogneiss. The banded gneiss is highly folded and contains a large proportion of light-colored igneous material, some of which seems to intrude and get deformed along with the rest of the rock. This type of rock, which is metamorphic but contains a significant amount of igneous material, is called a migmatite and represents rocks that were once partially molten. In this outcrop you can find small reddish garnet crystals that grew during metamorphism.

Between mileposts 125 and 130, WA 20 passes numerous exposures of the banded gneiss. The road to Diablo Dam offers more garnet-bearing outcrops. Just east of the turnoff, look for some light-colored granitic dikes. These rocks have been dated to 45 million years, which, aside from the Chilliwack Batholith, makes them some of the youngest rocks of the Cascades' Crystalline Core. Near milepost 129, look for an elongate inclusion of marble.

The Diablo Lake Overlook lies a quarter mile (400 m) west of milepost 132. Sourdough Mountain, the high ridge north of the lake, and the cliffs along the

Gorge Creek cuts a slot through orthogneiss of the Skagit Gneiss Complex. Closeup: Note the bands in the orthogneiss, formed by the metamorphism of igneous rock. Penny for scale.

Migmatite of the banded portion of the Skagit Gneiss. Note the predominance of light-colored igneous material.

lake's north shore consist of the banded gneiss. Davis Peak, which rises high to the northwest of Diablo Dam, consists of Skagit orthogneiss. Looking south over Thunder Arm, you can see the glacially eroded north face of Colonial Peak, also made of the orthogneiss. Looking across the highway, you can see a roadcut of the orthogneiss, cut by dikes of lighter-colored granitic rock.

The striking turquoise color of Diablo Lake comes from the effect of glacial sediment, or rock flour, that is so fine grained it remains suspended in the water almost indefinitely. Water absorbs long-wavelength light, such as reds and yellows, and reflects the remaining short- and medium-wavelength violets, blues, and greens. Rock flour absorbs much of the short-wavelength light, so it removes much of the violet and blue colors to reflect mostly greens. As a general rule, the higher the content of rock flour in a lake, the more blue it absorbs and the greener the color.

Pierce Falls flows beneath the roadway a half mile (0.8 km) east of milepost 133. The roadcut on the east side of the bridge displays rock of the banded gneiss, including a lot of deformed pegmatite. Within it are lenses of biotite gneiss, and everything is cut by a series of granitic dikes and sills. A number of researchers have argued that the biotite gneiss originated as a sedimentary rock that was

Diablo Lake, viewed from Diablo Lake Overlook, is turquoise because of the glacial rock flour suspended in the water. Davis Peak, formed of orthogneiss, is the prominent mountain the background.

heated to the point that it partially melted. Many of the structures in this outcrop indicate deformation while the rock was in this partially molten state.

Two viewpoints between mileposts 135 and 136 offer views of Ross Lake, which fills the upper Skagit River valley behind Ross Dam and extends some 19 miles (30 km) north into Canada. The Ross Lake fault zone, which forms the eastern edge of the Crystalline Core domain, cuts diagonally across the lake a few miles to the north. The high peaks on the east side of the lake belong to the Methow domain. They are part of the Hozomeen terrane, oceanic crust that formed from Permian through Jurassic time. Most researchers think it was the ocean floor on which the younger sedimentary sequence of the Methow domain was deposited, but the Hozomeen has since been metamorphosed and separated from the sedimentary rocks by a fault.

Up and Over Washington Pass

The road crosses the Ross Lake fault zone near milepost 140, east of which is phyllite of the Little Jack terrane, a low-temperature metamorphic rock marked by fine grains and a distinct sheen on its flat surfaces. The Little Jack terrane is difficult to understand in the context of the surrounding rocks. It is only found within the Ross Lake fault zone and seems to have attributes of both the Crystalline Core and Methow domains. Some of the best exposures of the phyllites are on the south side of WA 20 a quarter mile (400 m) west of milepost 141 and 0.1 mile (160 m) west of milepost 142. The fine-grained and platy nature of these rocks readily lend themselves to slope failure, as you can see by the many landslide scars along this stretch of road.

Just east of milepost 142, the road turns southeast to follow Granite Creek in a nearly straight line to Rainy Pass, about 16 miles (26 km) away. The straightness of Granite Creek implies that it follows a fault zone, along which the broken and crushed granitic rock erodes more easily. Between mileposts 144 and 156, WA 20 passes through granite of the Golden Horn Batholith, which intruded 48 million years ago, after the most recent period of metamorphism. About 100 square miles (260 km²) in area, this batholith forms most of the high mountains on the northeast side of the highway. At about milepost 156, the highway passes temporarily into the adjoining 90-million-year-old Black Peak Batholith. You can usually tell the rocks apart from a distance because the granite of the Golden Horn Batholith displays a distinct golden hue from oxidized iron in the rock.

The overlook at Washington Pass offers incredible views of the upper reaches of Early Winters Creek and the surrounding peaks, all made of granite of the Golden Horn Batholith. To the south you can see Liberty Bell Mountain and the Early Winters Spires. Kangaroo Ridge lies directly east, its jagged top the result of intense mechanical weathering along fractures in the rock. The rock at the overlook also consists of the Golden Horn granite. You can find patches of rock that are glacially polished, as well as small cavities in the rock formed from gas bubbles in the original magma. These gas bubbles typically form only when magmas intrude shallow levels of the crust. The Golden Horn Batholith likely intruded within the upper 2 miles (3.2 km) of Earth's crust.

East of Washington Pass along Early Winters Creek, the rocks of the Golden Horn Batholith show off their distinctive gold color especially well on weathered surfaces. An exposure across from the wide shoulder 0.6 mile (1 km) downhill from milepost 167 displays the rock's characteristic rapakivi texture, in which plagioclase feldspar forms whitish rims around pinkish orthoclase feldspar. Look also for small clots of mafic minerals.

Near milepost 168, you can see sedimentary rocks of the Methow domain on the ridge to the north. From this distance, you can detect a subtle layering to the rock and a difference in color compared to the granite to the west. The sedimentary rock that crops out just west of milepost 171 is part of the Harts Pass Formation, which was probably deposited in a submarine fan during the Cretaceous Period. If you are heading east, you pass into progressively younger Cretaceous rocks. First you encounter the Virginian Ridge Formation, which is also marine, and then between mileposts 177 and 178, near the mouth of the canyon, some cliffs of the Winthrop Formation, sandstone deposited in rivers. Above the Winthrop lies the Midnight Peak Formation, which consists of sedimentary and volcanic rocks. It frames both sides of the Methow Valley until the turnoff for Mazama, a half mile (0.8 km) southeast of milepost 179.

Liberty Bell Mountain and golden granite of the Golden Horn Batholith.

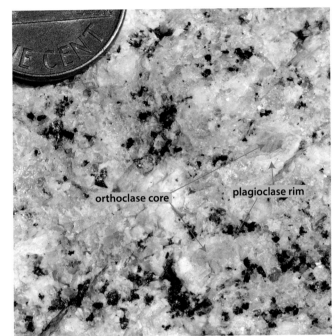

Rapakivi texture in granite of the Golden Horn Batholith.

orthoclase core · · plagioclase rim

Methow Valley

Between Mazama and southeast of milepost 184, the hills northeast of the highway consist of Cretaceous diorites that intruded the sedimentary sequence of the Methow domain. You can glimpse a great exposure of the Winthrop Formation through the trees between mileposts 186 and 187. A fault zone near milepost 187 brings sedimentary rocks of the Newby Group to the surface on both sides of the road. The Newby Group, which consists of deep-ocean sedimentary rocks as well as rhyolites and volcanic breccia, formed during the Jurassic Period, probably as a volcanic arc. On the west side of Winthrop, WA 20 passes along the side of a high river terrace composed of glacial outwash.

Between Winthrop and Twisp, WA 20 passes mostly glacial outwash and rocks of the Newby Group. An outstanding outcrop of steeply dipping and folded turbidites of the Newby Group shows up on the west side of the road about 1 mile (1.6 km) south of milepost 194 (at the time of this writing, there was no milepost 195). A prominent upright syncline graces the middle of the exposure. Just south of milepost 199, look for contorted mudstone of the Newby. A large pull-out immediately south of there offers a good view across the river to greenish volcanic rocks of the Newby below light-colored cliffs of

the Pipestone Canyon Formation. Consisting mostly of conglomerate and feldspar-rich sandstone, the Pipestone Canyon Formation was deposited near the
end of Cretaceous time after accretion of the Methow domain. Twisp sits on
gravel on the Methow River floodplain, surrounded by volcanic bedrock of the
Newby Group.

*The Midnight Peak Formation borders both sides of the Methow Valley.
View looking northwestward from Mazama.*

*The Pipestone Canyon Formation (high upper cliffs) was deposited on top
of volcanic rock of the Newby Group (lower, steep greenish slopes).*

WA 153
TWISP—PATEROS
30 miles (48 km)

WA 153 follows the Methow River to its confluence with the Columbia. At the junction with WA 20 east of Twisp is a good exposure of Jurassic volcanic rocks of the Newby Group, which likely formed in a volcanic arc. Regionally, these rocks lie beneath the sedimentary rocks of the Methow Basin. For the first 6 miles (10 km) south of WA 20, there are no bedrock exposures near the road, but the flat bench some 100 feet (30 m) higher to the east marks a former position of the floodplain, now stranded as a terrace. Remnants of this terrace, as well as older higher ones, follow the valley all the way to Pateros.

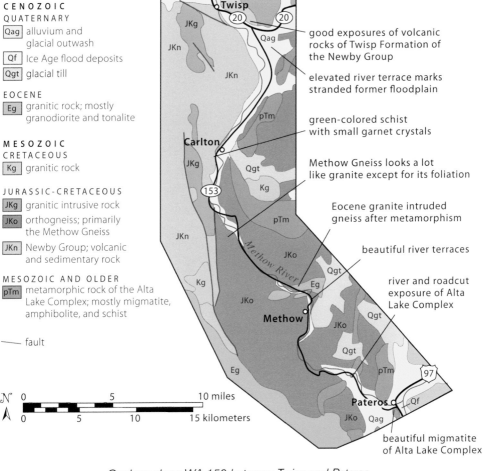

CENOZOIC
QUATERNARY
Qag alluvium and glacial outwash
Qf Ice Age flood deposits
Qgt glacial till

EOCENE
Eg granitic rock; mostly granodiorite and tonalite

MESOZOIC
CRETACEOUS
Kg granitic rock

JURASSIC-CRETACEOUS
JKg granitic intrusive rock
JKo orthogneiss; primarily the Methow Gneiss
JKn Newby Group; volcanic and sedimentary rock

MESOZOIC AND OLDER
pTm metamorphic rock of the Alta Lake Complex; mostly migmatite, amphibolite, and schist

—— fault

good exposures of volcanic rocks of Twisp Formation of the Newby Group

elevated river terrace marks stranded former floodplain

green-colored schist with small garnet crystals

Methow Gneiss looks a lot like granite except for its foliation

Eocene granite intruded gneiss after metamorphism

beautiful river terraces

river and roadcut exposure of Alta Lake Complex

beautiful migmatite of Alta Lake Complex

N 0   5   10 miles
   0   5   10   15 kilometers

Geology along WA 153 between Twisp and Pateros.

Just south of the small town of Carlton about halfway between mileposts 23 and 22, WA 153 crosses the Methow River and encounters a large outcrop of greenish schist that forms a small part of the Alta Lake Complex, which shows up again farther down the road near milepost 4. The green color of these rocks comes from the green mineral chlorite. Some of the schist also contains garnets. This exposure is too small to show on the road map.

The Methow Gneiss, which largely resembles a granite except for its foliation, shows up in a cliff a half mile (0.8 km) farther south near milepost 22. Many outstanding exposures of the gneiss show up until milepost 13, about 2 miles (3.2 km) west of the town of Methow. Its foliation, which is subtle in some hand specimens, shows up distinctly in some bedrock exposures from a distance. The Methow Gneiss was originally a granitic intrusion, about 100 million years ago, before it was metamorphosed in late Cretaceous time.

At milepost 13, just north of the town of Methow, you pass into a narrow belt of Eocene granite that intruded the gneiss. You can inspect this rock at a double

Outcrop of Methow Gneiss with otherwise faint, nearly horizontal foliation enhanced by fracturing.

River meanders and terraces near Methow. An older terrace sits at left center, higher than the terrace with the irrigated grass.

roadcut halfway between mileposts 13 and 12 and see that it lacks the foliation that characterizes metamorphic rocks. The river terraces here, former floodplains of the Methow River, are especially striking as the river makes several tight meanders in the narrow valley.

More good exposures of the Methow Gneiss and river terraces show up for another 8 miles (13 km) south of Methow. About a quarter mile (400 m) south of milepost 4, you cross the river onto the Alta Lake Complex, which shows up as the prominent outcrop at river level and next to the road. The dark color and distinct banding of this gneiss contrasts greatly with the Methow Gneiss, but the Alta Lake Complex also contains abundant amphibolite, schist, and migmatite. Here it forms a half-mile-wide (0.8 km) body surrounded by the younger Methow Gneiss, which intruded the Alta Lake Complex. Alta Lake, only 2 miles (3.2 km) south of the highway, fills a valley eroded into the Alta Lake Complex.

The river terraces between milepost 4 and Pateros are especially striking. Between milepost 1 and Pateros, WA 153 passes through the Alta Lake Complex. A spectacular exposure of migmatite forms a cliff along the road only a quarter mile (400 m) west of the junction with US 97.

WA 542
BELLINGHAM—ARTIST POINT AT MT. BAKER
60 miles (100 km)

WA 542 follows the North Fork of the Nooksack River for much of its length, ending in an awesome alpine landscape above 5,000 feet (1,500 m) in elevation between Mt. Baker and Mt. Shuksan. The route illustrates every element of the North Cascades west of the Straight Creek fault: the Northwest Cascades thrust system, the younger Chuckanut Formation that sits on top the thrust faults, the volcanism of the last several million years, and modern-day glaciation. The schematic cross section on page 111 illustrates these features.

For about the first half of the drive you pass along or between high ridges of mostly Chuckanut Formation. The most notable exception is Twin Sisters Mountain, which shows up at milepost 11 as the high peaks about 15 miles (24 km) to the southeast. The Twin Sisters consist mostly of dunite, an olivine-rich rock that forms in Earth's mantle. A good exposure of sandstone of the Chuckanut Formation lies near the junction of WA 542 and WA 9, about a quarter mile (400 m) south of milepost 14.

The landscape becomes noticeably more rugged by about milepost 16 as you head up the North Fork of the Nooksack River. Some visible scars in the

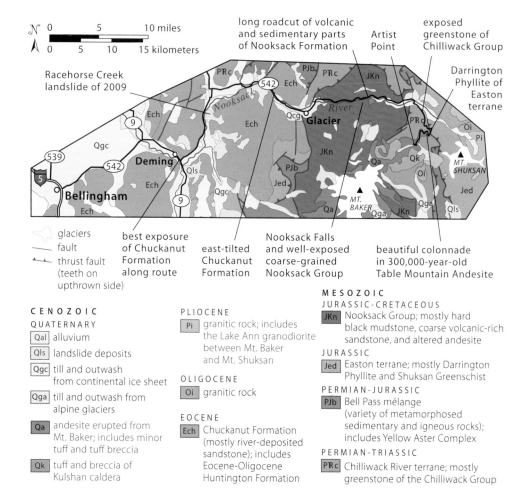

Geology along WA 542 between Bellingham and Artist Point at Mt. Baker.

mountainsides northeast of milepost 18 reflect the tendency of the Chuckanut Formation to erode by landslides, especially where the bedding dips in the same direction as the land's slope. A much larger landslide occurred a few miles to the north on Racehorse Creek in January 2009 during an unusual storm of warm rain that melted some of the existing snowpack. Some 1,500 slides formed during the storm, but the Racehorse slide stands out because it revealed fossil-rich beds of the Chuckanut Formation. One of these beds contained 10-inch-long (25 cm) footprints of diatryma, a flightless bird that reached 7 feet (2.1 m) in height. A rock bearing one of these prints is on display in the geology department of Western Washington University.

About halfway between mileposts 29 and 30, look for an outcrop of north-tilted Chuckanut Formation just before reaching a pull-out and view of Mt. Baker. The town of Glacier, just east of milepost 33, occupies a wider part of the river valley that is filled with landslide deposits. These deposits bury a fault zone that separates the Chuckanut Formation on the west from older thrust-faulted rock to the east. You cross the fault near the National Forest headquarters, just east of town.

East of the fault, WA 542 rises through rocks of the Jurassic-and-Cretaceous-age Nooksack Group, exposed here as a series of metamorphosed marine sedimentary rocks and andesitic lava flows. Above the Nooksack Group, you can see a high ridge straight ahead to the northeast between mileposts 34 and 35. These rocks consist of a thrust slice of the Chilliwack Group, which is mostly greenstone, metamorphosed basalt erupted on the seafloor during the Permian Period.

The first outcrops of the Nooksack Group show up on the north side of the road just west of milepost 38 and continue intermittently until the turnoff for Nooksack Falls, 0.7 mile (1.1 km) east of milepost 40. At Nooksack Falls, only 0.8 mile (1.3 km) down a hard-packed gravel road, Wells Creek spills over a cliff of coarse-grained, metamorphosed sedimentary rocks of the Nooksack Group. The water exploits vertical fractures in the rock to form two main falls with a narrow fin of rock between. The overlook, less than 100 yards (90 m) from the parking lot, rests on the same rock, with prominent bedding surfaces that dip toward the waterfall. On the rock behind the waterfall, you can see some horizontal lines marking traces of the bedding. The bedding, however, dips directly back into the falls at about 20 degrees.

A long bedrock exposure of both the volcanic and sedimentary parts of the Nooksack Group frames the north side of the road just east of milepost 41. Most of the western half of the exposure consists of the volcanic rock, whereas sedimentary rock makes up the eastern half. Several small pull-outs across from the roadcut allow easy inspection of the volcanic part and great views of the Nooksack River. You'll notice that the rock is highly fractured but otherwise pretty uniform. With a hand lens, you can see some crystals of plagioclase feldspar in a fine-grained, greenish-gray matrix, but the rock is pretty thoroughly altered. One pull-out lies across from some yellow sulfide mineralization. About 0.3 mile (480 m) east along the roadcut, the rock takes on a distinctly slaty appearance to indicate sedimentary origins. The best and safest place to

At Nooksack Falls, water spills over the Nooksack Group. The bedding appears horizontal because it is dipping directly away from the plane of the photograph.

The original sedimentary bedding of this slate of the Nooksack Group is defined by slight changes in grain sizes. The brown-colored bed (in the middle of the photo with the blue pen) is slightly coarser grained than the adjacent gray ones. Cleavage, which is the tendency of the rock to split, follows the bedding here.

visit the slate lies at the far east end of the roadcut. There, you can see that the original bedding, defined by some thicker and coarser-grained beds, dips very steeply toward the road. The rock breaks easily along closely spaced surfaces, called cleavage planes, that are approximately parallel to the bedding in these rocks. Cleavage typically forms at low temperatures of metamorphism.

Just west of milepost 47 you cross to the south side of the river and start the steep uphill climb that culminates at Artist Point. You pass numerous good bedrock exposures of Permian-age greenstone of the Chilliwack Group that have been pushed over the Nooksack Group along a thrust fault. Numerous pull-outs allow you access to these rocks, but an especially good outcrop lies 0.4 mile (640 m) uphill from milepost 48.

Just a quarter mile (400 m) uphill from milepost 49, you see an exposure of fresh-looking andesite lavas, marked by columnar jointing. Called the Table Mountain Andesite, these lavas erupted 300,000 years ago during the Pleistocene, flowing downhill and filling valleys eroded into the older metamorphic rocks. This exposure, almost 1,000 feet (300 m) below the elevation of the next exposures of the lava, demonstrates the steepness of the valleys that existed at the time of the eruption.

A quarter mile (400 m) uphill from milepost 51, you cross Razor Hone Creek and pass into a still higher thrust slice. It's composed of metamorphic rock of the Darrington Phyllite, which originated as shale and siltstone deposited in a deep ocean during the Jurassic Period. The road recrosses the creek and back into the Chilliwack Group about 1 mile (1.6 km) later.

You see more andesite between mileposts 52 and 53. Mt. Shuksan soars more than 1 mile (1.6 km) above directly to the east. The andesite displays beautiful, narrow columns, which might lead some to misidentify it as basalt. However, a close look at the rock shows numerous plagioclase crystals floating in a light-gray matrix, which is much more typical of andesite. The andesite dominates the bedrock for the remainder of the trip because it sits on top of and covers the greenstone of the Chilliwack Group. This point is demonstrated by the occasional reappearance of greenstone bedrock along the road, about a quarter mile (400 m) before milepost 53 and again at the first switchback above the ranger station, above milepost 55.

From Artist Point, you gain wonderful views of Mt. Baker and Mt. Shuksan and can take numerous hiking trails to even more views. Mt. Shuksan consists mostly of Jurassic-age Shuksan Greenschist, which belongs to a still higher thrust slice that was pushed over the top of the Darrington Phyllite. Shuksan Arm forms the ridge that extends northwest from the mountain on the other side of the valley. It contains rocks of three separate thrust slices.

A half mile (0.8 km) walk along Kulshan Ridge southeast to Huntoon Point allows views into the valley of Swift Creek and access to exposures of welded tuff of the Kulshan caldera. These rocks, which lie beneath the Table Mountain Andesite, erupted in a cataclysmic eruption 1.15 million years ago.

Mt. Baker

Mt. Baker forms northern Washington's preeminent landmark, with its graceful cone easily visible from the northern stretches of I-5 and from the San Juan Islands. You gain a particularly intimate view of the volcano from Artist Point at the east end of WA 542. At an elevation of 10,781 feet (3,286 m), it's the highest peak in the North Cascades and the third-highest peak in the state. Only 16 miles (26 km) south of the Canadian border, Mt. Baker is the northernmost active volcano in the conterminous United States and the only one to be shaped by both alpine and continental glaciation. Today, it hosts thirty-two alpine glaciers that form a radial pattern as they descend its flanks and drain into the Nooksack and Skagit Rivers.

Mt. Baker is the youngest volcano in the Mt. Baker volcanic field, which includes the Kulshan caldera and Black Buttes. Mt. Baker's most active period to date occurred between about 40,000 and 12,000 years ago, when it erupted thousands of feet of andesitic lava to build most of its base as well as its summit cone. Made of some two hundred steeply inclined lava flows, the summit cone hosts the ice-filled Carmelo Crater, formed by water-lava explosions during the latter part of this active period. The volcano also likely erupted abundant pyroclastic material, such as ash, pumice, and fragmental lava flows, although these materials, being less resistant to erosion, were largely removed by glacial ice. Some andesite and dacite lavas have yielded radiometric ages of 140,000, 90,000, and 80,000 years to reflect older, less well-defined, and probably much less active periods in its history.

During the last 10,000 years, Mt. Baker produced ash but no andesitic lava, although an eruption of basaltic lava took place a few miles south of the volcano between about 8,850 and 8,500 years ago. Its most significant recent eruption was andesitic ash about 6,700 years ago. This eruption likely accompanied the formation of Sherman Crater, just north of the summit, as well as several flank collapses, where portions of the mountain's sides suddenly failed in huge avalanches. These collapses spawned major lahars (volcanic mudflows) that flowed down the Nooksack Valley as well as into Baker Lake, west and south of the volcano, respectively. Some minor ash eruptions in 1843 and 1880 also spawned lahars.

In 1975, Sherman Crater showed an increase in gas emissions, interpreted as the result of magma intruding the volcano. Several years of intensive monitoring followed, although the monitoring tapered off through time with no eruptions. The US Geological Survey, however, continues to monitor the volcano's earthquakes, gas emissions, hydrology, and ground deformation. Similar to the other ice-clad Cascade volcanoes, the main hazards are lahars, which don't necessarily require an eruption to form.

View to the southwest of Mt. Baker and the Artist Point area from the Shuksan Arm. Table Mountain, on the right side of the photo, consists of Table Mountain Andesite, the same lava that forms the columns along the road. The flat-topped butte in the middle ground directly in front of Mt. Baker consists of rhyolite of the Kulshan caldera.

Other older elements of the Mt. Baker volcanic field have been reconstructed through geologic mapping, geochemistry, and radiometric age determinations. Andesite lavas that make the Black Buttes, which now form the rugged southwest ridge of Mt. Baker, erupted between about 500,000 and 200,000 years ago. Those eruptions produced some 7 cubic miles (30 km³) of material, about twice that of Mt. Baker. The Kulshan caldera, which adjoins Mt. Baker's northeast side, erupted between 12 and 19 cubic miles (50–80 km³) of rhyolitic material catastrophically about 1.15 million years ago. The caldera filled with volcanic material, so it is not a modern topographic feature. The andesite lava flows you see at Heather Meadows, Artist Point, and the Mt. Baker Ski Area erupted about 300,000 years ago, long after the caldera collapse. These lavas likely covered most of the caldera floor.

Near the Mt. Baker volcanic field, the Hannegan caldera and Lake Ann granodiorite attest to the long-lived nature of magmatism in the area. The Hannegan caldera, which lies about 12 miles (19 km) northeast of Mt. Baker, erupted about 3.7 million years ago, and the Lake Ann granodiorite, which adjoins the east side of the Kulshan caldera, intruded about 2.8 million years ago.

Mt. Rainier stratovolcano from Reflection Lakes

SOUTH CASCADES

The South Cascades are dominated by Oligocene and younger volcanic rocks generated in the magmatic arc above the Cascadia subduction zone. The arc lies just east of I-5, which forms a convenient western boundary to the region. Likewise, the northern boundary of the South Cascades coincides with I-90. The entire Cascade Range has been tilted up to the north, so although subduction-related volcanoes have erupted in the North Cascades, erosion has mostly stripped away the cover of volcanic rocks north of I-90. The pre-Cenozoic basement rocks of the North Cascades plunge southward beneath the cover of volcanic rocks in the South Cascades and probably underlie the entire region. Isolated exposures, such as on Manastash Ridge near Cle Elum and surrounding Rimrock Lake on the Naches River, consist of rock units very similar to some of those cropping out in the North Cascades and San Juan Islands.

Growth of the Cascade Magmatic Arc

Oligocene-age volcanic rocks in the South Cascades record the development of a magmatic arc that was established along the continental margin about 35 million years ago. Older sedimentary rocks of the coastal plain crop out beneath the volcanic cover in many places of the South Cascades. The rock units have local names, but the most familiar is the Puget Group. A few isolated volcanic centers developed in the broad coastal plain about 42 million years ago, but they may have been related to more widespread Eocene magmatism, known as the Challis episode, that occurred to the east.

Early work by the US Geological Survey established a sequence of three volcanic units in Mt. Rainier National Park: from lowest and oldest to highest and youngest, they are the Ohanapecosh, Stevens Ridge, and Fifes Peak Formations. Joe Vance of the University of Washington and coworkers found that the Stevens Ridge and Fifes Peak units are interlayered, and therefore they favor using only the Ohanapecosh and Fifes Peak as formal formations. We follow their usage but retain the Stevens Ridge designation for the rocks within the national park. Vance and coworkers dated the Ohanapecosh Formation as about 36 to 28 million years old and the Fifes Peak as about 27 to 20 million years old. The aggregate thickness of the units may be about 4.3 miles (7 km). The units consist of many rock types, including lava flows, ash flows, volcanic breccia, and sediments consisting largely of volcanic sand. The breccia likely originated as hot volcanic debris flows or lahars and now features coarse fragments

173

of volcanic rocks in a fine-grained matrix. As is typical in many subduction-related volcanic arcs, the composition of the lavas ranges widely from basalt to andesite and even rhyolite. The sequence of volcanic rocks at any one place is expectably complex, with abrupt changes in the thickness of the interlayered flows and breccias. In addition, hydrothermal activity in a volcanic system may alter the original volcanic rock. This alteration weakens the rock so that it more easily forms landslides and produces a range of new colors, most typically shades of green or yellow. In general, green colors result from higher-temperature alteration while yellow colors result from oxidation at lower temperatures. To add to the complexity, in most of the region the units are broadly folded in northwest-trending anticlines and synclines. The age of the folding is not well known but is probably mid- to late Miocene.

Intrusive rocks emplaced from late Oligocene to late Miocene time cooled from the magma that pooled at depth below the volcanoes. A typical example of the plutons along the axis of the Cascades arc is the Spirit Lake pluton north of Mount St. Helens. The texture and mineralogy of the granitic rocks indicate that the body of magma cooled close to the Earth's surface. Its age, about 21 million years old, is barely older than the 23-million-year-old remnants of the caldera and ash flows that the magma intruded as it rose toward the surface. Other well-known intrusive bodies are the Miocene Tatoosh pluton at Mt. Rainier and the Bumping Lake pluton at Chinook Pass. Geologists think there was a hiatus of intense magmatic activity in the arc in late Miocene and Pliocene time because there are few rocks from that time period.

By far the most familiar features in the entire Cascade Range are the majestic, isolated, ice-clad cones—the most recent, and active, manifestation of the Cascade volcanic arc. Rainier, St. Helens, and Adams in the South Cascades are stratovolcanoes. Other types of volcanoes, including shield volcanoes, cinder cones, plugs, domes, and calderas, occur in Washington, although the diversity and abundance of Cascade volcanic features are best represented in Oregon.

The type of volcano and volcanic eruption depends on the silica content of the magma as well as its water and gas content. In general, lower-silica magmas such as basalt tend to flow more easily and create volcanoes that are less steep than those formed of higher-silica, less fluid magmas. Magmas with a higher water and gas content tend to be much more explosive than those with a lower water and gas content. If higher-silica magmas also have high water and gas contents, they can create catastrophic eruptions, such as the May 18, 1980, eruption of Mount St. Helens—or even larger, caldera-forming eruptions, such as the one that formed the Kulshan caldera near Mt. Baker just over 1 million years ago.

A single eruption can produce a diverse set of volcanic features. For example, the May 18 eruption of Mount St. Helens, even though much smaller than many of its previous eruptions, produced a giant rock avalanche and numerous lahars and ash flows. Smaller subsequent eruptions and the growth of a spine of andesite in the cone attest to a still-active volcano. Today, the region bristles with seismometers monitoring the earthquakes beneath the cone that might herald another major eruption. The histories of Mt. Rainier and Mount St. Helens are detailed in later sections.

Alteration of Oligocene lavas near milepost 37 on WA 504 to Mount St. Helens.

SHIELD VOLCANO

Low gradient and large, consists of basaltic lava flows that were relatively fluid; may have cinder cones on flanks; resembles a shield lying on its side.

CINDER CONE

Steep, relatively small conical hill that consists of cinders erupted from a central vent.

STRATOVOLCANO / COMPOSITE VOLCANO

Steep and large, consists mostly of andesitic and pyroclastic material (explosive products such as ash and pumice); may consist of two or more coalesced volcanoes.

CALDERA

Steep-sided depression, typically 100s to 1,000s of feet deep and more than a mile across; formed by collapse of volcano into emptied magma chamber after major eruption.

DOME VOLCANO

Very steep gradient, but relatively small feature, made out of dacitic or rhyolitic lava, which tend to be very viscous (sticky) and so cannot travel far from the vent.

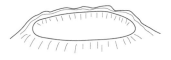

The type of volcano that forms depends largely on the type of magma (bars at left) that erupts. –Modified from Miller, 2014

basalt

andesite

dacite-rhyolite

Mt. Adams

Washington's second-highest peak, Mt. Adams reaches an elevation of 12,276 feet (3,742 m) above sea level, more than 10,000 feet (3,000 m) above Davies Pass at the edge of the Columbia River Gorge on US 97. Although nearly 2,000 feet (610 m) shorter than Mt. Rainier, lava flows from Mt. Adams cover a significantly larger area (about 230 square miles, 600 km^2) and occupy a larger volume (about 70 cubic miles, 300 km^3) than any other Cascade volcano, except for Mt. Shasta in California. The ice-capped stratovolcano hosts nine named glaciers that radiate outward from its summit. Today, these glaciers extend to elevations of about 7,000 feet (2,100 m), but during the end of the Pleistocene Epoch, they flowed to much lower elevations and covered almost the entire volcano.

Mt. Adams consists almost entirely of andesite lava flows erupted from a central vent at the summit as well as vents high on the mountain's flanks. Since the end of Pleistocene time, about 10,000 years ago, these eruptions built the summit area up from an elevation of about 8,000 feet (2,400 m). Its central vent, now marked by a steep summit cone, consists of andesite fragmented through its explosive interaction with ice. Much of this fragmental rock is also highly altered because of the effects of ongoing hydrothermal activity.

The basaltic Mt. Adams volcanic field encircles the stratovolcano. It contains more than sixty known vents whose history partly pre-dates but generally parallels that of Mt. Adams. The field covers an area of some 250 square miles (650 km^2)

View of Mt. Adams from near Davies Pass on US 97. The stratovolcano consists of a steep, ice-covered edifice and a broad base, both of which are built almost entirely of andesite lava flows.

and contains cinder and spatter cones and a few small shield volcanoes, as well as innumerable lava flows, although the heavy forest cover obscures most of the vents.

Two other Quaternary-age basaltic volcanic fields adjoin the Mt. Adams field. Immediately to the southwest lies the Indian Heaven volcanic field, and to the east lies the Simcoe volcanic field, both of which are mostly older—but partially contemporary with—than the Mt. Adams field. This part of the modern Cascades volcanic arc is unusually wide, reaching 90 miles (150 km) from the west side of Mount St. Helens to the eastern edge of the Simcoe volcanic field at the town of Goldendale.

Mt. Adams and its volcanic field date back some 520,000 years, although rocks as old as 900,000 years exist on the edges of the field. About 460,000 years ago, the stratovolcano's central vent shifted about 3 miles (5 km) northwest from its original location near upper Hellroaring Creek to its present location at today's summit. A well-dated series of lava flows shows a history of consistent activity since then, with only occasional quiet periods lasting for a few thousand to up to 30,000 years. Eruptions throughout this history were likely far less violent than nearby Mount St. Helens, which produced numerous ash-fall and pyroclastic flow deposits. At Mt. Adams, the eruption products consist almost entirely of lava flows.

Despite the presence of steam vents at the volcano's summit, today seems to be one of Mt. Adams's quiet periods. Neither the stratovolcano nor its surrounding volcanic field have erupted during historic times, and only one eruption might be younger than 3,500 years. Still, the volcano can produce dangerous debris flows sparked by a variety of causes, including minor eruptions or glacial outburst floods. The steep upper slopes are especially prone to landsliding because they are built on multiple thin lava flows, parts of which have been hydrothermally altered to clay.

Guides to the South Cascades

US 12
I-5—Yakima
137 miles (220 km)

US 12 crosses the Cascade Range at White Pass, with the Cowlitz River on its west side and tributaries of the Yakima River on its east side. For the 15 miles (24 km) between I-5 and Mayfield Reservoir, US 12 passes over a mostly low-relief surface developed on glacial outwash deposited by meltwater streams flowing from alpine glaciers. The outwash deposit is very thick here: the road abruptly drops more than 150 feet (45 m) where it enters the valley of Lakamas Creek between Marys Corner and Ethel but remains within the outwash.

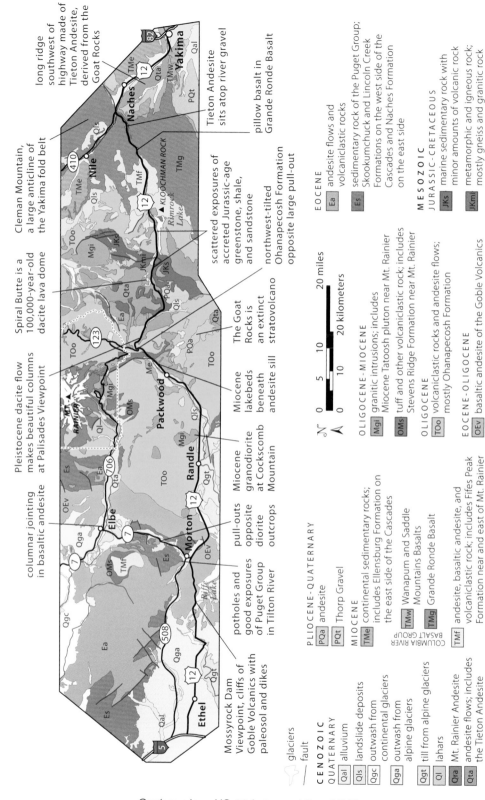

Geology along US 12 between I-5 and Yakima.

long ridge southwest of highway made of Tieton Andesite, derived from the Goat Rocks

Cleman Mountain, a large anticline of the Yakima fold belt

Tieton Andesite sits atop river gravel

pillow basalt in Grande Ronde Basalt

scattered exposures of accreted Jurassic-age greenstone, shale, and sandstone

northwest-tilted Ohanapecosh Formation opposite large pull-out

Spiral Butte is a 100,000-year-old dacite lava dome

The Goat Rocks is an extinct stratovolcano

Pleistocene dacite flow makes beautiful columns at Palisades Viewpoint

Miocene lakebeds beneath andesite sill

Miocene granodiorite at Cockscomb Mountain

columnar jointing in basaltic andesite

pull-outs opposite diorite outcrops

potholes and good exposures of Puget Group in Tilton River

Mossyrock Dam Viewpoint, cliffs of Goble Volcanics with paleosol and dikes

EOCENE

Ea — andesite flows and volcaniclastic rocks

Es — sedimentary rock of the Puget Group; Skookumchuck and Lincoln Creek Formations on the west side of the Cascades and Naches Formation on the east side

MESOZOIC

JURASSIC-CRETACEOUS

JKs — marine sedimentary rock with minor amounts of volcanic rock

JKmi — metamorphic and igneous rock; mostly gneiss and granitic rock

OLIGOCENE-MIOCENE

Mgi — granitic intrusions; includes Miocene Tatoosh pluton near Mt. Rainier

OMs — tuff and other volcaniclastic rock; includes Stevens Ridge Formation near Mt. Rainier

OLIGOCENE

TOo — volcaniclastic rocks and andesite flows; mostly Ohanapecosh Formation

EOCENE-OLIGOCENE

OEv — basaltic andesite of the Goble Volcanics

CENOZOIC

QUATERNARY

Qal — alluvium

Qls — landslide deposits

Qgc — outwash from continental glaciers

Qga — outwash from alpine glaciers

Qgt — till from alpine glaciers

Ql — lahars

Qra — Mt. Rainier Andesite

Qta — andesite flows; includes the Tieton Andesite

glaciers

fault

PLIOCENE-QUATERNARY

PQa — andesite

PQt — Thorp Gravel

MIOCENE

TMe — continental sedimentary rocks; includes Ellensburg Formation on the east side of the Cascades

COLUMBIA RIVER BASALT GROUP

TMw — Wanapum and Saddle Mountains Basalts

TMg — Grande Ronde Basalt

TMf — andesite, basaltic andesite, and volcaniclastic rock; includes Fifes Peak Formation near and east of Mt. Rainier

0   5   10   20 miles

0   10   20 kilometers

N

Outcrops of basaltic andesite lie on the south side of the road just east of milepost 83. These lavas, as well as the surrounding hills, are part of the Goble Volcanics, which erupted from the Cascades between about 45 and 32 million years ago. A truly impressive series of roadcuts in the basaltic andesite begins just west of milepost 89 and continues for a half mile (0.8 km) east of milepost 90. A large pull-out at milepost 90 for the Mossyrock Dam Viewpoint offers easy access to this long roadcut. A paleosol in the cliff across from the parking lot marks the boundaries of two flows and tilts gently eastward to ground level. The paleosol is cut by a vertical basaltic dike. You can see more paleosols in the roadcut less than a quarter mile (400 m) to the east; a third of a mile (530 m) to the east are some noticeably fragmental rocks. All of these rocks are inclined eastward.

Mossyrock Dam stretches 1,648 feet (502 m) across the Cowlitz River channel to back up Riffe Lake, which extends eastward about 13 miles (21 km). With a height of 606 feet (185 m), Mossyrock Dam is Washington's highest and provides about 40 percent of the electricity to Tacoma Power. The dam's abutments are set into the Goble Volcanics, which are nicely exposed along the short paved trail below the overlook. Along the trail, you'll notice zeolite crystals filling fractures and holes in the rock. Between the viewpoint and Morton, US 12 rises over a low pass along scattered outcrops of the Goble Volcanics.

At Morton, the Tilton River flows over good exposures of sandstone of the Puget Group, deposited in river and coastal environments during Eocene time. In many places, the sandstone displays potholes and other interesting shapes eroded by the river during high flows. To get to a bridge over the river, turn north on WA 7, turn left (west) on Main Street, and drive westward a half mile (0.8 km).

Sandstone of the Puget Group exposed along the Tilton River at Morton.

WA 7 to Elbe

North of Morton, WA 7 ascends an old river terrace and then, 0.6 mile (1 km) north of milepost 1, encounters an exposure of diorite, formed during either Miocene or Oligocene time. The surrounding rock consists of Puget Group sandstone but is poorly exposed, although some crops out between mileposts 2 and 3. About a quarter mile (400 m) south of milepost 7, two pull-outs on the east side of the highway lie across the road from exposures of more diorite. Just north of milepost 10, look west to see high cliffs of andesite flows, erupted during deposition of the Puget Group. Then, between mileposts 12 and 13, look northeastward to fabulous views of Mt. Rainier. Good exposures of Miocene-age basaltic andesite flows show up on both sides of the road as it takes a sharp bend, less than a quarter mile (400 m) south of milepost 14. An even better exposure lies on the north side of the highway at the T-junction with WA 706.

On clear days, you'll get a glimpse of Mt. Rainier from US 12 just west of Morton. More scattered exposures of the Goble Volcanics show up until about milepost 104. From there until Randle are beautiful views of Mt. Adams but sparse bedrock exposures.

Randle sits along the meandering Cowlitz River. The Cowlitz drains a watershed nearly 2,600 square miles in size (6,700 km²) that includes parts of Mount St. Helens, Mt. Rainier, and Mt. Adams. A glacier occupied the Cowlitz Valley here until as recently as 15,000 years ago, and the river is still fed largely by glaciers. Minute particles of rock flour, created through the abrasion of rock at the base of a glacier, absorb short-wavelength light, giving the water its soft greenish-gray color. You can get a nice view of the river from a bridge by driving less than a quarter mile (400 m) south on WA 131 toward Mount St. Helens.

Cockscomb Mountain, the prominent peak to the north and east of milepost 119, is Miocene granodiorite that intruded the surrounding volcanic Ohanapecosh Formation. Many similar intrusions dating from late Eocene to Pliocene time extend north-south along the Cascade Range and likely fed earlier episodes of volcanic activity. A somewhat older intrusion of diorite with numerous inclusions of andesite is exposed adjacent to the road about 0.1 mile (160 m) west of milepost 121. You can find a safe place to park on Davis Creek Road, which borders the outcrop's east side. An exposure of Ohanapecosh Formation lies along the highway just west of the Cora Bridge across the Cowlitz River, near milepost 123. This area is mapped as a landslide complex, however, so the outcrop, though pretty large, was probably carried here by the slide.

Cockscomb Mountain, made of Miocene granodiorite, intrudes the surrounding Ohanapecosh Formation.

For a few miles northeast of Packwood, some good exposures of an andesite sill crop out along the road. In some places, such as at milepost 134, you can see dark-colored, laminated shales beneath the sill. These shales were deposited in lakes during Miocene time. Beginning just east of milepost 135, the road passes into welded tuffs of the Stevens Ridge Formation, with some great roadcuts just west of milepost 137; about a quarter mile (400 m) east of milepost 137, you'll see volcaniclastic rocks of the Ohanapecosh Formation, which continue eastward almost until milepost 143.

Probably the best place to inspect the Ohanapecosh Formation is at the junction with WA 123, which heads northward into Mt. Rainier National Park. There, you can see multiple beds of these rocks, inclined moderately to the northwest (left, as you face the outcrop), overlying a thick layer of tuff, which is intruded by a basaltic sill. If you look closely, you'll see that these rocks are made entirely of volcanic particles, most of which originated in explosive eruptions. A recent study argues that soon after the eruptions, the particles were redeposited as sediments, mostly as thick gravity-driven flows in a large, deep lake.

Just east of milepost 141, the Palisades Viewpoint looks southwest across the Clear Fork of the Cowlitz River to some spectacular columns of a thick, Pleistocene-age dacite flow. The columns exceed 400 feet (120 m) in height, the likely result of the flow ponding against some kind of barrier, such as a tongue of glacial ice. The flow has been traced more than 5 miles (8 km)

The Ohanapecosh Formation at the junction with WA 123. The light-colored rock on the right side of the exposure is tuff that is intruded by a dark basaltic sill.

up the canyon to the south-southeast, where it erupted from a vent on the side of the Goat Rocks.

The Goat Rocks are a rugged series of peaks and glaciated valleys derived from a deeply eroded, extinct stratovolcano built atop the Ohanapecosh Formation. During Pleistocene time, the volcano likely attained a size similar to Mt. Hood in Oregon. The bedrock consists of numerous shallow intrusive bodies of probable Pliocene to Pleistocene age that may have fed early eruptions, as well as several large flows of Pleistocene andesite. One of these flows, the Tieton Andesite, rims the south side of US 12 from Naches to the outskirts of Yakima, an exceptional distance for an andesite flow.

At milepost 143, US 12 passes through a double roadcut of Pleistocene basalt, which contains a host of small and large air bubbles, called vesicles. Most of the exposures are netted to keep falling rocks off the roadway, although a pull-out at the east end of the roadcut allows one the chance to see some of the rock. The pull-out also gives a nice view of Lava Creek Falls as it plunges over the tall columns of the Pleistocene dacite flow seen at the Palisades Viewpoint.

From east of milepost 146 to a half mile (0.8 km) east of White Pass are highly fractured sedimentary rocks of Jurassic to Cretaceous age. Called the Russell Ranch Formation, these rocks were deposited in the ocean and then scraped off a subducting oceanic plate while being accreted to the edge of North America during the end of Cretaceous time. At the Mt. Rainier Viewpoint, you gain a good view southwest to the Goat Rocks and a fabulous view northeast to Mt. Rainier.

East of White Pass at milepost 151 is Pleistocene andesite that fractured into thin plates while cooling. Spiral Butte forms the prominent flat-topped peak to the northeast. The butte is a lava dome that erupted dacite about 100,000 years ago; the lava flowed in a clockwise direction to give the butte a spiral shape in plan view. Clear Creek Falls, halfway between mileposts 153 and 154, spills over andesite and into the long, glaciated, U-shaped valley below. Some of the dacite of Spiral Butte crops out at the sharp bend in the road just east of milepost 154.

Clear Creek Falls drops 228 feet (69 m) over Pleistocene-age andesite into the glaciated Clear Creek valley.

US 12 descends back into accreted rock 0.75 mile (1.2 km) east of milepost 156 at a roadcut of highly fractured gneiss of Jurassic age. East of there are some poor exposures of greenstone, which forms by metamorphism of basalt. Near milepost 160, eastbound travelers start getting views of Rimrock Lake. Look for outcrops of deformed sandstone and shale of the Russell Ranch Formation along the highway. On either side of a dirt road 0.6 mile (1 km) east of milepost 163, you can see a range of sedimentary rocks, from mudstone and siltstone to sandstone and lenses of chert. By contrast, the outcrop 300 feet (90 m) to the east is highly deformed, with innumerable veins and small shear zones of scaly mudstone with broken lenses of chert and greenstone. More greenstone, but this time showing remnant pillow structures, forms the prominent cliff about 0.75 mile (1.2 km) east of milepost 163. In many ways, this accreted rock along Rimrock Lake resembles that in the western North Cascades and San Juan Islands. It is exposed here and near White Pass because erosion carved its way through the volcanic cover.

Outcrop of accreted Jurassic rock near Rimrock Lake. Inset shows
scaly mudstone and broken pieces of chert.

Highly fractured Jurassic granitic rock, just east of milepost 164, marks the
eastern edge of the accreted rock's exposure. Its degree of deformation contrasts
markedly with the much more intact and younger intrusive rock that forms the
high cliffs just east of the Tieton Dam Viewpoint. Called the Westfall Rocks,
this younger rock consists of finely crystalline diorite that intruded at a very
shallow level in the crust. The diorite forms both abutments of Tieton Dam, as
well as Goose Egg Mountain and the tooth-shaped Kloochman Rock, visible
on the southeast side of Rimrock Lake. Look for interesting patterns in the
cooling fractures near the contact with the older granitic rock. The ridgeline
that rises gently behind Kloochman Rock is made of the Grande Ronde Basalt
of the Columbia River Basalt Group. The forested area between the two consists
mostly of landslide deposits.

East of Westfall Rocks and Tieton Dam, US 12 follows the Tieton River
through successively younger volcanic rocks, beginning with Oligocene and
early Miocene rock, and eventually into the Grande Ronde Basalt. Watch for
outcrops of nicely bedded, volcanic-rich sedimentary rock at milepost 167, and
then andesite flows and breccias of the Miocene-age Fifes Peak Formation near
milepost 169. An outstanding exposure of a lahar frames the north side of the
road for a quarter mile (400 m) between mileposts 169 and 170. More lahars

A lahar (volcanic mudflow) in the Fifes Peak Formation is exposed between mileposts 169 and 170.

and andesite flows of the Fifes Peak Formation show up sporadically along the road for the next several miles eastward and form beautiful ledges in the canyon walls near milepost 173.

Between milepost 176 and the intersection with WA 410, at the confluence with the Naches River, US 12 passes through a deep canyon defined by multiple lava flows adorned by colonnade and entablature. Except for a 0.75-mile (1.2 km) stretch of the Tieton Andesite on either side of milepost 183, the lava flows are entirely of the Grande Ronde Basalt. Look for pillow basalt and palagonite in the several good rock exposures on either side of the road near milepost 177. Palagonite is an orange-yellow mineral that is an alteration product of glassy basalt, the glass and pillows having formed as the lava poured into a lake. A good exposure of gravel, deposited by an early Tieton River, lies on the north side of the road about 0.25 mile (0.4 km) east of milepost 178.

An instructive exposure of the Tieton Andesite lies on the north side of the road at a sharp curve, 0.75 mile (1.2 km) east of milepost 182. The andesite, which erupted from the now-extinct Goat Rocks volcano about 1 million years ago, rests on top of river gravels deposited by an earlier incarnation of the Tieton River. The andesite filled a river channel that cut through the Grande Ronde Basalt.

Between the WA 410 junction and Yakima, US 12 follows the Naches River. On the north side of the road, south-dipping flows of the Grande Ronde Basalt mark the south limb of the Cleman Mountain anticline, which continues to the northwest for another 10 miles (16 km). The canyon at milepost 187 reveals a

The Tieton Andesite flowed down a valley about 1 million years ago, burying river gravels.

beautiful overturned anticline in its eastern wall. River terraces, which mark former positions of the river's floodplain, occupy the low area between the highway and the mountain. Near milepost 188, the basalt and the anticline end abruptly at a fault zone, which is obscured by landslide deposits. Southeast of there, the road follows well-bedded deposits of the Ellensburg Formation, deposited in rivers and lakes at about the same time that lava of the Columbia River Basalt Group was erupting in Miocene time.

If you turn north on Allan Road just east of milepost 191, you can take a short detour to inspect the Ellensburg Formation. Follow Allan Road 1 mile (1.6 km) to Old Naches Highway and turn right (east). Good exposures of ash deposits and volcanic-rich sedimentary rocks show up along the road within the first half mile (0.8 km). You also gain some nice views of the Naches River floodplain. At 2 miles (3.2 km) you can turn right (south) on Locust Lane to return to US 12.

Cliffs of the Wanapum Basalt, which is part of the Columbia River Basalt Group, form the lower part of the hills southeast of milepost 193. In some places, you can see the light-colored Ellensburg Formation interbedded with the basalt. Southwest of the highway, a lumpy ridge made of the Tieton Andesite continues all the way to the Naches River crossing west of milepost 199. The irregularity of the ridgeline originated from pressure ridges that developed in the flowing lava.

Between the Naches River crossing and I-82, US 12 passes numerous cliffs of the Columbia River Basalt Group. For the first mile to the south, you can see nice colonnade in the Wanapum Basalt. To the north, Lookout Point sits on the Wanapum Basalt; cliffs beneath it belong to the Grande Ronde Basalt.

Columbia Gorge National Scenic Area

Columbia Gorge National Scenic Area, which includes both sides of the Columbia River, was established by Congress in 1986. Given its location astride the Cascades, its bedrock is mostly volcanic in origin. In ascending order, the main bedrock units consist of volcanic conglomerate and lahars of the Miocene-age Eagle Creek Formation and overlying Columbia River Basalt Group, as well as the Miocene-Pliocene Troutdale Formation, which was deposited by the ancestral Columbia River. Younger rocks include Pliocene to Quaternary basaltic lava flows as well as intrusive rocks. Below the Eagle Creek Formation, but generally not exposed, lies the Oligocene-age Ohanapecosh Formation, which consists mostly of volcanic ash-flow tuffs. Near the top of this formation, most of the rock has been altered to an impermeable clay, a key unit to understanding the landscape of the Columbia River Gorge.

The Columbia River Gorge is asymmetrical: the Oregon side rises steeply above the river, whereas the Washington side rises much more gradually. Washington drivers on WA 14 look across the river to spectacular cliffs and waterfalls, while Oregon drivers on I-84 look across the river to an interesting yet comparatively subdued landscape marked largely by landslides. The gorge's asymmetry results from a shallow southward tilt of its bedrock, which promotes landsliding and decreases the overall gradient on the north side. At the same time, the southward inclination inhibits large landslides from forming on the south side of the river, allowing that side to maintain steep slopes.

Water passes easily downward through fractures in basalt of the Columbia River Basalt Group and through the permeable rock of the Eagle Creek Formation but tends to accumulate at the clay at the top of the Ohanapecosh Formation. This water increases buoyancy forces that effectively lift the overlying rock and

View eastward up the Columbia River over the Skamania landslide complex from Cape Horn. Large landslides such as the Skamania cover most of the Washington side of the river, whereas smaller slides and rockfalls dominate the steeper Oregon side.

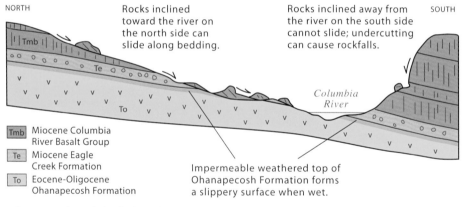

NORTH

Rocks inclined toward the river on the north side can slide along bedding.

Rocks inclined away from the river on the south side cannot slide; undercutting can cause rockfalls.

SOUTH

Tmb

Te

To

Columbia River

Tmb Miocene Columbia River Basalt Group

Te Miocene Eagle Creek Formation

To Eocene-Oligocene Ohanapecosh Formation

Impermeable weathered top of Ohanapecosh Formation forms a slippery surface when wet.

Cross section of the Columbia River Gorge showing how the southward dip of the rock units encourages large landslides to form on the Washington side. –Modified from Miller, 2014

allow it to slide down the slope of its depositional surfaces toward the Columbia River. Several large landslide complexes, each traversed by WA 14, line most of the river between mileposts 25 and 53. The same effect does not happen on the Oregon side, however, because there the rock tilts away from the river.

The rocks are tilted because they lie on the north side of a large syncline that lies a few miles south of the river in Oregon. The syncline is part of the Yakima fold belt, a series of folds and thrust faults that extends from about Hood River to the Tri-Cities. In several places along WA 14, especially east of Hood River, you can see how the rocks are folded. This fold belt is discussed in more detail in the Columbia Basin chapter of this book.

A common misconception about the Columbia River Gorge is that it was carved by the Ice Age floods, between about 18,000 and 15,000 years ago. These events did influence the shape of the valley by scouring and steepening some of its sides and removing loose debris, but the floods poured down an already existing valley. The Columbia River dates back to before 15 million years ago, when it influenced the path of the Columbia River Basalt Group. The unusually thick flows of Wanapum Basalt at Crown Point on the Oregon side, for example, fill a river channel that likely belonged to the former Columbia River. At Mitchell Point, visible on the Oregon side across from WA 14 milepost 58, the Troutdale Formation, deposited by the early Columbia River, lies between flows of the Grande Ronde and Wanapum Basalts. The Ice Age floods, however, did erode some spectacular scablands and deposit large accumulations of gravel along the river.

WA 14
VANCOUVER (I-5)—US 97
102 miles (164 km)

East of I-5, WA 14 follows the Columbia River through an urbanized landscape built on Ice Age flood deposits for the first 9 miles (14 km), with frequent beautiful views of Mt. Hood. At an elevation of 11,239 feet (3,426 m), Mt. Hood is Oregon's highest point. It's also one of the Cascade's most active stratovolcanoes, having had at least two major eruptions in the last 1,500 years, one of which occurred between 1780 and 1800. The eruption produced a lahar that swept down the Sandy River, through what is now Troutdale. The lahar added to older lahars of the Sandy River delta, which creates a prominent northward deflection of the Columbia River across from the town of Camas in Washington.

Some good roadcuts and quarry exposures of basaltic andesite lie on the north side of the highway between mileposts 9 and 10. The quarries use the rock for road base, landscaping, and even jetty construction. These rocks erupted between 730,000 and 590,000 years ago as part of the Boring Volcanics, named for the town of Boring, about 10 miles (16 km) to the south in Oregon. The Boring Volcanics are a series of basaltic lava flows, shield volcanoes, and cinder cones throughout the Greater Portland–Vancouver area that began erupting about 3 million years ago and continued until 57,000 years ago. Although related to the Cascade magmatic arc, they are unusual because they lie substantially west of the main Cascade volcanoes.

Between mileposts 12 and 13, WA 14 crosses Camas Slough of the Columbia River onto Lady Island at the mouth of the Washougal River. Lady Island is one of many islands within the channel of the lower 150 miles (240 km) of the Columbia River. These islands formed when rising sea levels at the end of the last glacial advance caused the Columbia River to deposit huge amounts of gravel in its channel. On the map, you can see that the islands all have streamlined shapes, parallel to the river channel. After recrossing onto the mainland at milepost 14, the highway passes over the floodplain until milepost 19. Look

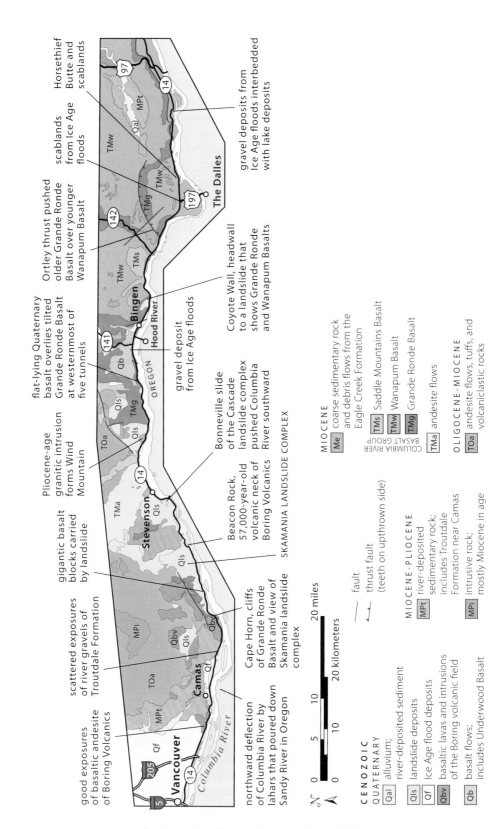

Geology along WA 14 between I-5 and US 97.

Gravel of the Troutdale Formation was deposited by the Columbia River in Miocene to Pliocene time.

eastward for good views of Larch Mountain, a gently sloping shield volcano in northern Oregon. Larch Mountain consists of andesite of the Boring Volcanics that erupted about 1.4 million years ago.

The highway rises off the floodplain just west of milepost 19 onto a steep hillside made of the Miocene-Pliocene Troutdale Formation. These rocks were deposited by the Columbia River; in some places, they date back as far as 15 million years ago. You can see scattered exposures of the Troutdale Formation between mileposts 20 and 23, with the best ones near milepost 22. Numerous pull-outs on the south side of the highway provide opportunities to view the outcrops as well as the river.

Cape Horn, at milepost 25, provides sweeping views of Beacon Rock, 11 miles (18 km) upriver. Directly below the overlook and extending eastward on the north side of the river, you can see part of the Skamania landslide complex, the westernmost of three large landslide complexes along the river. It consists of low-lying, uneven ground that is bounded by cliffs on its north side and the river on its south. In some places, cliffs form arcuate edges to the slide because they mark the locations where the landslide broke away. For the most part, the cliffs are made of the Grande Ronde Basalt of the Columbia River Basalt Group, as are the netted roadcuts on the west side of the highway opposite the pull-out. This particular landslide complex extends east from Cape Horn almost as far as Beacon Rock. The large rock in the river consists of basalt that was carried downward in the slide.

Just east of Cape Horn, the road descends along the cliffs and into the land-slide complex. Notice the beautiful entablature in the basalt. Below the cliffs, you pass by numerous outcrops of basalt, but these are blocks that were trans-ported with the rest of the slide. Some of these rock exposures are huge: a double roadcut exists near milepost 31.

Beacon Rock, just east of the Skamania landslide complex, overlooks the Columbia River at milepost 35. This landmark is a 57,000-year-old volcanic neck of the Boring Volcanics. Conglomerate of the Eagle Creek Formation makes up the roadcut on the north side of the road across from the trail-head parking lot. The Eagle Creek Formation lies beneath the Columbia River Basalt Group and contains material shed in early Miocene time from the early Cascade Range. Elsewhere in the Eagle Creek Formation, you can find lahars and ash deposits, as well as fossil plant fragments. Eagle Creek, the locality for which this formation is named, lies about 4 miles (6.4 km) to the east on the Oregon side of the river.

From immediately east of the high power lines near milepost 39 to about milepost 42, WA 14 passes over the Bonneville landslide, the most recent and largest landslide in the Cascade landslide complex. The landslide complex covers about 10 square miles (26 km^2) and consists of four overlapping land-slides. The Bonneville slide, which covers about 5.5 square miles (14 km^2), most likely occurred between AD 1425 and AD 1450. It pushed the channel of the Columbia River about 1 mile (1.6 km) southward, as can be seen from the map, which shows the river and the highway in a prominent curve around the land-slide's toe. A variety of lakes occupy the many topographically low areas on the hummocky surface of the slide.

The slide dammed the Columbia River to give rise to a temporary lake that backed up as far as Wallula Gap, some 150 miles (240 km) upriver. The lake eventually broke through the dam and flooded the gorge below, filling the area of today's downtown Portland with water to a depth of about 30 feet (9 m). In comparison, some of the Ice Age floods, which originated from failed ice dams near the Montana-Idaho border between about 18,000 and 15,000 years ago, inundated the Portland area to depths exceeding 400 feet (120 m)! The Bonn-eville slide also likely gave rise to a variety of Native American legends for the Bridge of the Gods, the temporary dam that allowed easy passage across the river. In the most prevalent legends, the bridge failed because of a jealous argu-ment between the gods Wy'east (Mt. Adams) and Pahto (Mt. Hood) over the beautiful Loowit (Mount St. Helens).

Bonneville Dam and Lock, at milepost 39, was built in 1938 to generate hydroelectric power. At least part of its foundation is grounded in material of the landslide complex. Looking southward across the Columbia, you gain a good view of the valley of Eagle Creek and cliffs of Eagle Creek Formation. Ashes Lake, just northeast of milepost 42, marks the northeastern edge of the Bonn-eville slide. Between Ashes Lake and Stevenson at milepost 44, WA 14 follows older parts of the slide complex. Good views north and west show the promi-nent headwall on the northwestern edge of the landslide complex; it consists mostly of the Grande Ronde Basalt overlying the Eagle Creek Formation.

Aerial view of the Bonneville slide of the Cascade landslide complex, looking upriver toward the east. Notice how it pushed the Columbia River channel southward.

Roadcuts from just west of milepost 46 to about milepost 47 consist of dark-colored andesite of early Miocene age that pre-dates both the Eagle Creek Formation and the Grande Ronde Basalt. Outcrops on both sides of milepost 48 consist of Quaternary-age basalt that flowed down the Wind River valley from its source at Trout Creek Hill, about 20 miles (32 km) to the northwest.

The easternmost landslide complex lies between the Wind River, just east of milepost 50, and Dog Mountain, near milepost 53. The slide formed two arms as it spread around either side of Wind Mountain, a Pliocene-age granitic intrusion that rises some 1,800 feet (550 m) above Home Valley. An imposing roadcut of the granite lies about 0.1 mile (160 m) east of milepost 51; more good outcrops at milepost 52 mark the edge of the intrusion. Similar to the other landslides, this one impinged on the river channel at Collins Point, where the road deflects slightly. Dog Mountain rises more than 2,800 feet (850 m) above the river and consists of the Grande Ronde Basalt. A good roadcut lies just east of milepost 53, where the rock comes right to the river's edge.

To the west of Dog Mountain, cliffs of the Grande Ronde Basalt frame the north side of the highway for about 7.5 miles (12 km). The basalt dips southward because it is part of some large folds of the Yakima fold belt. Above the inclined Grande Ronde, you can see the flat-lying Underwood Basalt, erupted in Quaternary time from Underwood Mountain, a small volcano just a couple miles to the north. Its horizontal nature indicates that the folding in the underlying rock occurred before the eruption of the younger rock. Numerous good exposures of the Grande Ronde Basalt exist where the road passes through a series of five tunnels between milepost 58 and just past milepost 60, but probably the best view of the angular relation between the Grande Ronde and the younger basalt is from the heritage marker just west of milepost 58 at the westernmost tunnel.

About halfway between mileposts 61 and 62, a side road to Spring Creek Fish Hatchery provides access to the river, good views of Mt. Hood, and the chance to watch paragliders and wind surfers dance on the waves of the Columbia River. This stretch of river seems to be especially windy, as it is confined by high ridges and occupies the thermal transition zone between the wet western side of the Cascades and the dry eastern side.

Just west of milepost 63 is more Underwood Basalt. Immediately east of the milepost you can see the basalt high on the cliff face, above a spectacular deposit of material left by the Ice Age floods. Some boulder-rich flood deposits interlayered with gravel beds probably reflect multiple events. The flood material is younger than the basalt, having been deposited against it. Both the flood deposits as well as the Underwood Basalt show up nicely in the first mile of WA 141 Alternate, which intersects WA 14 on the east side of the White River bridge. Just east of Bingen, at milepost 67, you cross a fault, east of which is the Grande Ronde Basalt, with cliffs on the north side of the road.

The bedrock for the next 35 miles (56 km) between Bingen and the intersection with US 97 consistently belongs to the Columbia River Basalt Group. Between mileposts 67 and 70 is the Grande Ronde Basalt. Coyote Wall, a prominent cliff at milepost 70, forms a landslide headwall with the Grande Ronde Basalt at its base and the southeast-dipping Wanapum Basalt at the top of the cliff. East of Coyote Wall is primarily Wanapum Basalt, with the still younger Saddle Mountains Basalt in a few places. You can see many outstanding examples of entablature and colonnade, as well as differently inclined flow surfaces

The flat-lying Underwood Basalt of Quaternary age forms the peak on the left, whereas the tunnels pass through the tilted Grande Ronde Basalt of Miocene age.

The flat-lying Wanapum Basalt and the Columbia River at the Chamberlain Lake Rest Area.

where the rocks lie on different parts of folds. From milepost 70 to 72, for example, the flows dip southeastward, but they are approximately flat-lying at the Chamberlain Lake Rest Area, immediately west of milepost 74. In addition to taking in the view and gazing at cliffs of the Wanapum Basalt at the rest area, you can watch trains emerge from the nearby tunnel. The older Grande Ronde Basalt makes an appearance on the east side of the Ortley thrust fault a quarter mile (400 m) east of milepost 78.

Scablands appear near milepost 82 and are particularly dramatic at the intersection of US 197, which crosses the Columbia River to The Dalles, Oregon. Look for irregular, bare bedrock exposures that rise above the otherwise low-relief surface. They were eroded by the Ice Age floods, which stripped off all the vegetation and surface deposits and scoured the bedrock between about 18,000 and 15,000 years ago.

WA 14 passes through more scablands at Horsethief Lake and Columbia Hills State Parks, beginning near milepost 84. Horsethief Butte, which consists of Wanapum Basalt, marks the western end of the ridge that separates the scablands from the river. It is accessible by a short hiking trail near milepost 86.

Gravel, deposited by the Ice Age floods, also shows up along the highway, most notably between mileposts 90 and 91 and between mileposts 94 and 95. About a quarter mile (400 m) west of milepost 95, look for gravels that are interbedded with lake deposits, suggestive of multiple flood events separated by periods of slack water. A large pull-out on the south side of the highway allows for easy access. For the 5 miles (8 km) west of US 97, the highway follows a bench some 500 feet (150 m) above the river that was carved by the floods and is now decorated with scablands.

Gravel and slack-water sediments deposited by the Ice Age floods. View looking west toward Mt. Hood in Oregon.

MOUNT ST. HELENS
NATIONAL VOLCANIC MONUMENT

Until 1980, Mount St. Helens looked like a typical volcano of the High Cascades: tall, steep, covered in glaciers, surrounded by a forested landscape, beautiful. It was unusually symmetrical, forming a near-perfect circle in plan view, and reached an elevation of 9,677 feet (2,950 m), the sixth-highest peak in Washington State. Like many of the other volcanoes, the surrounding area contained recently formed deposits to indicate the volcano was potentially active, but to most people an eruption seemed more of an abstract idea than a real possibility. On May 18, 1980, after only two months of warning signs, the volcano erupted catastrophically. Fifty-seven people lost their lives, as did countless wildlife. Airborne ash from the eruption turned day into night in Spokane, almost 250 miles (400 km) away.

Today, Mount St. Helens stands more than 1,300 feet (400 m) lower than before the eruption and displays an enormous crater that opens to the north. The volcano offers an unprecedented opportunity to see the effects of a major volcanic eruption, to watch the landscape recover, and to look inside a volcano to better understand its internal structure and history. It is one of the most closely studied volcanoes on Earth. Interested readers should consult the outstanding description and road guides by Patrick Pringle, and also the US Geological Survey website, for more details than can be offered here.

Swarms of small earthquakes, steam eruptions, and a prominent bulge on its upper slopes preceded the May 18 eruption. The bulge, formed by magma intruding high into the volcano, grew more than 5 feet per day (1.5 m/d) to protrude more than 450 feet (137 m) northward by May 17. The following morning at 8:32 a.m., a 5.1 magnitude earthquake triggered the eruption: the bulge collapsed as a gigantic landslide, which in turn released pressure on the magma below, much as a cap removed from a shaken bottle of soda.

The landslide, or debris avalanche, was Earth's largest in recorded history, with a volume of 0.67 cubic miles (2.8 km³). At speeds ranging from 70 to 150 miles per hour (113–241 km/hr), it spread north and northwest from the volcano, overtopped the 1,150-foot-high (350 m) Johnston Ridge 6 miles (10 km) away, and buried the upper 14 miles (23 km) of the North Fork Toutle River to an average depth of 150 feet (46 m). The landslide materials left behind a telltale hummocky topography of irregular hills and depressions across the area, much of which can now be visited along the Hummocks Trail near Coldwater Lake or along the Boundary Trail near Johnston Ridge Observatory.

Because the landslide released pressure on its north side, the initial blast aimed northward. Scalding gas and rocks exceeding 600 miles per hour (970 km/hr) blew down entire forests more than 10 miles (16 km) away. Fallen tree trunks still show a preferred orientation depending on their locations with respect to the volcano and the local topography. According to US Geological Survey estimates, enough trees were blown down to build some three hundred thousand two-bedroom homes. Less than fifteen minutes after the initial blast, a vertical column of rock, ash, and gas had risen 15 miles (24 km) into the atmosphere. More explosions within the hour added to the plume, which deposited measureable amounts of ash over 22,000 square miles (57,000 km²) and had circled the globe within fifteen days.

The eruption also produced pyroclastic flows and lahars. The pyroclastic flows, consisting mostly of superheated gas, ash, pumice, and small rock fragments, poured down the northern flanks of the volcano at speeds of 50 to 80 miles per hour (80–130 km/hr) as far as 5 miles (8 km) north of the crater. The lahars, made from a mixture of ice, water, and rock debris, had a consistency of wet concrete; they moved out from the volcano in all directions at 10 to 25 miles per hour (16–40 km/hr). The largest lahar was a clay-rich slurry that oozed from a landslide deposit. It flowed all the way down the North Fork Toutle River to the Cowlitz River, some 55 miles (90 km) away, destroying bridges and homes along the way. It added so much sediment to the river system that at a point almost 20 miles (32 km) downstream from the mouth of the Toutle River, the channel depth of the Columbia River was reduced from 40 to 14 feet (12–4.3 m).

Mount St. Helens continues to evolve. Besides dozens of smaller eruptions and frequent bursts of seismicity since the main eruption, the crater now hosts two lava domes as well as an active glacier. Even the blast zone, which was stripped of visible life during the eruption, is coming back with new vegetation and wildlife.

The later eruptions mostly occurred during two periods of time: June 1980 until about 1986 and September 2004 until 2013. Some of these eruptions were significant in their own right. In the months after the May 1980 eruption, for example, Mount St. Helens erupted five more times, forming eruption columns that reached 8 to 9 miles (13–14 km) high, as well as pyroclastic flows. Each eruptive period also produced separate lava domes, which together fill about 7 percent of the crater. The domes consist of dacite, a viscous lava intermediate between andesite and rhyolite that tends to plug vents rather than flow out of them. The youngest dome reaches a height of about 1,500 feet (460 m) above the crater floor.

The crater, with its high walls and near-constant shade, proved an ideal place for snow to accumulate. Crater Glacier was first recognized in 1996 and by 2004 had grown into a horseshoe around the first dome. With growth of the new dome, the glacier split into two arms and doubled its thickness in some places.

Aerial view eastward of Mount St. Helens and the North Fork Toutle River drainage, with Mt. Adams in background. The 1980 eruption blasted through crater walls northward to create a zone of devastation that extended about 10 miles (16 km) to the north. The debris avalanche and pyroclastic flows overtopped Johnston Ridge (JR) and dammed Spirit Lake (SL) and Coldwater Lake (CL). Lahars poured down and filled the channel of the North Fork Toutle River. Photo taken in 2002.

Although the glacier now completely encircles both lava domes, its extent is not immediately apparent because frequent rockfalls off the crater rim keep it covered almost entirely with rock.

Without question, Mount St. Helens is, and has been, the most active volcano in the Cascades for at least the past 4,000 years. Its history dates back about 300,000 years with eruptions of several dacite lava domes that formed just west of the present volcano. Since then, its activity has been sporadic, with intervals of frequent eruptions separated by dormant times that sometimes lasted tens of thousands of years. Most of these eruptions produced dacite domes and pyroclastic material, attesting to typically violent eruptions. The many geologists who have studied Mount St. Helens, particularly those of the US Geological Survey, identify six separate intervals of eruptions, called stages.

The most recent stage, called the Spirit Lake stage, began 3,900 years ago. It is further divided into six eruptive periods, during which Mount St. Helens built its edifice from a generally small to midsize volcano to a large stratovolcano. During the Castle Creek period, between about 2,200 and 1,600 years ago, the volcano produced a wider variety of lava, including several andesite and basalt flows. One of these basalt flows formed the lava tube now known as Ape Cave (see page 204).

Despite its spectacle, the 1980 eruption was nowhere near as large as some of Mount St. Helens's earlier eruptions. For example, during the Cougar stage, between 23,000 to 17,000 years ago, an eruption produced a debris avalanche about twice the size of the 1980 landslide. More recently, an eruption during the Smith Creek eruptive period, between 3,500 and 3,300 years ago, produced about four times the volume of ash that was produced by the 1980 eruption; lahars from the eruption likely reached as far as the Columbia. An eruption in 1479 similarly dwarfed the 1980 eruption, and only three years after that, Mount St. Helens erupted again, with a size comparable to the 1980 blast.

WA 504

Castle Rock—Johnston Ridge Observatory

51 miles (82 km)

WA 504 follows the North Fork Toutle River, which channeled the devastating lahar during the May 18, 1980, eruption of Mount St. Helens. The valley is cut into bedrock of lavas and some intrusions that formed during Cascade eruptions from Eocene to Miocene time. Traveling east from I-5, between mileposts 2 and 3 you'll pass some cliffs of basaltic andesite flows that are part of the Eocene-age Goble Volcanics. The rocks weather a rusty red but are dark gray and typically very fine grained on a fresh surface.

Silver Lake first comes into view near milepost 6. It formed about 2,500 years ago when a gigantic lahar dammed Outlet Creek, which flows northward into the Toutle River. The lahar's source was Spirit Lake, which broke through its own dam of debris avalanche material deposited by an earlier eruption. Similar

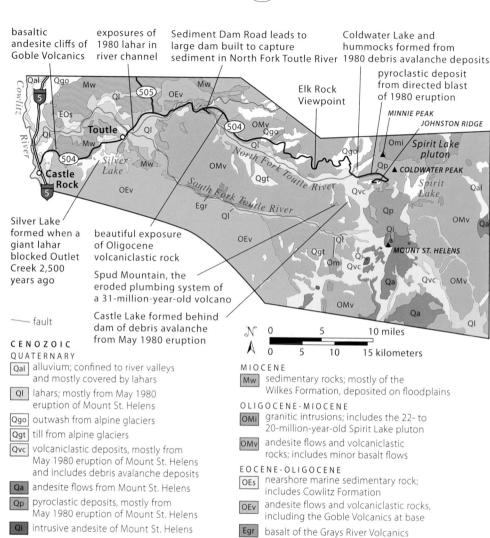

basaltic andesite cliffs of Goble Volcanics

exposures of 1980 lahar in river channel

Sediment Dam Road leads to large dam built to capture sediment in North Fork Toutle River

Coldwater Lake and hummocks formed from 1980 debris avalanche deposits

pyroclastic deposit from directed blast of 1980 eruption

Elk Rock Viewpoint

Silver Lake formed when a giant lahar blocked Outlet Creek 2,500 years ago

beautiful exposure of Oligocene volcaniclastic rock

Spud Mountain, the eroded plumbing system of a 31-million-year-old volcano

Castle Lake formed behind dam of debris avalanche from May 1980 eruption

—— fault

CENOZOIC
QUATERNARY

Qal — alluvium; confined to river valleys and mostly covered by lahars

Ql — lahars; mostly from May 1980 eruption of Mount St. Helens

Qgo — outwash from alpine glaciers

Qgt — till from alpine glaciers

Qvc — volcaniclastic deposits, mostly from May 1980 eruption of Mount St. Helens and includes debris avalanche deposits

Qa — andesite flows from Mount St. Helens

Qp — pyroclastic deposits, mostly from May 1980 eruption of Mount St. Helens

Qi — intrusive andesite of Mount St. Helens

N   0   5   10 miles
     0   5   10   15 kilometers

MIOCENE

Mw — sedimentary rocks; mostly of the Wilkes Formation, deposited on floodplains

OLIGOCENE-MIOCENE

OMi — granitic intrusions; includes the 22- to 20-million-year-old Spirit Lake pluton

OMv — andesite flows and volcaniclastic rocks; includes minor basalt flows

EOCENE-OLIGOCENE

OEs — nearshore marine sedimentary rock; includes Cowlitz Formation

OEv — andesite flows and volcaniclastic rocks, including the Goble Volcanics at base

Egr — basalt of the Grays River Volcanics

Geology along WA 504 between I-5 and Mount St. Helens.

lake breakouts are an important part of the history of Mount St. Helens, but this was the biggest. Its initial discharge of more than 325,000 cubic yards per second (250,000 m³/s) exceeded that of the Amazon River's! Just west of milepost 9 at its northeastern edge, the lake shallows into wetlands. The lahar is generally inconspicuous along the highway, but you can see some of its deposits by driving south about a quarter mile (400 m) on South Toutle Road, between mileposts 10 and 11. Upon crossing the Toutle River at milepost 11, WA 504 leaves the old lahar behind and follows the much younger 1980 lahar. A quarter mile (400 m) east of milepost 12, look for some uneroded exposures of the 1980 lahar in the river channel.

For the next 15 miles (24 km), the road passes occasional bedrock exposures of Oligocene and Eocene volcanic rocks and good views of the lahar-filled channel of the North Fork Toutle River. The Goble Volcanics of mostly Eocene age make up the roadcuts between mileposts 16 and just east of 17; the younger rocks make up the exposures beyond. The North Fork Toutle River passes through a beautiful meander bend at milepost 20; notice the large gravel bar on its inside bend, deposited by slow-moving water, and the steep cutbank and bedrock cliffs along its outside bend, eroded by fast-moving water.

Sediment Dam Road at milepost 21 leads 1 mile (1.6 km) southward to a structure completed in 1989 to capture some of the huge volumes of suspended sediment carried by the North Fork Toutle River since the eruption. Its basin, which can hold more than 250 million cubic yards (190 million m³) of sediment, filled much earlier than expected, rendering the structure no longer effective, although the height of the dam has since been raised slightly. The plain of accumulated sediment is accessed by a 1-mile-long (1.6 km) trail from the parking lot. Had this sediment not been trapped, its deposition farther downstream would likely have contributed to flooding in Kelso and Longview.

Three-quarters of a mile (1.2 km) east of milepost 24, a well-stratified exposure of volcanic-rich sedimentary rock, ash beds, and lahar deposits of Oligocene age borders the north side of the highway. Some small faults cut these rocks, and ancient petrified tree stumps poke through the lahars.

The Hoffstadt Bluffs Viewpoint, about a half mile (0.8 km) east of milepost 26, offers the chance to relax and take in a view of Mount St. Helens, now

Oligocene rocks between mileposts 24 and 25 contain ash and lahar deposits erupted from much older volcanoes than Mount St. Helens.

only 17 miles (27 km) away. Between here and Coldwater Lake near milepost 45, exposures of the Oligocene andesite lavas and volcanic-rich sedimentary rocks become more frequent and bold, many with pull-outs for easy access. They were also invaded by numerous dikes as well as some shallow andesite intrusions, such as near milepost 33. In many places, hydrothermal activity has altered the rocks to yellows and greens. Just east of milepost 36, look for broad zones of broken yellow rock cutting across more intact rock with a greenish cast. The green color derives from the mineral chlorite, which forms from high-temperature alteration; the yellow color comes from iron oxide, which typically forms at lower temperatures.

In addition to its outstanding view of Mount St. Helens and the blast zone, the Elk Rock Viewpoint near milepost 37 offers probably the best view of Mt. Adams, 30 miles (48 km) farther east. Spud Mountain, considered the eroded plumbing system of a 31-million-year-old volcano, forms the prominent peak south and slightly east across the valley. Dikes from Spud Mountain invaded some of the rocks in this area, causing some of the alteration. Eastward, you can see the south edge of the granitic Spirit Lake pluton in jagged Minnie Peak, the northernmost of the rugged peaks that extend north from Mount St. Helens. The pluton intruded the older volcanic rocks between 23 and 20 million years ago. Directly behind the overlook is a broad zone of yellow-altered andesite; look for a greenish dike across the road from the east exit.

Castle Lake, best seen from the viewpoint near milepost 40, formed when the debris avalanche from the May 18 eruption dammed its drainage. To prevent a possible outburst flood, engineers built a spillway in 1981 and continually monitor the lake level.

Although generally not exposed, glacial till covers much of the area between the Castle Lake Viewpoint and the Mount St. Helens Science and Learning Center at Coldwater Lake. An especially interesting bedrock exposure just east of milepost 42 shows a basalt flow overlying stratified air-fall tephras and thin lava flows. Near the middle of the exposure, the basalt fills a small channel cut into the older rocks. Some small faults visibly offset the rocks, and a dike intrudes the rocks on the west side of the exposure. Numerous basaltic dikes cut bedrock exposures along the road between mileposts 43 and 44 as the road descends toward Coldwater Lake.

Slightly past milepost 45, you can access the Coldwater Lake picnic area and boat launch. The lake formed after the 1980 debris avalanche dammed a valley. If you walk along the boardwalk from the picnic area, you can also see a delta to the east that protrudes northward into the south end of the lake. The delta is now heavily vegetated, but you can still see its general shape. Notice the islands of debris avalanche deposits within the lake. The whole area south of Coldwater Lake is hummocky from the debris avalanche, parts of which are accessible by the well-maintained Hummocks Trail, a short distance farther up the road. The trail makes a 2-mile (3.2 km) loop and accesses the North Fork Toutle River. If you don't have time for the entire loop, you can still see part of the debris avalanche up close by hiking about a quarter mile (400 m) in from the eastern-most trailhead. Looking up the lake to its headwaters, Minnie Peak of the Spirit Lake pluton weathers in a noticeably different style than the adjacent peaks made of Oligocene and Miocene volcanic rock.

The stretch of road from the Coldwater Lake area to Johnston Ridge initially passes through some of the hummocks left behind from the debris avalanche

Panoramic photo of Mount St. Helens and its blast zone from the Loowit Viewpoint. The dome is visible in the crater. Spud Mountain is the peak on the far right (west) side of the photo.

of 1980 and then ascends the valley of South Coldwater Creek. The valley is cut into Oligocene to Miocene basalt flows and volcaniclastic rocks, but its bottom contains well-exposed deposits of the debris avalanche and directed blast. A small pull-out at milepost 49 gives the chance to view a tree trunk protruding from the deposit. Loowit Viewpoint, just beyond milepost 51, offers a wonderful view of the volcano and blast zone, away from the crowds of the Johnston Ridge Observatory.

The Johnston Ridge Observatory, in the middle of the blast zone, offers unparalleled views of the volcano and surrounding peaks, including the basalt-capped Coldwater Peak just 2 miles (3.2 km) to the northeast. In addition to a bookstore and unusually informative displays, movies, and ranger-led activities, an easy hiking trail leads eastward along the top of Johnston Ridge. It winds through hummocks left by the debris avalanche and eventually connects with other trails.

APE CAVE AND TRAIL OF TWO FORESTS

This side trip to the south flank of Mount St. Helens leads to one of the world's longest lava tubes, as well as molds of tree trunks from an ancient forest. These features formed in a basaltic lava flow that erupted about 2,000 years ago. To get there, follow Lewis River Road (WA 503) for 36 miles (58 km) east from I-5 at Woodland. A mile (1.6 km) north of the Swift Dam overlook, go north on Forest Service Road 83 for about 1.5 miles (2.4 km) and follow the signs to Ape Cave and Trail of Two Forests. Bring warm clothing and flashlights.

Lava tubes can form when a lava flow continues flowing beneath its cooling exterior. As the outside of the flow hardens, it insulates the remaining lava on the interior, which continues to flow away from the vent. Because of lava tubes, lava flows can travel some distance from their vents before actually emerging at the surface. After the lava supply becomes depleted, the tube empties and becomes a long, narrow cave. Ape Cave, which extends nearly 2.5 miles (4 km) underground, formed in a pahoehoe lava flow that erupted during the Castle Creek eruptive period of Mount St. Helens, about 2,000 years ago. Pahoehoe lava has a smooth, undulating, or ropy, surface formed by the movement of fluid lava under a congealing surface crust. A nice example of the lava's ropy texture can be seen just outside the cave entrance. The cave shows different levels at which the lava flowed, as well as cave formations more typical of limestone caves. Downward-growing stalactites formed where lava dripped down from the ceiling, and upward-growing stalagmites formed where lava drips accumulated on the floor. Remnants of a lahar that entered the cave in the late 1400s also can also be found in the cave.

The nearby Trail of Two Forests winds through old-growth Douglas fir and western red cedar past more than a dozen molds of trees that were growing at the time of the eruption that formed Ape Cave. The molds, up to about 4 feet (1.2 m) across, appear as vertical or horizontal hollow cylinders through the lava. Upon encountering the trees, the lava caused them to burn as it cooled and solidified around the trunks. After any remaining charcoal or wood had disintegrated, a hollow shell remained.

Inside Ape Cave, a lava tube.

MT. RAINIER NATIONAL PARK

Mt. Rainier, the highest volcano in the Cascade chain, practically soars to its elevation of 14,410 feet (4,393 m). It reaches some 8,000 feet (2,400 m) above the tops of the surrounding peaks. With a base that extends well below 3,000 feet (900 m) in some places, it displays the greatest topographic relief in the conterminous United States. The mountain offers outstanding alpine scenery, deep glacial valleys, and at least twenty named glaciers. The history of volcanism in the vicinity of Mt. Rainier reaches back about 36 million years, with the modern volcano developing during the last half million years of the story.

Aerial view of Mt. Rainier, looking southeastward up the South Mowich Valley. The summit cone was built mostly during the Summerland eruptive period between 2,700 and 2,000 years ago. Steeply dipping lava flows on either side of the cone (best seen on the right in this photo) indicate the volcano was substantially higher earlier in its history.

At least three major volcanic and sedimentary units and one major intrusive unit pre-date the modern volcano. The oldest and most extensive of these rock units, the Ohanapecosh Formation, was deposited between 36 and 22 million years ago. It consists mostly of volcanic-rich sedimentary rocks, lahars, and basaltic andesite. Above that, the Stevens Ridge Formation formed about 26 million years ago and consists mostly of ash-flow tuffs. The youngest rock unit, the Fifes Peak Formation, was deposited from 26 to 22 million years ago and consists mostly of basaltic to andesitic lava flows and volcanic-rich sedimentary rocks. Granitic rocks of the Tatoosh pluton (mostly granodiorite) intruded those volcanic rocks during multiple events from 26 to 14 million years ago.

Rocks of the modern volcano are primarily andesite and dacite that date back about 500,000 years. They show that the volcano's growth took place in several different eruptive stages of intense magma production, separated by quieter times of fewer and smaller eruptions. These quieter times were dominated largely by erosion. During the first eruptive stage, which lasted from about 500,000 to 420,000 years ago, the volcano erupted both pyroclastic flows and andesite lavas, including the lava that forms Burroughs Mountain and the Sunrise area.

Granitic rock of the Tatoosh pluton is exposed at Christine Falls on Van Trump Creek north of Longmire.

Because some of the pyroclastic flows are preserved high on the modern volcano at Steamboat Prow and slope upward toward today's summit, we know that this earlier rendition of the volcano reached an elevation comparable to today's.

In a similar way, 200,000-year-old andesite perched high on Mt. Rainier's upper west side suggests the volcano reached an even greater height during its second main eruptive period between 280,000 and 180,000 years ago. Numerous flows from this time erupted from dikes that intruded the western and eastern flanks of the volcano.

Beginning about 40,000 years ago, Mt. Rainier rebuilt its upper reaches, which had been reduced by erosion since the end of last eruptive period. Most flows originated high on the volcano, with the exception of the large one at Ricksecker Point. At 40,000 years old, the Ricksecker flow is Rainier's youngest large ridge-forming flow. Mt. Adams and Mount St. Helens also experienced important growth periods during this time.

The most detailed record of Rainier's activity, however, comes from the last 11,000 years, after the glaciers reached their maximum extent. Although the volcano did not produce any substantial lava flows, it did produce more than thirty separate tephra deposits, which allow for a detailed chronology. Ten to twelve of these eruptions occurred over the last 2,600 years. The Summerland eruptive period, between 2,700 and 2,000 years ago, is especially notable because it marks when Rainier built a major portion of its modern summit cone. The most recent fully documented eruption occurred about 1,100 years ago. Eyewitness reports of a possible eruption in the nineteenth century have never been verified.

Lahars

Mt. Rainier produced numerous lahars (volcanic debris flows) in the last 11,000 years. Lahars can form whenever broken rock mixes with water and flows down the volcano's sides and into its river valleys as a slurry, much like flowing wet concrete. Lahars can be triggered by volcanic eruptions, when hot rock mixes with and melts glacial ice and snow, or by large avalanches that transform into volcanic debris flows. Many lahars actually grow in size as they pick up rock material and water downstream but eventually shrink as stream gradients decrease and they lose energy.

Mt. Rainier unleashed innumerable lahars during its half-million-year history, but as with the ash deposits, we can sort through the more recent record in much greater detail. Most notably, many large lahars in the past 6,000 years reached the Puget Lowland and greatly altered the landscape. An eruption 5,600 years ago caused an enormous section of Rainier's eastern side to collapse; the resulting Osceola Mudflow poured down the White River valley, burying the areas now occupied by the towns of Enumclaw and Kent and continuing all the way to Puget Sound. Deposits of this lahar line sections of the road to Sunrise, above the turnoff to the White River Campground, and occupy much of the White River Valley to the north and northwest, as well as portions of the lower Puyallup River valley. The National Lahar, triggered by whole-scale melting during an eruption about 2,500 years ago, flowed down the Nisqually River on the south side of the volcano and into Puget Sound. Another lahar, at roughly the same time as the National, left deposits along the Puyallup River as far downstream as the town of Puyallup. In the White River drainage, lahars triggered by moderately explosive eruptions about 2,700, 1,500, and 1,100 years ago reached as far as Tacoma and Seattle.

About 500 years ago the Electron Mudflow, triggered by a collapse of the southwestern side of the volcano, flowed more than 50 miles (80 km) down the Puyallup River as far as Sumner and Puyallup. The collapse of the mountain's southwest side was caused by Rainier's ongoing hydrothermal activity. Hot water, circulating through the volcano, caused the alteration and consequent weakening of the rock. Eventually, a zone of rock became so weak that a large section of the mountain failed catastrophically and fell onto the Nisqually Glacier. A concurrent eruption may have triggered the lahar but was not the major cause.

Glaciers and Ridges

In map view, Mt. Rainier presents a near-circular pattern, with its many glaciers radiating outward like spokes on a wheel. The glaciers now occupy the upper reaches of long, deep glacial valleys that were filled with ice near the end of Pleistocene time. Most of the ridges between these valleys are capped by andesite lavas erupted from the volcano over the course of its 500,000-year history.

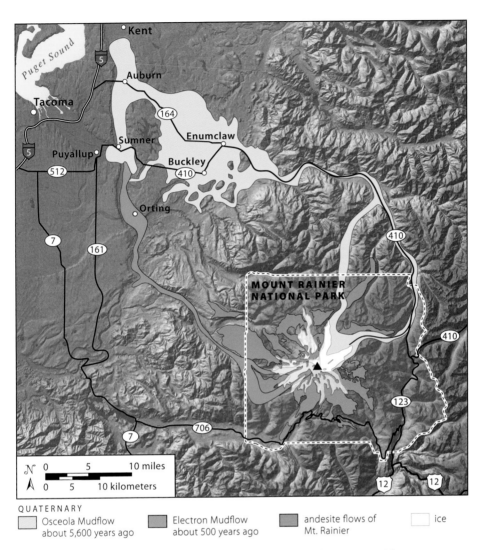

QUATERNARY

| | Osceola Mudflow about 5,600 years ago | | Electron Mudflow about 500 years ago | | andesite flows of Mt. Rainier | | ice |

The Osceola and Electron Mudflows, also called lahars, erupted from Mt. Rainier and extended into areas now heavily populated.

Lava tends to flow down valleys rather than ridgelines, so the lava-capped ridges pose an enigma. For decades, geologists surmised that they represented inverted topography, in which the lava once flowed down valleys eroded into softer material. Subsequent erosion removed the valley walls and left the more resistant lava as a high-standing ridge. At Mt. Rainier, this hypothesis is hard to support because some lava flows are quite young, and extraordinary rates of erosion would have been required to remove the surrounding rock, some of which is resistant granitic rock.

It turns out that glaciers controlled the flow of lava. As Mt. Rainier grew, so did its glaciers, and they grew to be much thicker and longer than they are today, easily filling and overtopping the valleys. Lava, pouring out of vents along the volcano's flanks, flowed down the ridgelines separating the ice-filled valleys. The margins of many of the lava flows show the effects of having been in contact with ice, such as glassy zones from extremely fast cooling. In addition, cooling fractures are horizontal or nearly horizontal at the flow margins rather than vertical.

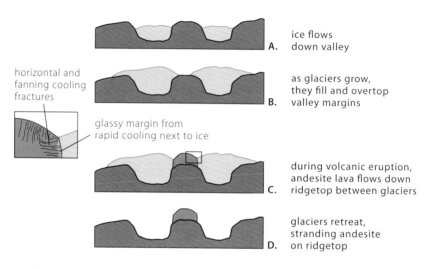

A. ice flows down valley

horizontal and fanning cooling fractures

B. as glaciers grow, they fill and overtop valley margins

glassy margin from rapid cooling next to ice

C. during volcanic eruption, andesite lava flows down ridgetop between glaciers

D. glaciers retreat, stranding andesite on ridgetop

Sequence of events that led to the ridge-capping lava flows of Mt. Rainier.

WA 706/PARADISE ROAD/STEVENS CANYON ROAD
ELBE—STEVENS CANYON ENTRANCE
48 miles (77 km)

At Elbe, Miocene basaltic bedrock shows up beautifully at the intersection of WA 7 and WA 706 on the north side of the road. A pull-out allows parking on the south side of the road, where you can see well-defined columns, especially on the east side of the exposure. A close inspection reveals easily visible, blocky plagioclase crystals scattered throughout the rock. More exposures of the rock show up sporadically for the next half mile (0.8 km) to the east. Farther east, the bedrock consists mostly of tree-covered Eocene-Oligocene volcanic rocks that show up as cliffs and peaks on the north side of the road.

For the 13 miles (21 km) from Elbe to the national park boundary, WA 706 follows the valley of the Nisqually River, which issues from the mouth of the Nisqually Glacier, high on Mt. Rainier. Much of the valley floor is covered by outwash deposits. During its maximum extent about 17,000 years ago, the Nisqually Glacier advanced about 7 miles (11 km) west of the park boundary.

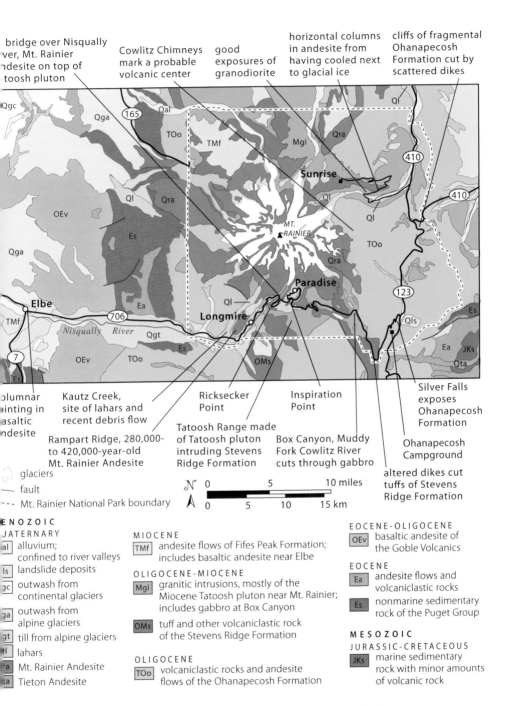

bridge over Nisqually ver, Mt. Rainier ndesite on top of toosh pluton

Cowlitz Chimneys mark a probable volcanic center

good exposures of granodiorite

horizontal columns in andesite from having cooled next to glacial ice

cliffs of fragmental Ohanapecosh Formation cut by scattered dikes

olumnar ointing in asaltic ndesite

Kautz Creek, site of lahars and recent debris flow

Rampart Ridge, 280,000- to 420,000-year-old Mt. Rainier Andesite

Ricksecker Point

Tatoosh Range made of Tatoosh pluton intruding Stevens Ridge Formation

Inspiration Point

Box Canyon, Muddy Fork Cowlitz River cuts through gabbro

Silver Falls exposes Ohanapecosh Formation

Ohanapecosh Campground

altered dikes cut tuffs of Stevens Ridge Formation

glaciers
— fault
- - - Mt. Rainier National Park boundary

0   5   10 miles
0   5   10   15 km

CENOZOIC

QUATERNARY
al alluvium; confined to river valleys
ls landslide deposits
gc outwash from continental glaciers
ga outwash from alpine glaciers
gt till from alpine glaciers
l lahars
ra Mt. Rainier Andesite
ta Tieton Andesite

MIOCENE
TMf andesite flows of Fifes Peak Formation; includes basaltic andesite near Elbe

OLIGOCENE-MIOCENE
Mgi granitic intrusions, mostly of the Miocene Tatoosh pluton near Mt. Rainier; includes gabbro at Box Canyon
OMs tuff and other volcaniclastic rock of the Stevens Ridge Formation

OLIGOCENE
TOo volcaniclastic rocks and andesite flows of the Ohanapecosh Formation

EOCENE-OLIGOCENE
OEv basaltic andesite of the Goble Volcanics

EOCENE
Ea andesite flows and volcaniclastic rocks
Es nonmarine sedimentary rock of the Puget Group

MESOZOIC

JURASSIC-CRETACEOUS
JKs marine sedimentary rock with minor amounts of volcanic rock

Geology along roads that access Mt. Rainier National Park.

Just west of the park boundary, good exposures of the Ohanapecosh Formation form cliffs on the north side of the road. Between the park boundary and Longmire, the road passes through 6 miles (10 km) of deep forest and scattered outcrops and views of cliffs of the Ohanapecosh Formation to one side of the road, and views of the braided Nisqually River to the other. The river flooded in November 2006 and destroyed a campground at the park entrance. The nearly abandoned channel of Kautz Creek, 3 miles (4.8 km) east of the entrance station, offers a glimpse of Mt. Rainier. Numerous snags in the channel are trees killed by a debris flow triggered by heavy rains and water released from Kautz Glacier in 1947. Deposits from this debris flow and older lahars fill most of the channel. During the storm in November 2006, Kautz Creek shifted its course to the east; the road crosses the new channel less than a quarter mile (400 m) to the east.

Longmire, which marks the site of the original park headquarters, sits atop lahar deposits estimated to have formed between AD 1479 and AD 1570. Cliffs of the western Tatoosh Range rise high to the east, and Rampart Ridge forms the prominent cliffs to the northwest. Rampart Ridge is made of the Mt. Rainier Andesite, which erupted between 420,000 and 280,000 years ago. The lava flow was channeled between the Kautz and Nisqually Glaciers, which completely filled the valleys on either side of the ridge. It is estimated that during this period of time, the Nisqually Glacier was more than 1,000 feet (300 m) deep at Longmire Meadows. The two glaciers merged where the two valleys now come together and blocked the lava flow. Beneath the cliffs, vegetation obscures the Ohanapecosh Formation, although a couple good exposures of it lie along the road within the next 1.5 miles (2.4 km).

Rampart Ridge, viewed from Longmire, is made of the Mt. Rainier Andesite.

About 2 miles (3.2 km) east of Longmire, a pull-out offers easy access to the boulder-strewn channel of the Nisqually River. The boulders give a good idea of what's to come along the road: andesite from Rainier lava flows, granitic rock of the Tatoosh pluton, and ash-flow tuff of the Stevens Ridge Formation of the Tatoosh Range.

Christine Falls cuts a beautiful cleft in bedrock of the Tatoosh pluton, and Van Trump Creek flows beneath the roadway about 4 miles (6.4 km) north of Longmire. Just to the east, WA 706 crosses the Nisqually River. From the bridge, you can look up the once-ice-filled channel toward the glacier, and you can also look downstream along the long Rampart Ridge toward Longmire. Bedrock on both sides of the bridge is granodiorite, although you can see vertical columns in overlying andesite as you cross the bridge. About a half mile (0.8 km) past the bridge, the road rises into the Mt. Rainier Andesite, called the Ricksecker lava flow here. Erupted 40,000 years ago, this flow is the youngest large ridge-capping lava flow on the mountain. The side road to Ricksecker Point offers even more views and the opportunity to inspect the andesite and overlying glacial till away from the main road.

Narada Falls spills over the edge of the Ricksecker lava flow onto granodiorite of the Tatoosh pluton. You can see where the two rocks come together by taking

Narada Falls pours over andesite of the Ricksecker lava flow, which erupted 40,000 years ago and flowed over the Tatoosh granodiorite. The granodiorite crops out in the hillside on the right side of the photo.

the short (0.1 mile; 160 m) trail down to the overlook. The contact between the two rock types here is unusually steep, reflecting some of the topography over which the lava flowed. Lane Peak, made of Stevens Ridge Formation, forms the prominent mountain directly south of the parking lot.

It's sometimes difficult to distinguish the light-colored andesite flows from the light-colored granodiorite without pulling off the roadway and looking closely at the rock. Their different fracturing characteristics, however, usually give them away. The andesite tends to break into thin plates or, in some places, vertical columns. These fractures form as the lava cools into rock. By contrast, the granodiorite is broken by more widely spaced fractures and tends to form large angular blocks. Many of the granitic rocks also contain dark-colored inclusions of older intrusive rock that was incorporated into the magma. The stretch of roadway above Narada Falls shows distinctively platy andesite for the first half mile (0.8 km); an outcrop of fractured granodiorite with inclusions lies at the junction of the Paradise and Stevens Canyon Roads.

Paradise Loop

The road to Paradise winds upward past numerous outcrops of the Ricksecker lava flow until it reaches the main visitor area. The entire area was inundated by a lahar, as indicated by scattered boulders and exposures of clay-rich, orange-yellow sediments, most of which are obscured by the prolific huckleberries. Called the Paradise Lahar, it flowed down the Nisqually River to a few miles past the park's western boundary, probably about 5,600 years ago. Its timing suggests it possibly occurred during the same event as the much larger Osceola Mudflow that came off the volcano's northeast side.

The Tatoosh Range as viewed southward from the trail above Paradise Lodge. Granodiorite (Mgi) generally appears light colored in this photo, whereas Stevens Ridge Formation (OMs) appears dark colored. With the exception of Unicorn Peak, which is entirely Tatoosh granodiorite, all the summits are made of Stevens Ridge Formation. From Denman Peak eastward (left), however, their bases consist of granodiorite.

The one-way road leading northward from the visitor center descends past outcrops of andesite into the upper reaches of the U-shaped Paradise Valley. The road crosses into granodiorite just before crossing Paradise Creek; you'll see the slotted canyon of Edith Creek carved into the granodiorite on the west. As you descend the east side of the valley to the intersection with the Stevens Canyon Road, you'll pass numerous outcrops of the Tatoosh pluton and gain spectacular views of both the Tatoosh Range and Mt. Rainier.

The rugged Tatoosh Range, immediately to the south, consists of the Tatoosh pluton and ash-flows of the older Stevens Ridge Formation. From Lane Peak westward, the range consists of Stevens Ridge Formation, which weathers to a dark gray. East of Lane Peak, the high points mostly consist of the dark-colored Stevens Ridge Formation while the light-colored areas between them consist of the granitic rock. The magma intruded from below, and the high points of Stevens Ridge Formation are all that remain of the rock that existed above the intrusion.

Stevens Canyon Road

East of the Paradise Road junction, Mt. Rainier completely dominates the view to the north of Stevens Canyon Road, rising more than 9,000 feet (2,700 m) above the smaller but jagged Tatoosh Range immediately to the south. The large pull-out at Inspiration Point offers an awesome view down the edge of the Tatoosh Range and the deep Paradise Valley. The point is perched on granitic rock, but directly above, you can see cliffs of andesite showing beautiful columns. Less than 1 mile (1.6 km) from there, the road descends to Reflection Lakes, with a lahar deposit forming hummocks on the south and west sides. Another half mile (0.8 km) eastward, the road descends into a glacial cirque occupied by Louise Lake.

Just east of Louise Lake, the road enters the top of Stevens Canyon, a glacial valley, which the road descends via a long switchback past alternating exposures of Tatoosh granitic rock and Mt. Rainier Andesite. Shortly after crossing Stevens Creek, choked with logs from a recent flood, an avalanche chute marks the edge of nearly continuous cliff exposures until the tunnel at Box Canyon, about 3.2 miles (5 km) down the road. You see granitic rock for the first 1.1 miles (1.8 km) and then, across from a long pull-out, ash-flow tuffs of the Stevens Ridge Formation. These rocks superficially resemble the granitic rock, although they exhibit smoother surfaces and, in some places, show gently inclined bedding surfaces. The tuff consists primarily of welded ash and pumice. Mt. Adams rises high to the south-southeast nearly 40 miles (64 km) away.

The easternmost highway tunnel, 0.2 mile (320 m) west of Box Canyon, shows what looks to be a light-colored sill intruding dark-colored rock. In fact, it's the other way around. The light rock is part of the Stevens Ridge Formation,

Ash-flow tuff of the Stevens Ridge Formation along the road in Stevens Canyon.

and the dark rock on either side of it is gabbro, an intrusive rock that chemically resembles basalt; it has intruded the surrounding Stevens Ridge Formation, of which the light-colored rock is a remnant. Box Canyon is a deeply incised gorge cut into the gabbro by the Muddy Fork of the Cowlitz River. The highway bridge allows a view into the canyon, some 180 feet (55 m) straight down to the stream. You can get an even better view by walking a short distance upstream to a hiking bridge; the trail passes glacially streamlined and striated bedrock. A good exposure of gabbro sits directly across the road from the parking lot.

For the 10 miles (16 km) between Box Canyon and the Stevens Canyon Entrance Station, the road makes a giant switchback, heading south-southeast to the crest of Backbone Ridge, and then descending into the Ohanapecosh River valley by heading north-northeast. Numerous blocky roadcuts of Ohanapecosh Formation line the uphill side of the road south of Box Canyon; these give way to Stevens Ridge Formation after about 4 miles (6.4 km). The large pull-out just west of Backbone Ridge gives an unobstructed view of Mt. Rainier and easy access to the Stevens Ridge Formation across the road, where the tuff is cut by several greenish dikes. The green color comes from chemical alteration of the original andesitic rock. East of Backbone Ridge, the road passes entirely through many exposures of volcanic-rich rocks of the Ohanapecosh Formation. Note the bedding surfaces that show the rocks are inclined toward the west. Unfortunately, this narrow stretch of road is practically devoid of safe pull-outs.

WA 123
US 12—CAYUSE PASS
16 miles (26 km)
See map on page 211.

WA 123 gives access to Mt. Rainier National Park's southeast corner. The highway rises from an elevation of 1,200 feet (365 m) in the south to 4,694 feet (1,431 m) at Cayuse Pass as it follows the Ohanapecosh River upstream nearly to the headwaters of Chinook Creek, its largest tributary. Bedrock exposures for the entire length of the drive consist of volcanic and sedimentary rocks of the Ohanapecosh Formation, deposited during Oligocene time. Even with the heavy vegetation at lower elevations, you can still see many exposures of the Ohanapecosh Formation, some of which clearly show bedding moderately inclined to the west.

Between mileposts 3 and 4, the Ohanapecosh Campground marks the former site of the Ohanapecosh Hot Springs, which grew from a backcountry destination in the early 1900s to a privately operated resort in the 1950s. Today, you can find warm seeps and deposits of travertine along the Hot Springs

Silver Falls flows over outcrops of the Ohanapecosh Formation.

Nature Trail connecting the visitor center to loop B of the campground. Silver Falls, reached by a short trail 1.6 miles (2.6 km) north of the campground, or a quarter mile (400 m) south of Stevens Canyon Road, plunges 60 feet (18 m) over cliffs of the Ohanapecosh Formation.

As you head north on WA 123, you rise above the moss-covered outcrops of the lower elevations. Look northwest near milepost 9 to cliffs of the Ohanapecosh Formation dipping westward. Roadcuts in this vicinity have a greenish tint to them, created by fine particles of the mineral chlorite disseminated throughout the rock. The chlorite originated from high temperatures generated by the intrusion of the Tatoosh pluton. A quarter mile (400 m) north of milepost 10, glacial striations adorn the outcrop on the east side of the highway, although they are more easily seen by southbound travelers.

On both sides of Deer Creek, which WA 123 crosses between mileposts 12 and 13, look toward Mt. Rainier and the Cowlitz Chimneys, a series of sharp peaks eroded from shallow, erosion-resistant volcanic plugs and other intrusive bodies of the Ohanapecosh Formation. They likely mark an eruptive center of the Ohanapecosh. Glacial till, forming an exposure of mixed boulders and fine sediment, shows up just north of milepost 13, and the view southward from near milepost 14 shows the U-shaped profile of the Chinook Creek valley.

The best exposures of the Ohanapecosh Formation along this route consist of a quarter-mile-long (400 m) roadcut just north of milepost 15. Look carefully at these rocks to see that they are largely made of volcanic fragments and display some bedding surfaces inclined gently northward. Some dikes cut the rocks as well. In the remaining mile south of Cayuse Pass, look for exposures of glacial till on the east side of the road.

The Cowlitz Chimneys, the eroded remnants of the Oligocene-age eruptive center of the Ohanapecosh Formation.

WA 410
SUMNER—NACHES
112 miles (180 km)

WA 410 crosses the Cascade Range over Chinook Pass on the east side of Mt. Rainier National Park. The west end of the road begins in Sumner, where the White River joins the Puyallup, which flows northwestward into Commencement Bay at Tacoma. For the first 2.5 miles (4 km) eastward, you traverse a broad floodplain carved into glacial sediments by rivers that flowed beneath glacial ice. Here, the Puget Lobe reached about 1,800 feet thick (550 m). Just east of milepost 11, however, you rise above the floodplain onto the broad glacial plain that you'll follow all the way to Enumclaw. Halfway between mileposts 12 and 13, and at milepost 13, notice the glacial outwash gravels on the north side of the road. Lake Tapps, slightly more than 1 mile (1.6 km) north of the highway on 214th Ave. East, was created by a dam built in 1911 for hydroelectric power. Elongate islands and peninsulas in the lake show how the glacial ice streamlined the till into a series of northwest-southeast drumlins.

Just west of milepost 17, WA 410 passes imperceptibly from deposits of glacial till to those of the Osceola Mudflow, which the highway follows all the way to Enumclaw. The lahar originated high on Mt. Rainier and flowed down the White River as far as Puget Sound. The road crosses the White River at milepost 22, between Buckley and Enumclaw. Some giant boulders, carried by the lahar, lie in the river channel on both sides of the bridge.

For the 25 miles (40 km) between Enumclaw and the national park boundary, WA 410 passes intermittent exposures of volcanic-derived rocks that pre-date the modern Mt. Rainier volcano. Until milepost 54, nearly all the exposed bedrock consists of andesite flows and lahars of the Miocene-age Fifes Peak Formation; these rocks include the quarry exposure near milepost 28. South of milepost 54, the bedrock consists of pyroclastic and volcanic-rich sedimentary rocks of the Oligocene-age Ohanapecosh Formation intruded by the granitic Tatoosh pluton. You will also cross several fairly flat, forested areas of the Osceola Mudflow.

Two side trips are worth checking out. Just west of milepost 30, the road to Mud Mountain Dam leads 2.5 miles (4 km) south to an overlook, exhibit, and trailhead, accessible only on weekdays between 6 a.m. and 4 p.m. The dam was built for flood-control purposes between 1939 and 1948 by the US Army Corps of Engineers. Halfway between mileposts 49 and 50, Forest Road 73 (hard-packed gravel) leads 1.7 miles (2.7 km) southwest up the valley of Huckleberry Creek to some well-defined but tree-covered hummocks of the Osceola Mudflow. At 1.3 miles (2.1 km), make sure you bear right.

Halfway between mileposts 51 and 52, the Skookum Falls Viewpoint offers a look to a terrific waterfall that spills over a 22-million-year-old tuff at the top of the Fifes Peak Formation. The same rock unit exhibits columnar jointing across the road from the parking lot.

In the northeast corner of Mt. Rainier National Park, you pass exposures of the Tatoosh pluton between just north of milepost 60 and the turnoff to

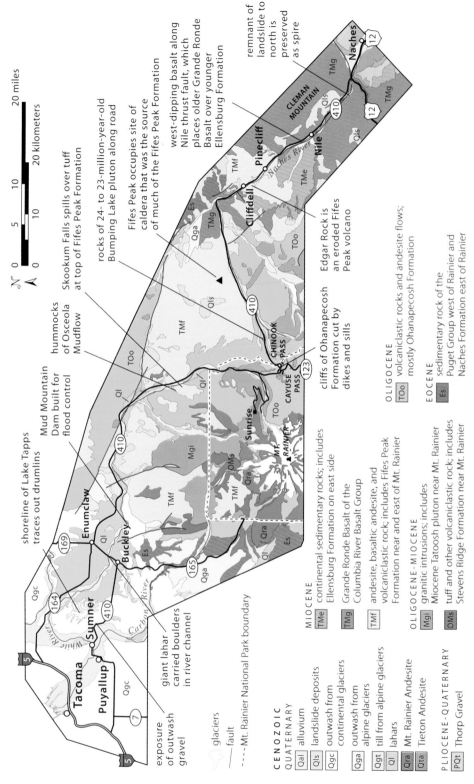

Geology along WA 410 between Sumner and Naches.

N
0   5   10   20 miles
0   10   20 kilometers

remnant of landslide to north is preserved as spire

west-dipping basalt along Nile thrust fault, which places older Grande Ronde Basalt over younger Ellensburg Formation

Fifes Peak occupies site of caldera that was the source of much of the Fifes Peak Formation

rocks of 24- to 23-million-year-old Bumping Lake pluton along road

Skookum Falls spills over tuff at top of Fifes Peak Formation

Edgar Rock is an eroded Fifes Peak volcano

cliffs of Ohanapecosh Formation cut by dikes and sills

hummocks of Osceola Mudflow

Mud Mountain Dam built for flood control

shoreline of Lake Tapps traces out drumlins

giant lahar carried boulders in river channel

exposure of outwash gravel

Naches
Cleman Mountain
Pinecliff
Nile
Cliffdell
Naches River
Chinook Pass
Cayuse Pass
Sunrise
Mt. Rainier
Enumclaw
Buckley
Sumner
Tacoma
Puyallup
White River
Carbon River

- - - - Mt. Rainier National Park boundary

glaciers
— fault

CENOZOIC
QUATERNARY
Qal alluvium
Qls landslide deposits
Qgc outwash from continental glaciers
Qga outwash from alpine glaciers
Qgt till from alpine glaciers
Ql lahars
Qra Mt. Rainier Andesite
Qta Tieton Andesite

PLIOCENE-QUATERNARY
PQt Thorp Gravel

MIOCENE
TMe continental sedimentary rocks; includes Ellensburg Formation on east side
TMg Grande Ronde Basalt of the Columbia River Basalt Group
TMf andesite, basaltic andesite, and volcaniclastic rock; includes Fifes Peak Formation near and east of Mt. Rainier

OLIGOCENE-MIOCENE
Mgi granitic intrusions; includes Miocene Tatoosh pluton near Mt. Rainier
OMs tuff and other volcaniclastic rock; includes Stevens Ridge Formation near Mt. Rainier

OLIGOCENE
TOo volcaniclastic rocks and andesite flows; mostly Ohanapecosh Formation

EOCENE
Es sedimentary rock of the Puget Group west of Rainier and Naches Formation east of Rainier

Road to Sunrise

Only 15 miles (24 km) long, the road to Sunrise climbs nearly 3,000 feet (900 m) up a long ridge of Miocene granodiorite that is capped in part by a flow of the Mt. Rainier Andesite. Set your odometer to zero because this road does not have mileposts.

For the first 5 miles (8 km) to the bridge over the White River, you drive through heavy forest with sparse rock exposures. Good examples of the granodiorite show up just past the bridge at the intersection with the White River Campground Road. The granodiorite has an age of 14.1 million years, which makes it one of the youngest intrusions in the Tatoosh pluton. A short drive down the campground road passes numerous other outcrops of the granodiorite.

The road to Sunrise climbs steeply past the campground road, passing more good outcrops of the granodiorite for about the next half mile. Just past mile 6 (10 km), you can see several yellowish exposures of the Osceola Mudflow. This mudflow originated high on Mt. Rainier about 5,600 years ago and flowed all the way to Puget Sound.

Near mile 7.2 (11.6 km), the road crosses into a flow of the Mt. Rainier Andesite. Called the andesite of Burroughs Mountain, this flow erupted about

Deposit of Osceola Mudflow along the road to Sunrise.

496,000 years ago and reaches thicknesses that exceed 1,000 feet (300 m). At mile 7.6 (12.2 km) you pass a stunning exposure of columnar jointing in a cliff. The columns are nearly horizontal, so what you see is a cross-sectional view of polygons. The horizontal orientation of the columns shows up regularly along the lava flow's margins and likely originated from the flow cooling against the side of a glacier.

Outcrop of near-horizontal columns in andesite flow of Burroughs Mountain, expressed as polygon-shaped cross sections, about 7.6 miles (12.2 km) from WA 410. Photo is about 30 feet (9 m) across.

You've crossed back into the granodiorite by the time you reach a sharp right turn in the road near mile 8 (13 km). From there to Sunrise Point, near mile 12.4 (20 km), the road stays within the granodiorite. Sunrise Point offers spectacular views in all directions, as well as a chance to inspect fresh surfaces of the granodiorite in the walls and paving stones of the overlook. Near mile 13 (21 km), you pass into pyroclastic rocks of the Ohanapecosh Formation, which crop out along the north side of the road. At mile 15 (24 km), Sunrise offers a wide variety of hiking opportunities as well as a visitor center and ranger station.

Sunrise. From the Sunrise turnoff to Cayuse Pass, all exposures are part of the Ohanapecosh Formation. Look for some outstanding views of Rainier and the ridge-like Burroughs Mountain between mileposts 64 and 65. Burroughs Mountain, easily accessed from the Sunrise area, formed after andesite flowed down the narrow ridge between two glacial valleys that were overfilled with ice at the time.

Between Cayuse Pass and Chinook Pass, WA 410 rises nearly 750 feet (230 m) in elevation as it switches back through cliffs of east-tilted Ohanapecosh Formation. You'll notice how variable this rock unit can be, from thin-bedded mudstones to volcanic-rich sandstones to extremely thick bodies of volcanic breccia. Just below the second hairpin turn near the top of the steep grade, look for both felsic and mafic dikes and sills that cut through the rocks. Tipsoo Lake occupies a small glacial cirque just below the pass, and on clear days it provides a stunning foreground to a photo of Mt. Rainier. The prominent rugged peaks in front of Rainier are the Cowlitz Chimneys, a series of erosionally resistant volcanic plugs. They probably mark vents for the Ohanapecosh Formation.

Chinook Pass occupies a low spot in the regional divide between the Cowlitz River watershed on the west and the American and Naches Rivers watershed on the east. The bedrock consists of the Ohanapecosh Formation intruded, in some places, by fingers of the Tatoosh pluton; Yakima Peak, to the northwest, and Naches Peak, to the southwest, are both made largely of granitic rock. The view eastward down the Rainier Fork of the American River shows it to be a classic U-shaped glaciated valley. Look for glacially polished surfaces on the exposed bedrock. Almost all the way to milepost 76, the cliffs consist of beautifully exposed Ohanapecosh Formation intruded in two places by granodiorite of the Bumping Lake pluton. This intrusion, named for Bumping Lake about

Mafic sill (dark-colored rock) intruding lighter-colored fragmental rocks of the Ohanapecosh Formation on the west side of Chinook Pass.

Mt. Rainier and Emmons Glacier viewed from near Chinook Pass. The craggy peaks in the foreground are the Cowlitz Chimneys, which probably originated as an eruptive center for the Ohanapecosh Formation.

7 miles (11 km) to the southeast, is probably between 24 and 23 million years old, so it is older than most of the Tatoosh pluton. You can inspect rocks of the Bumping Lake pluton at milepost 71 and on both sides of milepost 73. At both places you can find dark-colored, angular inclusions of andesite in the granodiorite. Look for glacial till exposed above the bedrock immediately east of milepost 71.

The road descends into the forested valley of the American River near milepost 76. The valley has a broad, glacially carved profile until the canyon narrows between mileposts 84 and 85. The easternmost good exposure of the Ohanapecosh Formation is a double roadcut of porphyritic andesite about 0.75 mile (1.2 km) northeast of milepost 78.

The Fifes Peak Viewpoint, between mileposts 80 and 81, offers a great view of the vertical south face of Fifes Peak, formed of palisade-like columns of pyroclastic rock. Paul Hammond of Portland State University mapped this area in detail and showed that the pyroclastic rock fills a caldera that erupted much of the Fifes Peak Formation. Between milepost 84 and a quarter mile (400 m) east of milepost 90, you'll see scattered outcrops that show the variability of the Fifes Peak Formation. Besides andesite and basalt flows, look for lahars, such as the one 0.5 mile (0.8 km) east of milepost 85, and ash-flow tuffs, such as the one immediately east of the intersection with Bumping Lake Road, between mileposts 88 and 89.

The high cliffs on both sides of the river consist of the Grande Ronde Basalt of the Columbia River Basalt Group; they reach road level near milepost 91. For the next several miles to the Naches River confluence, you can't miss the spectacular basaltic cliffs, with beautiful columns and entablature. Between the confluence and Cliffdell, the exposures aren't nearly as exciting.

South of Cliffdell, WA 410 passes through the Fifes Peak Formation. Edgar Rock, the prominent peak across the Naches River to the south, is a deeply eroded remnant of a Fifes Peak volcano with lava flows and lahars steeply inclined to the northwest. Along the road, look for good exposures of lahars in the cliffs, especially at milepost 97, which is directly across the river from Edgar Rock. Gold Creek Road at milepost 99 hosts a roadcut with several steeply inclined dikes that were likely part of this volcano; a half mile (0.8 km) farther south on US 12, more dikes invade the Fifes Peak Formation.

From just north of milepost 101 to US 12, the Grande Ronde Basalt forms the main outcrops, some with beautiful columns where the road narrows. Look for some river deposits with large boulders stuck to the canyon side directly across from milepost 101. Cleman Mountain, a large anticline in the Columbia River Basalt Group, rises to the northeast. The road to Nile, which intersects

Ash-flow tuff in the Fifes Peak Formation is exposed between mileposts 88 and 89. The light-colored particles in the rock are fragments of pumice.

WA 410 between mileposts 105 and 104, crosses the river and accesses the Ellensburg Formation, which forms the hills south of the river until halfway between mileposts 107 and 108. The Nile thrust fault places the Columbia River Basalt Group to the northeast over the Ellensburg Formation to the southwest. The best roadside expression of the fault is the west-dipping Grande Ronde Basalt about a quarter mile (400 m) north of milepost 106.

A huge landslide complex becomes apparent south of milepost 110, where the canyon opens up and the high basaltic cliffs recede into the background. The hummocky area below the cliffs and extending almost to the road is the landslide. The main mass of the landslide extends to about milepost 113, where you start seeing river gravels along the north side of the canyon.

At milepost 114, it's worthwhile to park in the large pull-out and take in the view. On both sides of the river, you can see some spectacular river terraces, stranded as the Naches River cut deeper into the canyon. The quarry behind the pull-out exposes Ellensburg Formation sandwiched between overlying gravel and underlying basalt. The spire-like rock across the river is cemented material of the landslide complex you passed between mileposts 110 and 113!

OKANOGAN HIGHLANDS

The Okanogan Highlands, a gently mountainous, till-mantled upland, extends east from the Pasayten fault to the Idaho border, and south to the Spokane and Columbia Rivers. Grass and sage land in Okanogan County merge eastward with more heavily forested uplands dissected by mostly southward-flowing rivers. During the Pleistocene Epoch, the southern edge of the great Cordilleran Ice Sheet extended south to within a few tens of miles of the present courses of the Columbia and Spokane Rivers. Four fingerlike lobes of the ice sheet roughly followed the present river valleys for which they are named: Okanogan, Columbia, Colville, and Pend Oreille. The Okanogan Lobe temporarily dammed the Columbia River near the location of the modern Grand Coulee Dam. The Pend Oreille River flows north today, having reversed its direction as the lobe of valley ice receded.

Compared to the other regions in the state, the Okanogan Highlands preserves the longest record of geologic history, from Precambrian events at the former edge of North America to Pleistocene glaciation. Metamorphic and igneous rocks of Archean age—more than 2.5 billion years old—crop out in westernmost Idaho, about 20 miles (32 km) east of Spokane. These old rocks, which may be present in easternmost Washington but have not yet been identified, are the basement, or core, of Laurentia, an ancient continent. Overlying the basement rock is the Belt Supergroup, exposed mostly in Montana and Idaho. It consists of sedimentary rock, up to 12 miles (19 km) thick, deposited in a subsiding basin about 1.4 billion years ago. Although why the basin subsided is still controversial, the rock record is clear. In Washington, Belt rocks include the Deer Trail Group and the Prichard Formation, the lowest unit in the supergroup, and the Prichard's metamorphosed equivalent, the Hauser Lake Gneiss.

Laurentia was part of the more extensive supercontinent Rodinia. About 750 million years ago, the western part of Rodinia rifted from Laurentia and drifted away. Determining its whereabouts today—possible locations include Australia, Antarctica, and Siberia—is an ongoing challenge for geologists interested in reconstructing supercontinents through geologic time. When continents rift and new ocean basins form between them, the new continental margins subside and layers of sediment accumulate upon them, thickening toward the ocean.

In Washington, the Proterozoic-age Windermere Supergroup, which is younger than 700 million years old, and the overlying early Paleozoic units define the rifted margin of Laurentia. This collection of westward-thickening sediments is a world-class example of sediments deposited on a rifted continental margin. Some of the formations provide additional insights into the

227

Chiliwist Butte, made of Mesozoic orthogneiss of the Okanogan Batholith, rises southwest of Malott on US 97.

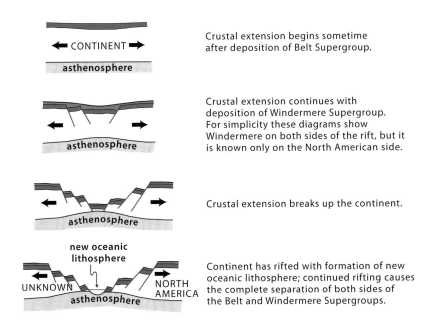

CONTINENT

asthenosphere

Crustal extension begins sometime after deposition of Belt Supergroup.

asthenosphere

Crustal extension continues with deposition of Windermere Supergroup. For simplicity these diagrams show Windermere on both sides of the rift, but it is known only on the North American side.

asthenosphere

Crustal extension breaks up the continent.

new oceanic lithosphere

UNKNOWN

asthenosphere

NORTH AMERICA

Continent has rifted with formation of new oceanic lithosphere; continued rifting causes the complete separation of both sides of the Belt and Windermere Supergroups.

The rifting of Rodinia during late Precambrian time formed the western margin of Laurentia (North America).

history of the rifted margin. The Leola and Huckleberry Formations represent lavas erupted from the newly formed ocean crust. The Shedroof Conglomerate in Washington and the correlative Toby Conglomerate in British Columbia are glacial sediments formed during Snowball Earth, when the globe was partly or completely shrouded in ice. East of the Columbia River, the well-defined early Paleozoic sequence overlying the Windermere includes the Cambrian Addy Quartzite and Maitlen Phyllite, the limestone of the Cambrian-Ordovician Metaline Formation, and the Ordovician-Silurian Ledbetter Slate. The Metaline limestone and dolomite host important deposits of lead and zinc sulfides that have been mined for nearly one hundred years. North and south of Kettle Falls on the Columbia River, poorly dated early and possibly late Paleozoic rocks of the Covada Group largely consist of black shale, phyllite, and slate but include minor quartzite and volcanic rocks. These units probably accumulated west of the better-known, contemporaneous strata east of the Columbia River.

Accretion of the Quesnellia Terrane

In mid-Jurassic time an extensive terrane of sedimentary and volcanic rocks was added to the margin of the continent. These oceanic rocks are best exposed in eastern British Columbia, but scattered remnants in the Okanogan Highlands suggest they were widespread in Washington, too. In British Columbia, the rocks are divided into two terranes: the Slide Mountain terrane, consisting of

Permian volcanic and ultramafic rocks and chert, and the overlying Quesnellia terrane, consisting largely of Triassic and early Jurassic volcanic and sedimentary rocks. These terranes formed in an island arc and were thrust eastward onto the margin of North America in early or mid-Jurassic time.

In Washington, these Mississippian, Pennsylvanian, Permian, Triassic, and Jurassic rocks are collectively called the Quesnellia terrane, although we

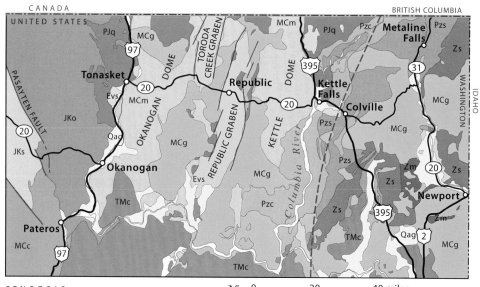

CENOZOIC
QUATERNARY
Qag alluvium and glacial deposits

MIOCENE
TMc lava flows of the
Columbia River Basalt Group

EOCENE
Evs volcanic and nonmarine
sedimentary rocks

MESOZOIC-CENOZOIC
MCg intrusive rocks; mostly granitic
or dioritic in composition and
mostly Cretaceous or Eocene in age
(many of the Cretaceous granites east
of Kettle Falls locally contain both
muscovite and biotite mica)

MCc metamorphic and igneous rock of the
Cascades Crystalline Core domain; contains
Alta Lake Complex and Methow Gneiss

MCm metamorphic and igneous rock
of the Okanogan and Kettle domes;
includes metamorphic rock formed
from older sedimentary rock

0   20   40 miles
0   30   60 kilometers

——— fault
– – – Kootenay arc (edge of Paleozoic North America)
MESOZOIC
JURASSIC-CRETACEOUS
JKo Okanogan Batholith; mostly
granitic (tonalite) and orthogneiss

JKs sedimentary rock of the Methow Basin

PERMIAN-JURASSIC
PJq rocks of Quesnellia terrane; includes a variety of
metamorphosed volcanic and sedimentary rocks,
most notably marble and greenstone

PALEOZOIC
Pzc sedimentary rocks of the Covada Group

LATE PROTEROZOIC–ORDOVICIAN
Pzs sedimentary rocks of the Laurentian margin

PROTEROZOIC
Zs sedimentary rocks of the
Windermere and Belt Supergroups
Zm metamorphic rocks of sedimentary origin

Generalized geologic map of the Okanogan Highlands.

suspect that some late Paleozoic and undated greenstone and ultramafic rocks are part of the Slide Mountain terrane. Perhaps because of the discontinuous nature of the remnants, the names of many rock units are local rather than province-wide. Nevertheless, Eric Cheney of the University of Washington has provided a useful nomenclature based partly on correlative rocks in Canada. In the Chesaw area (east of Oroville), late Paleozoic ultramafic rocks and diverse metamorphosed sedimentary and volcanic rocks constitute the Knob Hill and Attwood Groups. The unconformably overlying late Triassic Brooklyn Formation consists of sedimentary rocks, limestone, and greenstone. Rocks of similar age and type west of the Okanogan River and north of Omak are called the Cave Mountain Formation in some literature. Jurassic strata and volcanic rocks of the Rossland Group are present in places near Oroville, Chesaw, and Curlew and are more widely exposed north of Kettle Falls. Scraps of Quesnellia rocks, most notably serpentinite, are preserved along the major normal faults bounding the Republic graben. The terrane was probably more extensive in the central Okanogan Highlands but has been intruded by granitic plutons and covered by Eocene volcanic rocks.

The thrusting of the Quesnellia terrane onto North America folded the sedimentary rocks of the continental margin. The late Proterozoic and early Paleozoic rocks east of the Columbia River, from south of Colville to the Canadian border, are replete with northeast-trending, upright and overturned folds. These rocks are in the southern part of the Kootenay arc, so named because the trend of the folds curves from northeast to northwest across the Canadian border. Eric Cheney mapped two fold-and-thrust belts northeast of the Columbia River in the deformed late Proterozoic and early Paleozoic units of the Kootenay arc. The Colville River fold-and-thrust belt at Chewelah consists of west-dipping thrust faults; the southeast-dipping thrust faults in the Pend Oreille fold-and-thrust belt between Northport and Metaline Falls deformed rocks as young as the Jurassic-age Rossland Group.

Cretaceous Magmatism

In mid-Cretaceous time, magma intruded the crust, forming granitic plutons across the whole of the Okanogan Highlands. Using the mineralogy of the plutons, geologists have recognized two distinct groups. The first group, typified by the 114- to 111-million-year-old Okanogan Batholith on the western edge of the province, has the geochemical character of the plutonic roots of a magmatic arc. Oceanic crust subducting beneath a continent typically forms magmatic arcs, but the plate tectonic setting during mid-Cretaceous time is controversial. The second group of plutonic rocks, in the east near Spokane, has unusual mineralogy. Instead of the typical humdrum minerals found in arc-related plutons—quartz, feldspar, minor biotite, and hornblende—the Cretaceous granites in the east contain not just one but two micas, biotite and muscovite, and the granites curiously lack hornblende. Moreover, geochemistry indicates that continental crust, not just oceanic crust, was partly melted. Why this occurred is still hotly debated, but it produced the unusual mineralogy and chemistry of plutons such as the Mt. Spokane Granite and the Newman Lake Gneiss.

The Crazy Eocene Epoch

The mid-Cretaceous magmatism was followed by a quiet 50-million-year-long period, but then, about 51 million years ago in the Eocene Epoch, all hell broke loose. A wave of magmatism swept southwest from the northeast corner of the state, and andesite and rhyolite erupted in the north-central part of the Okanogan Highlands. At the same time, the crust was being extended and stretched, not only by conventional, steeply dipping normal faulting but also by profound slip on gently dipping normal, or detachment, faults that gave rise to metamorphic core complexes. These dramatic events may have been caused by the accretion of the Siletz terrane to the western margin of North America about 250 miles (400 km) to the west. Jeff Tepper of the University of Puget Sound proposed that when Siletzia was accreted, the Farallon Plate that was being subducted beneath the continent broke off and allowed hot mantle to rise and melt the crust, causing the magmatism.

As the crust stretched, basins (grabens) formed where rock dropped down between steeply dipping normal faults. Volcanics and sediments accumulated in the grabens and are now preserved in the Republic and Toroda Creek grabens near the town of Republic. Geologists at the US Geological Survey subdivided the Eocene volcanic and interlayered sedimentary rocks into three formations: the oldest and largely sedimentary O'Brien Creek Formation; the Sanpoil Volcanics, consisting of rhyolitic flows and dikes; and the youngest Klondike Mountain Formation, rhyolitic flows and diverse sedimentary rocks. Mining in the Republic mineral district, which began in the late nineteenth century, exploited quartz veins laden with gold and silver in the Klondike Mountain Formation. Thanks to the work of Jeff Tepper and others, we know that the volcanic rocks in the graben are flanked on both sides by granitic plutons of the same age. We infer that the plutons were the subsurface magma chambers that fed the volcanic rocks erupting at the surface. The volcanic and sedimentary units were likely much more extensive in Eocene time.

The crustal stretching also formed metamorphic core complexes: the Okanogan and Kettle core complexes in Washington and the Priest River Complex along the Washington-Idaho border. To illustrate how a core complex forms, we encourage you to try a simple analog model using your hands. Place your right hand beneath the left at eye level and hold them tilted about 20 degrees to the left. Now let your right hand slide to the right, out from under your left. In the real world, your left hand is the hanging wall and your right the footwall separated by a gently dipping normal, or detachment, fault. You have demonstrated crustal extension, because your hands and elbows have moved apart. Furthermore, you have drawn your right fingertips up from deep below your left hand, and in so doing you have brought formerly deeply buried rocks up closer to Earth's surface. By slightly arching your hands during slip, you can see why core complexes are also referred to as domes.

In the Okanogan core complex, the normal fault is the Okanogan detachment fault. Slip along this fault strongly deformed the upper kilometer or so of the footwall and converted the metamorphosed sedimentary rocks and igneous rocks into mylonites—rocks that have strong layering imparted by deformation

Cross section from Tonasket to Kettle Falls, across the Okanogan core complex, Toroda Creek and Republic grabens, and the Kettle core complex. The Okanogan detachment fault does not appear on the cross section because it was eroded off the top of the mylonite zone. It crops out just west of the cross-section line.

along the fault at high temperatures. The Okanogan footwall consists of medium- to high-grade metamorphic rocks, whereas the hanging wall consists of slightly metamorphosed parts of Quesnellia overlain by unmetamorphosed Eocene volcanic and sedimentary rocks. The Okanogan and Kettle core complexes are north-south-trending domes that reflect the originally curved geometry of the detachment faults. The metamorphic rocks in the footwalls included interlay- ered sedimentary units, now metamorphosed to paragneiss, but there is no consensus on the rocks' age, original sedimentary rock type, or origin.

GUIDES TO THE OKANOGAN HIGHLANDS

US 2
SPOKANE—NEWPORT
38 miles (61 km)

US 2 between Spokane and Newport crosses forested lands developed on gravel deposits of the Ice Age floods, ones that most certainly originated from Glacial Lake Missoula. Poking through these deposits are occasional bedrock outcrops. North of Spokane, the highway mostly follows the top of a terrace overlooking the Little Spokane River. Exposures of gravel along the roadside consist of flood gravels. The most prominent gravel deposits show up on the side of the terrace

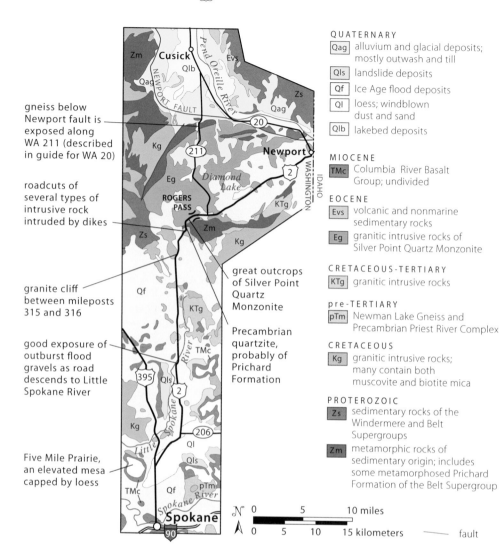

QUATERNARY
Qag alluvium and glacial deposits; mostly outwash and till
Qls landslide deposits
Qf Ice Age flood deposits
Ql loess; windblown dust and sand
Qlb lakebed deposits

MIOCENE
TMc Columbia River Basalt Group; undivided

EOCENE
Evs volcanic and nonmarine sedimentary rocks
Eg granitic intrusive rocks of Silver Point Quartz Monzonite

CRETACEOUS-TERTIARY
KTg granitic intrusive rocks

pre-TERTIARY
pTm Newman Lake Gneiss and Precambrian Priest River Complex

CRETACEOUS
Kg granitic intrusive rocks; many contain both muscovite and biotite mica

PROTEROZOIC
Zs sedimentary rocks of the Windermere and Belt Supergroups
Zm metamorphic rocks of sedimentary origin; includes some metamorphosed Prichard Formation of the Belt Supergroup

gneiss below Newport fault is exposed along WA 211 (described in guide for WA 20)

roadcuts of several types of intrusive rock intruded by dikes

granite cliff between mileposts 315 and 316

good exposure of outburst flood gravels as road descends to Little Spokane River

Five Mile Prairie, an elevated mesa capped by loess

great outcrops of Silver Point Quartz Monzonite

Precambrian quartzite, probably of Prichard Formation

0   5   10 miles
0 5 10 15 kilometers   —— fault

Geology along US 2 between Spokane and Newport.

as the highway descends to cross the Little Spokane River near milepost 304. The road passes some Grande Ronde Basalt at milepost 308. A more distinct exposure of the overlying Wanapum Basalt forms a cliff that caps a hill to the east about halfway between mileposts 310 and 311.

Between mileposts 312 and 313, look for some small exposures of granitic rock on both sides of the road. An even better exposure forms a cliff on the west side of the road halfway between mileposts 315 and 316. With few exceptions, all the granitic rock in northeastern Washington intruded during Cretaceous or Eocene time. The granites between mileposts 315 and 316 are Cretaceous and, in some places, contain both white (muscovite) and black (biotite) micas that indicate a high level of aluminum in the magma.

Just south of milepost 318, Cretaceous granitic rock intrudes an older, darker-colored igneous body. Both these rocks are cut by basaltic dikes. About 0.4 mile (640 m) north of the milepost, cliffs on the east side of the road are made of Precambrian quartzite of the Prichard Formation, deposited near the bottom of the Belt Supergroup. On close inspection, you can find bedding defined by small variations in grain size. The bedding dips steeply northward, in contrast to the many closely spaced fractures in the rock that dip southward. Immediately north, you see more granite; good outcrops of this granite continue for the next mile to Rogers Pass and then sporadically for most of the way to Newport. Called the Silver Point Quartz Monzonite, this granitic rock has radiometric ages that typically fall between 52 and 47 million years, which places it in the Eocene Epoch. It intruded at about the same time the crust in this area was being stretched along the shallowly dipping Newport detachment fault. The fault, which WA 20 crosses just north of Newport, carried rocks to the east and exposed metamorphic rocks to its west.

Although Eocene granites lie just beneath the surface between Rogers Pass and Newport, the area is mostly forested flood deposits of Glacial Lake Missoula. Diamond Lake, at milepost 325, occupies a depression in the flood deposits on the edge of more of the Eocene granite. US 2 crosses the Little Spokane River and its alluvium near milepost 330. Between mileposts 331 and 332, the highway passes imperceptibly into the watershed of the Pend Oreille River, which flows north into Canada.

Roadcut of Silver Point Quartz Monzonite adjacent to the quartzite of the Prichard Formation (outcrop at far right).

US 97
PATEROS—OKANOGAN—CANADIAN BORDER
82 miles (132 km)

Between Pateros and the mouth of the Okanogan River, US 97 winds past spectacular river terraces of glacial outwash along the Columbia River. They are especially striking at Pateros, where they exceed 400 feet (120 m) in height! The scattered bedrock exposures are part of the southern margin of the Okanogan Batholith, a series of plutons that intruded between about 114 and 111 million years ago in Cretaceous time. After determining these ages, Hugh Hurlow and Bruce Nelson of the University of Washington suggested that the batholith formed the roots of a magmatic arc that extended north into Canada. Some especially good exposures of the granitic rocks show up along the highway between milepost 255 and just north of milepost 256.

North of milepost 264, US 97 follows the Okanogan River northward between hills and ridges that were buried by the Okanogan Lobe of the Cordilleran Ice Sheet. The Okanogan River rises some 115 miles (185 km) to the north in southern British Columbia. River terraces of glacial outwash deposits dominate much of the landscape. For at least the first 10 miles (16 km), the road follows the top of a low, recently formed terrace, but in some places older, higher ones rise high above the river. Several good exposures of granitic rocks of the Okanogan Batholith show up along the road near and north of milepost 267. Most of these rocks are gneisses, which formed through high-temperature deformation of the original granitic parent rock. Between mileposts 268 and 269, look for granitic rock intruded by mafic rock.

River terraces on the Columbia River near Pateros reach more than 400 feet (120 m) in height. Note the flat surface at the top, the former level of the river floodplain.

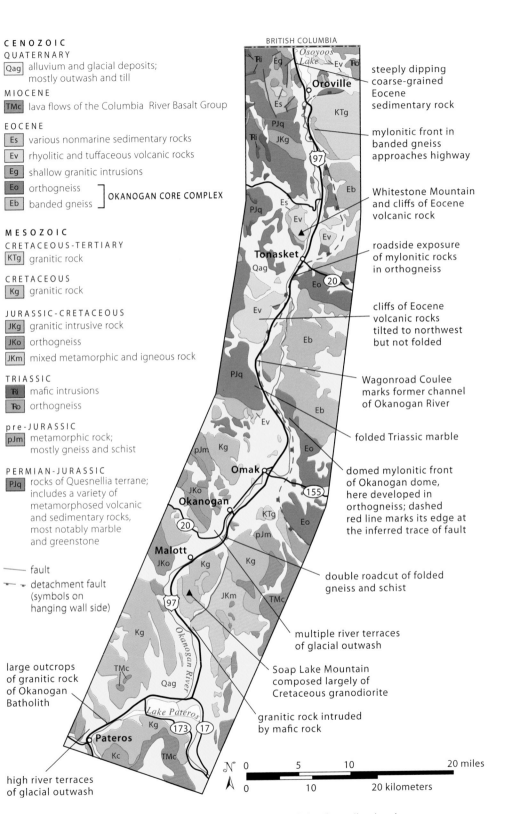

CENOZOIC
QUATERNARY
Qag alluvium and glacial deposits; mostly outwash and till

MIOCENE
TMc lava flows of the Columbia River Basalt Group

EOCENE
Es various nonmarine sedimentary rocks
Ev rhyolitic and tuffaceous volcanic rocks
Eg shallow granitic intrusions
Eo orthogneiss ⎫ OKANOGAN CORE COMPLEX
Eb banded gneiss ⎭

MESOZOIC
CRETACEOUS-TERTIARY
KTg granitic rock

CRETACEOUS
Kg granitic rock

JURASSIC-CRETACEOUS
JKg granitic intrusive rock
JKo orthogneiss
JKm mixed metamorphic and igneous rock

TRIASSIC
Ri mafic intrusions
Ro orthogneiss

pre-JURASSIC
pJm metamorphic rock; mostly gneiss and schist

PERMIAN-JURASSIC
PJq rocks of Quesnellia terrane; includes a variety of metamorphosed volcanic and sedimentary rocks, most notably marble and greenstone

— fault
⟶ detachment fault (symbols on hanging wall side)

large outcrops of granitic rock of Okanogan Batholith

high river terraces of glacial outwash

BRITISH COLUMBIA

Osoyoos Lake
Oroville

steeply dipping coarse-grained Eocene sedimentary rock

mylonitic front in banded gneiss approaches highway

Whitestone Mountain and cliffs of Eocene volcanic rock

Tonasket

roadside exposure of mylonitic rocks in orthogneiss

cliffs of Eocene volcanic rocks tilted to northwest but not folded

Wagonroad Coulee marks former channel of Okanogan River

folded Triassic marble

Omak

Okanogan

domed mylonitic front of Okanogan dome, here developed in orthogneiss; dashed red line marks its edge at the inferred trace of fault

Malott

double roadcut of folded gneiss and schist

multiple river terraces of glacial outwash

Soap Lake Mountain composed largely of Cretaceous granodiorite

granitic rock intruded by mafic rock

Okanogan River

Lake Pateros
Pateros

0   5   10   20 miles
0   10   20 kilometers

Geology along US 97 between Pateros and the Canadian border.

Folded gneiss and schist near milepost 283.

The road passes through some vegetated dune deposits hosting a stand of ponderosa pine between mileposts 273 and 274. The sand probably originated in the glacial outwash but became separated from the otherwise coarse gravel by wind.

Near Malott, halfway between mileposts 278 and 279, you pass Soap Lake Mountain to the east, with outcrops of Cretaceous granodiorite forming the steep hillside near milepost 281. A spectacular double roadcut of older highly folded metamorphic rock lies 0.3 mile (480 m) north of milepost 283.

At Omak, US 97 crosses the Okanogan River. From the broad river terrace north of Omak, you get increasingly better views of the domed upper surface of the Okanogan metamorphic core complex, or dome, to the northeast. The layering in this surface rock consists of a metamorphic foliation that forms in a high-temperature fault zone. Called a mylonite zone, it formed in already existing rocks of the middle crust, and here the zone reaches thicknesses of about 6,000 feet (1,800 m). Between Omak and Tonasket, just north of milepost 314, US 97 follows the western edge of this mylonite zone, which dips westward as the Okanogan detachment fault beneath the rocks of Quesnellia to the west. It formed during Eocene time when this part of the crust was stretched, allowing metamorphic rocks of the Okanogan core complex to rise and the rocks of Quesnellia to move westward and downward.

Beginning at about milepost 300, the road leaves the floodplain of the Okanogan River and heads over a low divide into Wagonroad Coulee, which served as the Okanogan River's channel prior to glacial advance in the Pleistocene Epoch.

West-dipping mylonitic gneisses mark the western edge of the Okanogan metamorphic core complex, or Okanogan dome.

Glacial outwash deposits are visible in many hillsides between milepost 300 and Tonasket. The cliffs on the west side consist of metamorphosed Triassic limestone, part of the Cave Mountain Formation of Quesnellia. Near milepost 302, you can see that they make a large overturned fold, although in poor light it can be difficult to see. This folding must have taken place prior to the Eocene extension because just a few miles north, at milepost 306, you can see cliffs of Eocene volcanic rocks that are not folded. These volcanic rocks consist mostly of tuffs and dacite flows and are tilted northwestward, directly away from the road. Their tilting probably occurred during movement on the Okanogan detachment fault. These volcanics erupted from the Challis volcanic field that extended from southern British Columbia all the way to southern Idaho. US 97 passes some mylonitic gneiss of the core complex just south of milepost 313.

North of Tonasket, the mylonite zone of the Okanogan detachment fault swings northeastward, so the road heads into the overlying Quesnellia terrane and the Eocene volcanic rocks. Whitestone Mountain, just north of town, exposes spectacular cliffs of the volcanic rocks on the west side of the road. At and north of Ellisforde, near milepost 321, are more hills made of metamorphosed volcanic and sedimentary rocks of Quesnellia, intruded in some places by granitic rock. These rocks come close to the road a quarter mile (400 m) north of milepost 323. Perhaps the most dramatic view, however, is northward, where a high ridge of the mylonite zone angles northwest to rejoin the highway near milepost 325.

The Okanogan River meanders through a wide floodplain between Ellisforde and Oroville. You'll see beautifully developed river terraces on the glacial outwash. Oroville sits at the confluence of the Okanogan and Similkameen

A large, overturned fold in limestone of the Cave Mountain Formation of Quesnellia. White lines trace some of the bedding planes.

The Eocene volcanic rocks of Whitestone Mountain are tilted eastward toward the mylonite zone along the front of the Okanogan core complex. Red lines on the photo trace some of the bedding.

Rivers, both of which flow south from Canada. A dam on the Okanogan at Oroville impounds Osoyoos Lake.

North of Oroville, US 97 passes alongside Eocene sedimentary and volcanic rocks. Near milepost 334, they are intruded by Eocene granitic rocks. You can inspect some steeply dipping Eocene sedimentary rocks adjacent to the highway just north of milepost 335. There, you'll find conglomerate, with angular fragments of granite and basalt, overlain by shale. Northward, you cross a fault and follow hills of Jurassic to Cretaceous granitic rock for the 1.5 miles (2.4 km) to the border.

US 395

SPOKANE—CANADIAN BORDER AT LAURIER

112 miles (180 km)

Less than 1 mile (1.6 km) north of I-90, US 395 crosses the Spokane River, which drains about 6,200 square miles (16,100 km²) of northeastern Washington and northern Idaho before emptying into the Columbia River about 45 miles (72 km) to the west-northwest. At Spokane Falls, in the heart of downtown Spokane, the river drops 146 feet (44.5 m) over a series of falls in Riverfront Park (see section on Spokane in the Columbia Basin chapter). Four miles (6.4 km) north of the river, look westward to the mesa of Five Mile Prairie, which rises nearly 500 feet (150 m) above the surrounding landscape. Five Mile Prairie consists mostly of Columbia River Basalt Group capped by loess.

Just south of milepost 168, a spectacular roadcut of Cretaceous granite lies along the southbound lanes. The rock is part of the Mt. Spokane Batholith, which makes up many of the mountains to the east, including Mt. Spokane. The rock is cut by innumerable Cretaceous dikes of pegmatite and aplite, as well as some mafic dikes of probable Eocene age. While both crystallize from silica-rich magmatic fluid, pegmatites have large, well-formed crystals, whereas aplite has small, granular crystals. A much smaller outcrop lies adjacent to the northbound lanes. In the valley of the Little Spokane River below, you can see exposures of gravel, deposited by the outburst floods from Glacial Lake Missoula.

For about the 30 miles (48 km) south of Clayton, US 395 crosses prairie established on Missoula flood gravels. Bedrock exposures are sparse. Look for some exposed Cretaceous granite about 0.75 mile (1.2 km) north of milepost 170 and some Columbia River Basalt Group about 0.2 mile (320 m) south of milepost 177.

North of Clayton, US 395 passes through the Silver Point Quartz Monzonite, an Eocene intrusion that is exposed as far south as the Spokane River and as far east as Newport. It has radiometric ages that typically fall between 52 and 47 million years, so it intruded while the region was being stretched apart along the Newport fault to the east. Probably the best exposure lies on the east side of the road just a quarter mile (400 m) north of milepost 188.

Loon Lake marks the southern edge of Pleistocene glacial deposits along this route. Although exposures of glacial gravel are generally poor, some decent

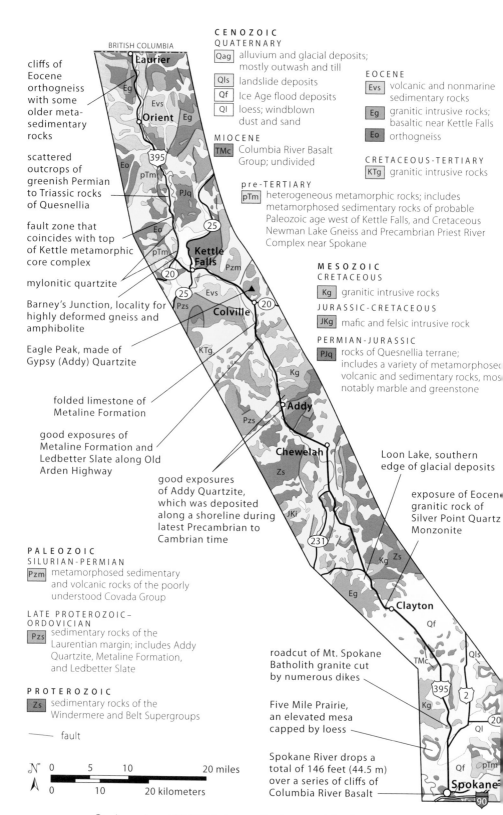

BRITISH COLUMBIA

cliffs of Eocene orthogneiss with some older meta-sedimentary rocks

scattered outcrops of greenish Permian to Triassic rocks of Quesnellia

fault zone that coincides with top of Kettle metamorphic core complex

mylonitic quartzite

Barney's Junction, locality for highly deformed gneiss and amphibolite

Eagle Peak, made of Gypsy (Addy) Quartzite

folded limestone of Metaline Formation

good exposures of Metaline Formation and Ledbetter Slate along Old Arden Highway

good exposures of Addy Quartzite, which was deposited along a shoreline during latest Precambrian to Cambrian time

Laurier

Orient

Kettle Falls

Colville

Addy

Chewelah

Clayton

Spokane

Loon Lake, southern edge of glacial deposits

exposure of Eocene granitic rock of Silver Point Quartz Monzonite

roadcut of Mt. Spokane Batholith granite cut by numerous dikes

Five Mile Prairie, an elevated mesa capped by loess

Spokane River drops a total of 146 feet (44.5 m) over a series of cliffs of Columbia River Basalt

CENOZOIC
QUATERNARY
Qag · alluvium and glacial deposits; mostly outwash and till
Qls · landslide deposits
Qf · Ice Age flood deposits
Ql · loess; windblown dust and sand

MIOCENE
TMc · Columbia River Basalt Group; undivided

pre-TERTIARY
pTm · heterogeneous metamorphic rocks; includes metamorphosed sedimentary rocks of probable Paleozoic age west of Kettle Falls, and Cretaceous Newman Lake Gneiss and Precambrian Priest River Complex near Spokane

EOCENE
Evs · volcanic and nonmarine sedimentary rocks
Eg · granitic intrusive rocks; basaltic near Kettle Falls
Eo · orthogneiss

CRETACEOUS-TERTIARY
KTg · granitic intrusive rocks

MESOZOIC
CRETACEOUS
Kg · granitic intrusive rocks
JURASSIC-CRETACEOUS
JKg · mafic and felsic intrusive rock
PERMIAN-JURASSIC
PJq · rocks of Quesnellia terrane; includes a variety of metamorphosed volcanic and sedimentary rocks, most notably marble and greenstone

PALEOZOIC
SILURIAN-PERMIAN
Pzm · metamorphosed sedimentary and volcanic rocks of the poorly understood Covada Group

LATE PROTEROZOIC–ORDOVICIAN
Pzs · sedimentary rocks of the Laurentian margin; includes Addy Quartzite, Metaline Formation, and Ledbetter Slate

PROTEROZOIC
Zs · sedimentary rocks of the Windermere and Belt Supergroups

— fault

N

0   5   10   20 miles
0   10   20 kilometers

Geology along US 395 between Spokane and the Canadian border.

Roadcut of granite cut by light-colored dikes of pegmatite and aplite. This granite is part of the Mt. Spokane Batholith.

examples are at milepost 193 and halfway between mileposts 196 and 197. Numerous closed depressions—potholes formed by melting of large ice blocks left by the retreating glacier—dot the landscape but are mostly obscured by trees. A good example lies just south of milepost 199.

Between mileposts 200 and 201, US 395 passes through several exposures of the McHale Slate, part of the Proterozoic Deer Trail Group. These rocks are generally described as argillites—shales that experienced low-grade metamorphism. The Deer Trail Group is lumped with the Belt Supergroup.

Between Chewelah and Colville, the road cuts across folded and faulted Proterozoic and Paleozoic rocks that are intruded by Cretaceous granitic rocks. Some of the most prominent exposures for the first few miles are in the Cambrian Addy Quartzite and greenstone of the Proterozoic Huckleberry Formation of the Windermere Supergroup. The Addy Quartzite, which was once beach sand on an ocean shore, forms the light-colored, bedded cliffs between mileposts 210 and 211. It also crops out halfway between mileposts 212 and 213 and forms the hills on both sides of the town of Addy, near milepost 217. A cut on the east side of the highway exposes the quartzite on the south side of town. A series of normal faults that run perpendicular to the highway cuts the quartzite and underlying greenstone, causing them to repeat.

About 2 miles (3.2 km) north of Addy, the highway passes through a Cretaceous granitic body named the Starvation Flat Quartz Monzonite. It is nicely exposed as a roadcut about a quarter mile (400 m) south of milepost 222 at the south intersection of US 395 and Old Arden Highway, which parallels US 395 through Arden. We know the granitic rock is younger than the sedimentary rock because it cuts across the structures in the Paleozoic sedimentary rocks.

Just north of milepost 225 at the north intersection of Old Arden Highway, you can see where the northern edge of the Starvation Flat Quartz Monzonite intrudes the Cambrian Metaline Formation and Ordovician Ledbetter Slate, both of which were deposited in a shallow sea. The Ledbetter Slate crops out closest to the highway, but within about 100 yards (90 m) farther north, you cross into limestone of the Metaline Formation and then into the granite. The Metaline Formation shows up in scattered roadcuts on the east side of US 395 between mileposts 226 and 228, with some beautiful folds halfway between mileposts 227 and 228.

Between Colville and Kettle Falls, US 395 passes by cliffs of Paleozoic-age metamorphosed and unmetamorphosed sedimentary and volcanic rock to the northeast, and hills of Eocene-age volcanic and volcanic-rich sedimentary rocks to the southwest. Near milepost 231, you can see a good exposure of the volcanic rocks about 1 mile (1.6 km) to the southwest. These rocks are part of the Sanpoil Volcanics, the same rock unit that fills much of the Republic graben, which WA 20 crosses about 30 miles (48 km) to the west. The high hills behind them are Paleozoic sedimentary rocks. Northeast of the highway, Eagle Peak is made of Cambrian-age Gypsy Quartzite, a different name but the same rock unit as the Addy Quartzite; younger rocks make up the hills along the road to the northwest. At milepost 237, US 395 crosses a fault and passes into Permian-age metasedimentary and metavolcanic rocks of Quesnellia. A mafic intrusive body, probably of Eocene age, intrudes the Permian rock to form the ridge immediately north of Kettle Falls.

Between Kettle Falls and the Columbia River, US 395 passes just north of Hawks Nest, a forested hill made of marine metasedimentary rock of Permian age intruded by mafic rock. Bluffs of Quaternary sediments line the east side of the river. West of the river, cliffs consist of highly deformed gneiss, amphibolite, and quartzite of probable early Paleozoic age. These rocks, which line the west side of the highway between the river crossing and the Canadian border, mark the fault zone along the edge of the Kettle metamorphic core complex. This fault zone, active during the stretching of the crust in Eocene time, deformed, uplifted, and exposed the metamorphic rocks to the west. Perhaps the best expressions of the fault zone are mylonites: fine-grained foliated rocks that were deformed deep in the crust at high temperatures in a ductile shear zone.

At milepost 242, look for a wonderful exposure of highly deformed gneiss and amphibolite on both sides of the highway. Called Barney's Junction, this exposure is worth taking some time to explore because it also contains mylonitic zones, abundant pegmatite, and lozenge-shaped bodies of green pyroxene-bearing gneiss. Some spectacular folds are exposed in the outcrop on the east side. A quarter mile (400 m) to the north, you can see mylonitic quartzite. Just north of milepost 243, the road passes several outstanding exposures of amphibolite.

Just north of milepost 245, Northport–Flat Creek Road leads northeast across the mouth of the Kettle River and follows the Columbia River through

Deformed gneiss at
the outcrop at Barney's
Junction. Top: amphibolite
gneiss (about 20 inches
across, 50 cm); bottom
right: a sheath fold, with a
three-dimensional geometry
similar to a sock; bottom
left: a recumbent fold.

Paleozoic metasedimentary and volcanic rocks. US 395 continues northward along the Kettle River, which meanders across a narrow floodplain partially filled with glacial outwash deposits. A beautiful roadcut of mylonitic quartzite exists on the west side of the highway just north of the junction, but for most of the next 8 or 9 miles (13–14 km), the road mostly crosses over glacial outwash deposits. Good exposures of greenish Permian to Triassic rocks show up, however, especially between mileposts 250 and 254. There is easy access to the Kettle River, a sandy beach, and a gravel bar halfway between mileposts 256 and 257. There, metamorphosed sedimentary rock in the riverbed has been scoured into potholes.

Between Orient and the border, the cliffs lining the road become increasingly dramatic. Similar to those farther south, they consist mostly of gneiss but include mylonitic quartzite and amphibolite. The fact that the three metamorphic rock types are found together suggests they originated as different types of sedimentary rock. A good example lies just a half mile (0.8 km) north of Orient, where the road passes an exposure of gneiss that contains layers of mylonitic quartzite.

WA 20
Twisp—Okanogan
32 miles (52 km)

Between Twisp and Okanogan, WA 20 passes over a subdued landscape of granitic gneiss (orthogneiss) and granitic intrusive rock of the Okanogan Batholith. Ice covered the region during the Pleistocene Epoch, rounding the bedrock and depositing glacial till in the valleys.

Immediately east of Twisp, WA 20 passes a series of outcrops of greenish and reddish volcanic rocks of the Newby Group, erupted in the Methow Basin in Jurassic to Cretaceous time. Just east of the intersection with WA 153, the road heads uphill for a half mile (0.8 km) through outwash gravels to a wide river terrace, more than 100 feet (30 m) above the modern floodplain. An outcrop of andesite of the Newby Group pokes through the gravel, its green color derived from the mineral chlorite disseminated throughout the rock.

Farther northwest, the Pasayten fault separates the Newby Group and younger sedimentary rocks of the Methow Basin to the west from the Okanogan Batholith to the east. About a quarter mile (400 m) west of milepost 205,

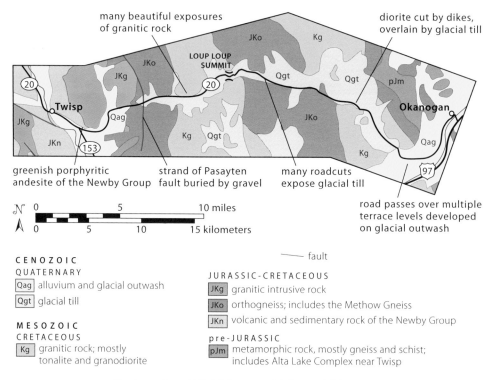

Geology along WA 20 between Twisp and Okanogan.

WA 20 crosses a southeastern strand of the fault, but it's buried by gravel. The next several miles farther west are mostly till covered and littered in some places with granitic boulders. Good exposures of granite appear about milepost 210.

Loup Loup Summit, between mileposts 214 and 215, marks the high point of the route at an elevation of 4,013 feet (1,223 m). The most interesting bedrock and till exposures lie along the long downgrade east of the summit. Just below milepost 221, look for glacial till overlying diorite across from a wide pull-out. The diorite is cut by a variety of pegmatite and granitic dikes. Just below there, you can see even more variety in the bedrock, which exhibits metamorphic layering in places.

Diorite cut by pegmatite dikes downhill from milepost 221.

Just east of milepost 227, the road crosses a wide river terrace partly covered by apple orchards. This terrace consists of glacial outwash, some of which makes a double roadcut at milepost 229. Meltwater streams emanating from the Okanogan Lobe deposited the outwash. As the Okanogan River deepened its channel, it left the terraces stranded on each side. To the east, you can see a series of three terraces on the east side of the Okanogan River. The highest of these terraces is the oldest, marking the level of the river's floodplain at that early time. The road follows the matching high terrace on the west side of the river until about milepost 230, where the highway cuts through more gravel and descends to a younger, lower level that it follows to the town of Okanogan. On crossing the river, WA 20 rises back up to this low terrace and intersects US 97. (See US 97 road guide for the section of WA 20 north of Omak.)

Outwash gravel with rounded cobbles exposed in the side of an old floodplain terrace of the Okanogan River.

WA 20
TONASKET—KETTLE FALLS
84 miles (135 km)

WA 20 between Tonasket and Kettle Falls displays Washington's Eocene crustal extension better than anywhere else in the state. The Okanogan and Kettle metamorphic core complexes illustrate the effects of faulting at depth in the hot middle crust, while the Toroda Creek and Republic grabens illustrate the effects of faulting in the cooler, upper part of the crust. Filling the grabens are Eocene-age volcanic and sedimentary rocks that host important gold deposits as well as fossil localities. Ice covered this part of the Okanogan Highlands during the Pleistocene, depositing glacial till as it receded.

Within the first mile east of Tonasket, WA 20 passes from outwash gravel into highly deformed gneisses of the front of the Okanogan metamorphic core complex. These gneisses are called mylonites because they are mostly very fine-grained and strongly layered, and they display lineations formed from elongated and entrained mineral grains—all a product of high-temperature deformation in which the rock flows instead of fractures. Along this mylonite zone, which reaches 6,000 feet (1,800 m) in thickness, the metamorphic core of the range rose upward to the east from depths of 6 to 12 miles (10–19 km).

These mylonites are especially resistant to erosion and weathering and control the shape of much of the landscape. The steepness of the canyon east

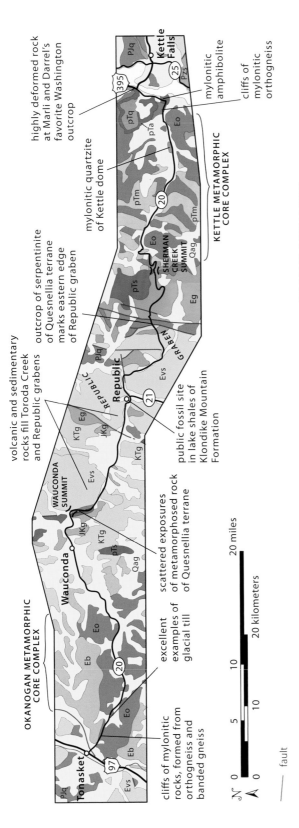

Geology along WA 20 between Tonasket and Kettle Falls.

Looking west down the valley toward Tonasket at the mylonite zone along the western edge of the Okanogan metamorphic core complex. Strongly foliated rocks are visible on the right on the left side in the background they control the slope of the land. Inset photo shows detail

of Tonasket, for example, mimics the dip of the layering in the mylonites. As the mylonite zone flattens to the east, so does the landscape. One outstanding locality where you can stop and inspect the mylonites is 0.3 mile (480 m) east of milepost 264. There, you can see interesting and complex relationships between the mylonites and pegmatite dikes, some of which probably intruded during the deformation.

The road passes by several great examples of glacial till starting only a half mile (0.8 km) east of milepost 264. These sediments, deposited directly by glacial ice, consist of anything and everything carried by the ice, from clay to boulders, all thrown together haphazardly with no stratification.

As the gradient flattens near milepost 267, notice that the mylonite layers also become less steep. They even dip eastward for a short interval near milepost 270. You continue to pass scattered outcrops of the mylonitic gneiss and till until east of milepost 279, where you encounter an imposing cliff of the mylonite, some of which is clearly folded. The mylonites are intruded by Eocene and Cretaceous granites that crop out halfway between mileposts 280 and 281, as well as between Wauconda and Wauconda Summit.

At Wauconda Summit, WA 20 crosses a fault, east of which is metamorphosed Triassic rock at the western edge of the Toroda Creek graben, an Eocene-age sedimentary basin that continues north of the road into Canada. You can see some of these rocks on the west side of the road as well as cliffs of Eocene lava flows to the north and east on the east side of the pass. These lavas, which consist of andesites, dacites, and rhyolites, fill this southern part of the graben. Cretaceous granodiorite, near and east of milepost 294, marks the southeastern edge of the Toroda Creek graben and forms the intervening area between it and the Republic graben a few miles to the east.

At about milepost 300, WA 20 crosses a fault and passes eastward into the Republic graben. The fault-bound basin is filled with a variety of Eocene-age rock, which here consists of mostly dacite lava flows of the Sanpoil Volcanics, dacite intrusions, and river and lake deposits of the Klondike Mountain Formation. A roadcut of the intrusive dacite lies on the north side of the road a quarter mile (400 km) west of milepost 302.

The Republic area was one of Washington's premier gold-mining regions, beginning with a rush of prospectors in 1896. The gold precipitated from hydrothermal fluids within the pre-Mesozoic rocks of Quesnellia as well as within the Klondike Mountain Formation.

Fine-grained lake deposits in the bottom third of the Klondike Mountain Formation host spectacular leaf fossils of ginkgo, larch, dawn redwood, and

Fossil-collecting site in lakebeds of the Klondike Mountain Formation. The Ginkgo leaf (inset) is a part of the collection at the Stonerose Interpretive Center.

sassafras, as well as some insect and fish fossils. Higher in the formation, the rocks become coarser grained, more typical of river deposits, and contain few fossils. The Stonerose Interpretive Center on the northwest side of Republic provides fossil-hunting advice and displays many beautiful specimens of fossil leaves from the area, as well as fossils from throughout the state. You can search for fossils in an exposure of the lakebeds a block north of the interpretive center.

On the south side of Republic, WA 20 passes lakebed deposits of the Klondike Mountain Formation on the east, but Sanpoil Volcanics occur on both sides of the highway shortly east of the intersection with WA 21. Along with the omnipresent glacial till, these rocks form nearly all the roadside exposures and cliffs until about milepost 313, where you cross the eastern fault of the Republic graben. East of the fault is greenish serpentinite of Permian or Triassic age. Serpentinite forms from the metamorphism of low-silica igneous rock, which probably originated as oceanic material in the Quesnellia terrane.

For the next several miles east of the serpentinite, WA 20 climbs through Eocene granitic rock to Sherman Pass, between mileposts 319 and 320. East of the pass is pre-Tertiary metamorphic rock of the Kettle metamorphic core complex. Some good exposures of these rocks occupy the area near the summit, but they give way to a forested landscape within the first couple miles. In many ways, the Kettle core complex is a mirror image of the Okanogan core complex. Just like the road rose eastward from Tonasket through a mylonite zone into the Okanogan core complex, the road now descends eastward through the Kettle metamorphic core complex into a mylonite zone along its front. The main

Greenish serpentinite in the fault zone at the eastern edge of the Republic graben.

A roadcut of mylonitic orthogneiss on the east side of the Kettle metamorphic core complex.

rocks consist of quartzite, gneiss, and amphibolite, a dark-colored metamorphic rock composed chiefly of the minerals amphibole and plagioclase. Many of the outcrops contain dikes of pegmatite. Toward the east side of the Kettle core complex, you pass several outstanding exposures of these rocks: mylonitic quartzite at 0.6 mile (1 km) east of milepost 333, mylonitic orthogneiss a half mile (0.8 km) east of milepost 338, and amphibolite just east of milepost 340.

At the intersection of WA 20 and US 395, a place informally called Barney's Junction, is authors Marli and Darrel's favorite Washington outcrop. It exposes an incredible array of mylonitic rocks, described in more detail in the US 395 road guide in this chapter.

WA 20
NEWPORT—COLVILLE
83 miles (134 km)

Between Newport and Tiger, WA 20 passes over deposits of till and glacial lake sediments along the Pend Oreille River, which reversed its course at the end of the Pleistocene Ice Age. Today, the river flows northward into Canada before looping westward and southward to join the Columbia just north of the Canadian border. However, from Newport to Canada, the river's tributaries nearly all flow southward along the slope of the regional land surface, opposite the present flow direction of the river. Moreover, the river's channel becomes narrower rather than wider in its downstream direction, indicating the river once flowed south. During the Pleistocene, the Pend Oreille ice lobe filled this valley to depths of several thousand feet, depressing the underlying crust with its weight. Because the ice thickness increased northward and lingered there longer during glacial retreat, it depressed the crust even more to the north,

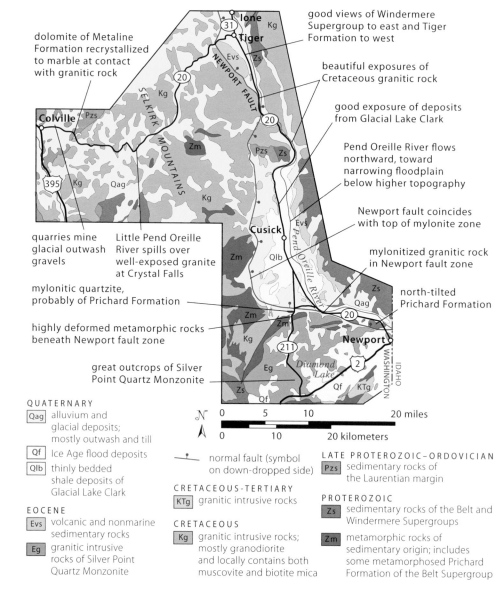

dolomite of Metaline Formation recrystallized to marble at contact with granitic rock

good views of Windermere Supergroup to east and Tiger Formation to west

beautiful exposures of Cretaceous granitic rock

good exposure of deposits from Glacial Lake Clark

Pend Oreille River flows northward, toward narrowing floodplain below higher topography

Newport fault coincides with top of mylonite zone

quarries mine glacial outwash gravels

Little Pend Oreille River spills over well-exposed granite at Crystal Falls

mylonitic quartzite, probably of Prichard Formation

highly deformed metamorphic rocks beneath Newport fault zone

mylonitized granitic rock in Newport fault zone

north-tilted Prichard Formation

great outcrops of Silver Point Quartz Monzonite

QUATERNARY

Qag alluvium and glacial deposits; mostly outwash and till

Qf Ice Age flood deposits

Qlb thinly bedded shale deposits of Glacial Lake Clark

EOCENE

Evs volcanic and nonmarine sedimentary rocks

Eg granitic intrusive rocks of Silver Point Quartz Monzonite

𝒩 0   5   10   20 miles

0   10   20 kilometers

normal fault (symbol on down-dropped side)

CRETACEOUS-TERTIARY

KTg granitic intrusive rocks

CRETACEOUS

Kg granitic intrusive rocks; mostly granodiorite and locally contains both muscovite and biotite mica

LATE PROTEROZOIC–ORDOVICIAN

Pzs sedimentary rocks of the Laurentian margin

PROTEROZOIC

Zs sedimentary rocks of the Belt and Windermere Supergroups

Zm metamorphic rocks of sedimentary origin; includes some metamorphosed Prichard Formation of the Belt Supergroup

Geology along WA 20 between Newport and Colville.

enough to cause a slight northward slope to the valley. As the river reestablished its channel after ice retreat, it flowed northward, cutting down faster than the crust was able to spring back.

Within the first mile north of Newport, WA 20 passes a quarry in the Eocene Silver Point Quartz Monzonite on the southwest side of the road, as well as an exposure of it at the crest of the first hill. At and northwest of milepost 435, you pass well-bedded, north-tilted rock of the Prichard Formation, which forms the base of the 1.5- to 1.4-million-year-old Belt Supergroup. The Prichard here exhibits its characteristic rust-colored weathering from the oxidation of sulfide minerals that formed throughout the rock. If you look closely, you can find brassy crystals of chalcocite and chalcopyrite. Even though these rocks have been slightly metamorphosed, you can also see original sedimentary layering and features. Alternating beds of sandstone and shale suggest these rocks, known as turbidites, were deposited from turbidity currents that flowed down slopes on the ocean floor, carrying sediment that settled to the bottom. In some places you can find individual beds in which the coarsest grains lie at the bottom and grade upward into finer material, a consequence of settling from suspension. Large pull-outs, 0.2 and 0.7 mile (320 and 1,100 m) northwest of milepost 435, allow close inspection of these layered rocks.

North-tilted turbidites of the Prichard Formation.

The rock ledges you can see to the south between mileposts 430 and 429 are part of the Prichard Formation, whereas the higher hills behind them consist of the Silver Point Quartz Monzonite. The granitic rock is faulted against the Prichard Formation along the Newport fault, a large normal fault that formed from crustal extension during the Eocene Epoch. Parts of the fault are exposed

Gneiss below the Newport Fault

WA 211 crosses the Newport fault zone. If heading north on WA 20, you can reach WA 211 by heading west on Westside Calispell Road at Dalkeena and then turning south on WA 211. If coming from the north, turn south on WA 211 at Usk. A good exposure of the north-dipping Prichard Formation sits on the west side of the road a half mile (0.8 km) south of milepost 11. If you look closely at these rocks, you can see that they are foliated, lineated, and fine grained, hallmarks of high-temperature deformation. Similar to the gneissic outcrops immediately to the south, these rocks were deformed at some depth in the crust by movement along the Newport fault.

A second roadcut, which continues for a good half mile (0.8 km) on the east side of Davis Lake, consists of gneissic rock mixed with fragments of metamorphosed sedimentary rock. Many of these inclusions consist of quartzite that belongs to the Precambrian Prichard Formation, the lowest part of the Belt Supergroup. A safe pull-out exists on the south side of the roadcut.

The roadcut of gneiss opposite Davis Lake. The dark bands near the center of the photo are metamorphosed sedimentary rocks.

in two railroad cuts beginning about halfway between mileposts 429 and 428. At the easternmost exposure, you can see somewhat deformed granite with large feldspar crystals, whereas at the exposure about 0.2 mile (320 m) farther west, you can see that the rock is much finer grained and highly deformed. These rocks are mylonites, which form during fault movement in the middle part of the crust where the temperatures are high. The best exposure of these mylonitic rocks is on the same side of the cut as the railroad tracks.

The Calispell-Cusick Valley, a glacially widened section along the Pend Oreille River, once formed the floor of a large glacial lake. Called Glacial Lake Clark, it formed from meltwater emanating from the retreating Pend Oreille ice lobe. You can see some of its light-colored deposits in the hillsides as well as an especially good exposure at Locke Cutoff Road near milepost 413. The lake deposits continue intermittently in low-lying areas as far north as the Canadian border, to indicate the great extent of the lake.

Conglomerate of the Tiger Formation forms the cliffs high on the hillside west of milepost 411 and on the east side of the river. The Tiger Formation accumulated in alluvial fans and the riverbeds of rivers flowing eastward off mountains that were actively rising along the Newport fault in Eocene time. Some outstanding and accessible exposures of these rocks lie along WA 31 about 7 miles (11 km) north of the community of Tiger.

Cliffs of the Tiger Formation, a conglomerate that formed in alluvial fans and riverbeds adjacent to highlands rising along the nearby Newport fault during the Eocene Epoch.

Cliffs high to the west between mileposts 407 and 406 consist of rocks of the Belt Supergroup. They are intruded by a large mass of Cretaceous granitic rock to the north. Between milepost 405 and just north of Lost Creek at milepost 395, all bedrock exposures are granites. A lovely spot to park and look at the granite as well as the river is at milepost 405, and an especially beautiful granite with large pink feldspar crystals crops out immediately north of milepost 403. Some excellent exposures of the lakebeds of Glacial Lake Clark appear between mileposts 400 and 399. Near milepost 392 you can look westward over the glacial lake deposits to cliffs of the Tiger Formation. Across the river to the east are cliffs of metamorphosed volcanic rock of the Windermere Supergroup. Formed toward the end of Proterozoic time, the rocks of the Windermere Supergroup were deposited during rifting of the supercontinent Rodinia.

Between Tiger (at the intersection with WA 31 to Ione) and Colville, WA 20 crosses the southern part of the Selkirk Mountains, which continue northward several hundred miles into British Columbia. This part of the range is of fairly low relief, having formed on Cretaceous granitic rock that is largely blanketed by forested glacial deposits. Although little granite actually shows up along this stretch of road, you can tell it's just beneath the surface because much of the soil consists of deeply weathered, disintegrated granite called grus. Because the weathering that forms grus tends to break the granite apart more than it alters its chemistry, grus retains the same look and mineral content as the original granite.

Grus, disintegrated granitic rock, blankets much of the landscape of the southern Selkirk Mountains.

West of Tiger, WA 20 rises steeply for about 4.5 miles (7.2 km) to Tiger Meadows, a small enclosed basin developed on the glacial deposits. By milepost 384, you've crossed into the headwaters of the Little Pend Oreille River, which flows south and westward into the Colville Valley. Good outcrops of granite show up just west of milepost 383 but remain sparse until the beautiful exposures at Crystal Falls, near milepost 370. Some excellent places to inspect grus are at milepost 374 and near the big pull-out just east of milepost 373. Between mileposts 373 and 372, the highway skirts the south edge of a couple of tree-covered glacial moraines. You can see a variety of rounded cobbles in the hillsides.

Crystal Falls spills over granite next to a small pull-out 0.75 mile (1.2 km) west of milepost 370. West of Crystal Falls, WA 20 leaves the Little Pend Oreille River's watershed, passing occasional exposures of the granite bedrock, grus, and glacial till for the next 10 miles (16 km). A high cliff of dolomite of the Cambrian-to-Ordovician-age Metaline Formation, about 0.1 mile (160 m) north of milepost 359, marks the granite's edge. The Metaline dolomite here was metamorphosed to marble because of the intrusion of the granite, some of which crops out along the road just south of the cliff. The Metaline Formation accumulated in a shallow sea at the beginning of Paleozoic time, long before the accretion of terranes during the Mesozoic Era. These rocks form part of an arcuate belt that was deposited along the continental margin of Laurentia. Called the Kootenay arc, the belt became highly deformed during the later accretionary events.

The 4 miles (6.4 km) of WA 20 east of Colville pass over glacial deposits, exposed as cobbles and sand in the hillsides. Quarries mine the gravel halfway between mileposts 357 and 356 and at milepost 355.

WA 31
Tiger—Metaline Falls— Canadian Border
27 miles (43 km)

WA 31 follows the Pend Oreille River north to the Canadian border. North from Tiger, WA 31 crosses a gentle surface developed on glacial lake deposits. Some high cliffs a few miles to the north consist of limestone and dolomite of the Metaline Formation. These carbonate rocks were deposited in a shallow sea during the Cambrian and Ordovician Periods at what was then the western edge of the Laurentian continent, ancient North America. The Metaline forms most of the large exposures between here and the border. Like the Metaline exposed along WA 20 near Colville, these rocks are part of the Kootenay arc, a belt of sedimentary rock that became strongly deformed by the accretion of terranes in Mesozoic time. Halfway between mileposts 4 and 5 you start seeing the Metaline Formation exposed along the highway.

The Box Canyon Viewpoint, immediately south of milepost 7, offers some impressive views of the Metaline Formation and the Box Canyon Dam. Just over 62 feet (18.9 m) high, the dam was built between the years 1951 and 1956 for hydroelectric power and has little or no storage capacity. Its abutments are in dolomite of the Metaline Formation. Northward from the overlook you can see tilted gravels of the Tiger Formation along the highway.

WA 31 passes the Tiger Formation just north of the viewpoint. A pull-out just south of milepost 8 provides a safe spot to park. The exposure consists of southwest-tilted beds of gravel full of limestone, dolomite, sandstone, and slate fragments that were derived from the Metaline Formation, Addy Quartzite, and Ledbetter Slate. The highly angular nature of the rock fragments indicates they were not transported very far from their original source. Not coincidentally, a northern extension of the Newport fault lies immediately to the west. Because the conglomerate formed in Eocene time at about the same time as the fault,

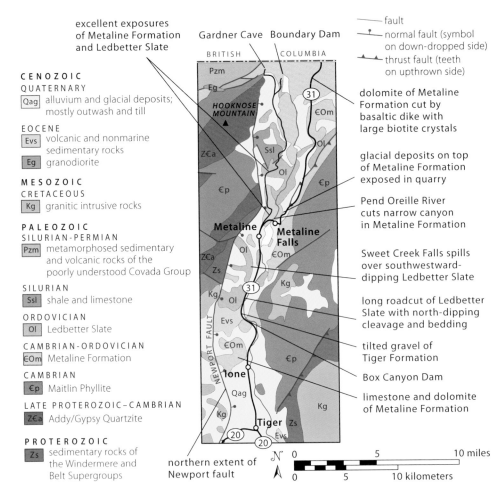

excellent exposures of Metaline Formation and Ledbetter Slate

Gardner Cave Boundary Dam

BRITISH COLUMBIA

—— fault
—•— normal fault (symbol on down-dropped side)
—▲▲— thrust fault (teeth on upthrown side)

CENOZOIC
QUATERNARY
Qag alluvium and glacial deposits; mostly outwash and till

EOCENE
Evs volcanic and nonmarine sedimentary rocks
Eg granodiorite

MESOZOIC
CRETACEOUS
Kg granitic intrusive rocks

PALEOZOIC
SILURIAN-PERMIAN
Pzm metamorphosed sedimentary and volcanic rocks of the poorly understood Covada Group

SILURIAN
Ssl shale and limestone

ORDOVICIAN
Ol Ledbetter Slate

CAMBRIAN-ORDOVICIAN
€Om Metaline Formation

CAMBRIAN
€p Maitlin Phyllite

LATE PROTEROZOIC–CAMBRIAN
Z€a Addy/Gypsy Quartzite

PROTEROZOIC
Zs sedimentary rocks of the Windermere and Belt Supergroups

dolomite of Metaline Formation cut by basaltic dike with large biotite crystals

glacial deposits on top of Metaline Formation exposed in quarry

Pend Oreille River cuts narrow canyon in Metaline Formation

Sweet Creek Falls spills over southwestward-dipping Ledbetter Slate

long roadcut of Ledbetter Slate with north-dipping cleavage and bedding

tilted gravel of Tiger Formation

Box Canyon Dam

limestone and dolomite of Metaline Formation

HOOKNOSE MOUNTAIN

Metaline Metaline Falls

Ione

Tiger

northern extent of Newport fault

N

0   5   10 miles
0   5   10 kilometers

Geology along WA 31 between Tiger and the Canadian border.

The coarse, angular nature of the cobbles in tilted conglomerate beds of the Tiger Formation indicate they were deposited close to their source.

many researchers suggest that slip on the fault created highlands, which in turn were eroded, with the angular gravel and rocks deposited in alluvial fans.

Between a quarter mile (400 m) north of milepost 8 and milepost 11 are outcrops of the Ledbetter Slate, deposited in a deep ocean during the Ordovician Period, after the deposition of the Metaline limestone. A long outcrop of the slate parallels the road near milepost 10. Most prominently, you see gently north-dipping slaty cleavage, which on close inspection is nearly parallel to the bedding in the rock. A near-vertical set of fractures forms steep faces parallel to the road, and numerous quartz veins and fractures cut the outcrop at a steep angle.

Sweet Creek Falls pours over ledges of Ledbetter Slate just south of milepost 11. There, the bedding dips southwestward, indicating the rock was folded or faulted between there and the exposure at milepost 10. If you walk to the upper falls (less than a quarter mile, 400 m), you can see numerous rounded cobbles and small boulders of Addy Quartzite in the streambed.

High cliffs of well-bedded glacial lake deposits show up on the west side of the highway just north of milepost 11. Named for the visible metals showing in

Sweet Creek Falls spills over southwest-dipping Ledbetter Slate. Addy Quartzite forms the bronze-colored cobbles in the stream channel.

some of the rock, the Metaline mining district was Washington's premier lead-zinc mining area from about 1928 until the mid-1950s. Most of the lead came from galena, and most of the zinc came from sphalerite, both of which formed by hydrothermal replacement of carbonate material in the Metaline Formation.

From the town of Metaline, you can head northward to Crawford State Park and Gardner Cave (see sidebar), or continue on WA 31 beneath cliffs of the Metaline Formation to cross the Pend Oreille River to Metaline Falls. At milepost 14, the bridge across the river is built on Metaline Formation. From there WA 31 switches back up to a high terrace built on glacial lake deposits. Look for a quarry that exposes the Metaline Formation beneath glacial deposits at the top of the grade.

As you drive northward, you gain nice views to the west for the first few miles. Some poor exposures 0.2 to 0.4 mile (329–640 m) south of milepost 18 belong to the Ledbetter Slate. It is lighter colored here because it has been metamorphosed to phyllite. If you stop at the heritage marker about a quarter mile (400 m) south of milepost 20, you gain a good view of the Selkirk Mountains. Hooknose Mountain, the most prominent peak, consists of the Cambrian Gypsy Quartzite, also called the Addy Quartzite.

Between milepost 22 and the Canadian border, the highway follows the Metaline Formation. An outstanding outcrop of the Metaline Formation lies about 0.75 mile (1.2 km) north of milepost 24. There, the dolomite was metamorphosed to a marble and is cut by a basaltic dike on its south side. The basalt contains large crystals of biotite scattered throughout its otherwise fine-grained matrix. It is highly weathered on its outer surfaces and so exhibits spectacular weathering rinds where you see cross sections of the rock. More good exposures of the Metaline line the first quarter mile (400 m) of the East-side Access Road to the Boundary Dam Vista, just south of milepost 26.

North-dipping marble of the Metaline Formation. The basaltic dike forms the steep brown-stained slope slanting down to the right, approximately perpendicular to the bedding, in the middle of the photo.

Gardner Cave

The carbonate rocks of the Metaline Formation host numerous small caves. Gardner Cave, within Crawford State Park, is the largest, extending as a single main passage about 1,000 feet (300 m). The guided tour goes about halfway and then returns along the same path (reservations via the park's website are recommended). Although mostly coated by flowstone, the original bedding surfaces can be spotted in places and show that the cave passage formed parallel to the bedding. Most of the cave decorations consist of a variety of flowstone deposited by water flowing evenly over various surfaces, but you can also see some stalactites growing down from the ceiling, stalagmites growing up from the ground, and small helictites growing sideways. The cave contains one unusually large column near the deepest level of the tour, where a stalactite joined a stalagmite.

Stalactites typically taper downward from the ceiling, so the unusual shape of the upper part of this large column in Gardner Cave invites speculation as to its origin.

Cave decorations form because slightly acidic groundwater dissolves limestone as it passes through the rock but then precipitates it as calcite when it reaches the interior of the cave. This precipitation occurs when carbon dioxide escapes from the water into the cave air, causing the water to lose some of its acidity. Water dripping from a cave ceiling, for example, leaves behind a small calcite deposit on the ceiling and on the floor where it lands. As these deposits grow, they form stalactites and stalagmites, respectively.

The 12-mile (19 km) drive to Gardner Cave and Crawford State Park provides nice views of the Selkirk Mountains, made of early Paleozoic sedimentary rocks. It passes some excellent exposures of the Metaline Formation in the first mile and then Ledbetter Slate between mileposts 4 and 6.

COLUMBIA BASIN

The Columbia Basin preserves world-class examples of two incredible geologic phenomena—flood basalts and megafloods. The basalts erupted in Miocene time and flooded a basin centered on the Tri-Cities of Pasco, Kennewick, and Richland. In most of the province, the basalt flows are monotonously flat, but west of the Tri-Cities prominent ridges define anticlines of the Yakima fold belt. The megafloods of late Pleistocene time originated when ice dams broke and released the water of giant lakes in northwestern Montana, as well as in northeastern Washington, northern Idaho, and southern British Columbia. The floods coursed generally southwest across the Columbia Basin, scouring the basalt into the Channeled Scabland. Blanketing much of the basalt in southeastern Washington is loess, windblown dust derived from glacial abrasion. The proximity of the continental ice sheets to the basin during the end of the Pleistocene created enormous amounts of this material. The floods scoured away the loess in places, but elsewhere it remained to provide fertile soil for agriculture.

Flood Basalts of Miocene Time

Stephen Reidel at Washington State University, who has studied the Columbia River Basalt Group for decades, notes that it constitutes the youngest and smallest of the flood basalt provinces on Earth. Nevertheless, the volume of the basalt in Washington, Oregon, and Idaho is impressive; it's estimated to be about 50,000 cubic miles (210,000 km³). The group consists of several formations, which have been classified using chemical composition, mineralogy, and the record of changes of Earth's magnetic field preserved in the basalt. More than 350 individual flows were erupted in the province, and a few experts can identify some flows simply by picking up a hand specimen and examining it with a hand lens.

Most of the basalt was erupted from thousands of north-northwest-striking feeder dikes in the eastern part of the province. Geologists favor the hypothesis that the Yellowstone hot spot, the head of a mantle plume, was responsible for the dike swarm. Perhaps the most remarkable aspect of the basalt group is the vast extent of individual flows. Some coursed down the Columbia River all the way to the coast. Basalt, which is already a fluid type of lava, can flow extremely far after emerging from a feeder dike. The surface of the flow cools immediately to form a crust insulating the interior, which remains hot and continues flowing as more magma is injected. Geologists still don't know the duration of a single

The Yakima River winds through multiple basalt flows of the Columbia River Basalt Group in the Yakima River Canyon along WA 821.

distribution of all
Columbia River Basalt Group
flows, exclusive of
Steens Basalt

Imnaha Basalt
(16.7–16.0 million years)

Grande Ronde Basalt
(includes Picture Gorge and
Prineville Basalts in Oregon;
(16.0–15.6 million years)

Wanapum Basalt
(15.6–15.0 million years)

Saddle Mountains Basalt
(15–6 million years)

*Distribution of the various units of the Columbia River Basalt Group,
excluding the Steens Basalt, which lies farther south in Oregon and
erupted about 16.8 million years ago.*

eruption and flow. Was it days to weeks or weeks to years? The question is difficult to answer because no flood basalts are erupting today on Earth.

In Washington the main formations in the Columbia River Basalt Group are, from lowest and oldest to highest and youngest, the Imnaha Basalt, about 16.7 to 16 million years old; Grande Ronde Basalt, 16 to 15.6 million years old; Wanapum Basalt, about 15.6 to 15 million years old; and Saddle Mountains Basalt, about 15 to 6 million years old. The Grande Ronde constitutes about 70 percent of the volume of the group, so a relatively short period of less than a half-million years witnessed enormous eruptive activity. The architecture of the basin becomes evident when you consider how the thickness of the basalt varies in the province: in the deepest part beneath the Tri-Cities, about 2.2 miles (3.5 km) accumulated, but the thickness decreases along the margins, where the basalt unconformably overlies diverse bedrock units. Sediments that were deposited in lakes and rivers and interlayered with the basalt flows are collectively called the Ellensburg Formation, or near Spokane, the Latah Formation. Some of the sediments consist of ash and volcanic mudflows derived from active Cascade volcanoes to the west.

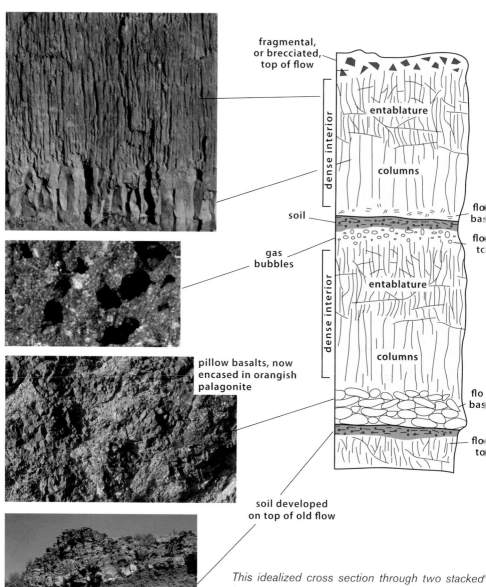

fragmental,
or brecciated,
top of flow

entablature

columns

dense interior

soil

gas
bubbles

entablature

dense interior

columns

pillow basalts, now
encased in orangish
palagonite

flo
ba

flo
to

flo
ba

flo
to

soil developed
on top of old flow

This idealized cross section through two stacked basalt flows displays typical internal features: a flow top, flow base, and a dense interior. The base of a flow may consist of glassy fragments, formed from the rapid cooling of hot magma on the cold land surface, and rubble littering the ground. In some parts of the province, the lava entered fresh-water lakes, and distinctive pillows formed when bulbous protrusions of magma from the front of a flow detached like gobs of toothpaste and tumbled down into the water to form a delta. The surfaces of the pillows may freeze to make a glassy rind, which then reacts with water to form yellow-brown palagonite. The dense interior of a flow typically features a lower colonnade capped by the entablature. Both consist of fractured basalt; the difference lies in the geometry and organization of the fractures. The distinctive, easily recognizable, and dominantly vertical columns in the colonnade formed as basalt contracted during cooling. The more disorganized and closely spaced fractures in the entablature, if present, probably record faster cooling. In your travels, you may see well-developed nonvertical or even curved columns, which are perhaps explained as places where magma flowed through tubes or irregular caverns filled with cold groundwater.

Megafloods of Pleistocene Time

All who have worked to develop the fascinating history of late Pleistocene floods in the Columbia Basin pay homage to J Harlen Bretz. As a young faculty member at the University of Chicago, he conducted research in his homeland of the Pacific Northwest in the 1920s. He recognized the evidence that a catastrophic flood of a magnitude hitherto unimaginable to earthbound geologists had swept southwest across the Columbia Basin from east of the Spokane area. For several decades, the scientific community rejected his hypothesis, but Bretz persevered because he saw and correctly interpreted the evidence. Subsequent work has shown that as many as forty and perhaps even ninety floods surged across the basin as two major ice dams were breached repeatedly between about 19,000 and 15,000 years ago. The most dramatic evidence for the floods, easily visible on satellite images, from the air, or when driving on the ground, is what Bretz called the Channeled Scabland. Here, the thin cover of loess—windblown silt derived from glacial drift—was scoured away to reveal the rocky outcrops of basalt below. The gently rolling landscape of the Palouse, where basalt is covered by a layer of loess, is ideal for growing wheat and other irrigated crops, while the barren scablands are agricultural wasteland.

During the maximum advance of the Cordilleran Ice Sheet, lobes of ice extended from the front of the sheet southward down major valleys. The Purcell Trench Lobe formed an ice dam that blocked the Clark Fork at the Montana-Idaho border, impounding Glacial Lake Missoula, a vast lake up to 2,000 feet (610 m) deep that extended about 155 miles (250 km) to the southeast. The southern extent of the Okanogan Lobe, marked by a terminal moraine in central Washington, dammed the Columbia River to form Glacial Lake Columbia. One likely explanation for how the ice dams were breached is based on simple physics. Ice cubes float, but only 10 percent of the volume of a cube is above the water because the density of ice is 90 percent that of water. When the level of a lake reached 90 percent of the height of the dam, the ice would lift off the bedrock at its base, begin to float, and open a channel through which the water could surge. The discharge rates at the breached dams are difficult to imagine: they probably ranged from 35 to 700 million cubic feet per second (1 to 20 million m³/s). By comparison, the greatest discharge measured on the Mississippi River during the devastating floods of 2011 was 2.3 million cubic feet per second (65,000 m³/s). Victor Baker and his colleagues at the University of Arizona estimate that the flow of some floods coursing across the scablands had average depths of 820 feet (250 m) and velocities of perhaps 50 miles per hour (80 km/hr). Individual floods probably continued for several days.

The floods ultimately reached the Pacific Ocean via the lower Columbia River, but they used diverse paths across the Columbia Basin, carving scablands and coulees, a local term for dry, rocky valleys and ravines. The earliest outbursts from Glacial Lake Missoula flowed southwest through present-day Spokane and channeled down the Columbia River. As the Okanogan Lobe advanced, it blocked this pathway and impounded Glacial Lake Columbia. Later floods overtopped the brim of Glacial Lake Columbia in the vicinity of Spokane to repeatedly spill southward over the Cheney-Palouse Scabland.

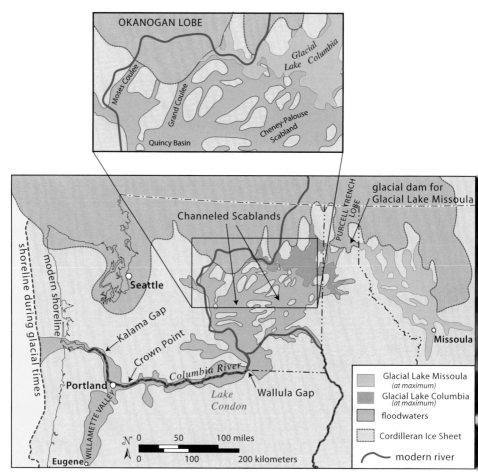

Generalized map of the Cordilleran Ice Sheet and glacial lakes that produced the Ice Age floods. Most broke through the dam at the north end of Glacial Lake Missoula and poured into Glacial Lake Columbia, which then overtopped its brim.

The floods also were diverted southward at the Okanogan Lobe dam into what became Moses and Grand Coulees. Eventually, after the last flood from Glacial Lake Missoula, the ice dam blocking Glacial Lake Columbia failed and released a flood of its own. Later floods, possibly from Lake Kootenay in Canada, passed unimpeded down the Columbia Valley.

Brian Atwater of the US Geological Survey investigated sediments in the Sanpoil River valley, which was an arm of Glacial Lake Columbia. He concluded that they are the deposits of fifteen back floods caused by Glacial Lake Missoula floods entering Glacial Lake Columbia about 15,000 years ago. Moreover, the character of the sediments interlayered with the Glacial Lake Missoula flood sediments indicated that a back flood entered Glacial Lake Columbia every thirty-five to fifty-five years. Atwater's novel analysis is one of the few that gives insight into the tempo of ice-dam floods.

Megafloods created a rich and startling assemblage of landforms. The coulees are strange valleys devoid of any modern rivers or traces of historic streams. Grand and Moses Coulees preserve excellent examples of hanging valleys. Along the rims of the coulees, high above their floors, are tiny V-shaped notches spaced every few hundred meters. The notches are the channels of small streams that flowed across the basalt plateau before the floods. The channels now end abruptly at the steep-walled cliffs of the coulees because their lower courses were stripped away by surging floods. Another landform is a colossal gravel bar deposited by a Missoula flood on the inner bank of a bend in the Columbia River south of Wenatchee. The grandest and most compelling of all the flood-created landforms, however, are the dry waterfalls, or cataracts, with scales that dwarf Niagara Falls today. The famous Dry Falls in Grand Coulee is a line of sheer cliffs about 3 miles (5 km) long and over 330 feet (100 m) high. They originally had formed about 25 miles (40 km) farther south at the mouth of the coulee; the cataract migrated upstream as flood after flood excavated the fractured basalt. Another cataract formed and migrated northward up Grand Coulee, but the waters of artificially dammed Banks Lake shroud most of the evidence. Palouse Falls is a smaller example of a cataract that still sports an active waterfall. Even the casual observer would realize that the modern Palouse River is merely a trickle incapable of carving the giant plunge pool at the base of the falls.

The floods encountered several impediments, or bottlenecks, on their way to the sea. The most significant was Wallula Gap south of the Tri-Cities, where the channel of the Columbia River narrows to a 0.6-mile-wide (1 km) cleft between confining cliffs. Here the river crosses the crest of an active fault system and

These rhythmically deposited lakebeds near Wenatchee, known as Touchet Beds, were deposited in water ponded behind Wallula Gap during the Ice Age floods. View is about 3 feet (0.9 m) high.

related anticlines in the basalt; the Columbia was able to continue to carve its channel even while the fold grew in late Cenozoic time. The gap is so narrow that the surging floodwaters could not rapidly exit through the restriction and instead temporarily ponded to form enormous ephemeral lakes north of the gap. Geologic evidence for the now-vanished lakes includes the famous Touchet Beds in the Walla Walla Valley, rhythmically deposited lake sediments representing as many as forty impounded floods. Some impounded lake waters back flooded into the Yakima Valley and up the Columbia River. Richard Waitt, of the US Geological Survey, interprets the sediments as far upriver as the mouth of Moses Coulee near Wenatchee as part of the record of back floods from the restriction at Wallula Gap.

Yakima Fold Belt

The Columbia River basalts in the western edge of the province have been deformed into an impressive system of folds and related thrust faults that are topographically expressed as ridges that rise up to 2,000 feet (600 m) above what would otherwise be a monotonous plateau. The shapes and geometry of the folds are remarkably different from those we ordinarily see in fold systems,

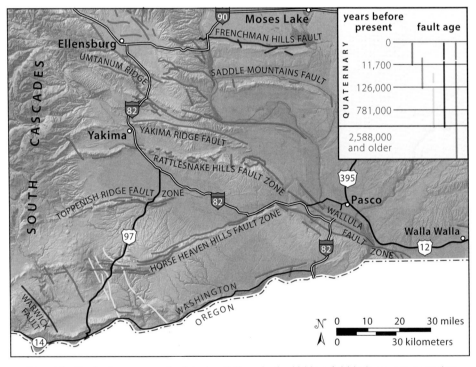

Thrust faults, fault zones, and related anticlines in the Yakima fold belt are expressed at the surface as basalt ridges. Colors indicate the approximate ages of individual structures. Field studies aimed at determining the most recent fault slips have shown that many of these faults have been active in Quaternary time.

where the alternating anticlines and synclines are more or less regularly spaced in map view and have smoothly rounded shapes. Yakima anticlines are asymmetric: the south limbs dip more gently away from the crests than the north limbs. Structural geologists ascribe this asymmetry to thrust faults beneath the ridges that also dip southward. The intervening synclines between the anticlines are broad basins in basalt.

Another tantalizing geometric feature of the Yakima fold belt is the apparent fanning of the ridges from the south to the north. The Horse Heaven Hills trend northeast-southwest, the Rattlesnake Hills are oriented northwest-southeast, and the major ridges to the north trend more easterly. The Rattlesnake Hills and their continuation in the Wallula fault zone are located in a topographic feature called the Olympic-Wallowa line (OWL), or, in the Columbia Basin, the Cle Elum-Wallula line (CLEW). For decades, geologists have speculated about why, in this narrow zone, structures would be oriented northwest-southeast. Perhaps a right-lateral strike-slip fault is hidden in the basement rocks below the basalt. Further research will provide a clearer picture of how the Yakima fold belt is developing.

The total north-south shortening recorded by the fold belt is only about 16 miles (26 km), but it demands an explanation. Only in the past several years

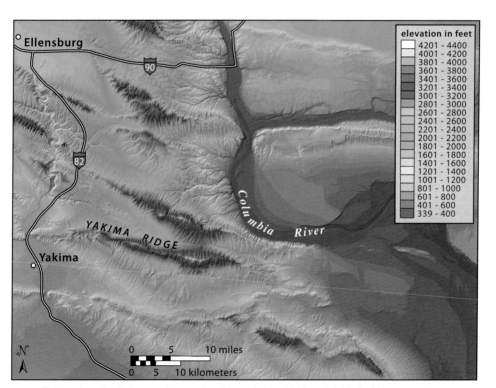

This colored relief map shows the ridges in the Yakima fold belt. Warmer colors indicate lower elevations, whereas cooler colors indicate higher elevations. –Image courtesy of Harvey Greenberg, University of Washington

have data from the Global Positioning System illuminated a possible answer. This part of Washington is rotating clockwise relative to a fixed reference point in western Idaho. The Earth's crust in the fold belt is moving northward, toward British Columbia, creating the fan shape.

The fold belt began forming contemporaneously with the eruption of the Grande Ronde Basalt about 16 million years ago and was active until at least 10 million years ago, the age of the youngest rock affected by the folding. The ridges continue to grow intermittently at much slower rates. The thrust faults, although uncommonly exposed, are of special interest to paleoseismologists, earth scientists who date the timing and recurrence of slip to assess the capability of faults to cause earthquakes. Results published in 2016 of a study by a group led by Brian Sherrod of the US Geological Survey indicate that slip on the fault responsible for lifting up the ridge at Wallula Gap generated three earthquakes in late Pleistocene time.

Guides to the Columbia Basin

SPOKANE

Spokane lies near the west end of the broad valley of the Spokane River just a few miles west of the Idaho state line. The area hosts unusually diverse bedrock geology, from Precambrian rocks to Miocene-age flows of the Columbia River Basalt Group. The Grande Ronde and Wanapum Basalts form benches and cliffs that frame much of the Spokane Valley, as well as numerous outcrops along the interstate. Across much of the area, sedimentary rocks of the Latah Formation are interlayered with the lava flows, indicating the development of rivers and lakes during quieter periods.

The combined thickness of the basalt and Latah Formation near Spokane reaches about 300 feet (90 m) but varies because of topography that existed before the eruptions. Where the lava filled ancient valleys, it formed thicker accumulations; where it crested hills and ridges, it formed thinner ones. As a thickness for the Columbia River Basalt Group, 300 feet (90 m) is extremely thin, but Spokane sits near the northeastern edge of the flood basalt province, where the flows would be expected to taper off. By contrast, the basalt flows reach a combined thickness of more than 12,000 feet (3,660 m) in the Pasco Basin, about 200 miles (320 km) to the southwest.

Some of the better places to see the Columbia River Basalt Group are at Riverfront Park in the heart of the city; Riverside State Park along the Spokane River, about 5 miles (8 km) downriver; and Palisades Park on the west side of the city. At Riverfront Park, you can walk a paved footpath along the Spokane River as it drops some 146 feet (44.5 m) over a series of spectacular falls developed in the Grande Ronde Basalt. Look for islands of hackly entablature in the river. Two hydroelectric facilities draw power from the falls. At the much

View northwest down the Spokane River at Riverside State Park. The Bowl and Pitcher are the large outcrops of the Grande Ronde Basalt on the left side of the photo. Note the river terrace that lines the northeast side of the river.

quieter Riverside State Park, you can explore the Bowl and Pitcher, two large outcrops of entablature of the Grande Ronde Basalt along the river. At Palisades Park, you can inspect flows of the Wanapum Basalt while overlooking the city. From there, you can see that the Wanapum Basalt forms a prominent rim that surrounds much of the Spokane area, which is shaped like a large bowl.

Toward the end of glacial times, between 18,000 and 15,000 years ago, Glacial Lake Columbia filled the Spokane Valley. The lake formed where the Okanogan Lobe of the Cordilleran Ice Sheet advanced across and dammed the Columbia River near the site of today's Grand Coulee Dam. From there, the lake extended eastward across the Spokane Valley to Coeur d'Alene, Idaho.

During roughly the same period, Glacial Lake Missoula repeatedly broke through its ice dam at the Montana-Idaho border and unleashed floodwaters that quickly poured over the Rathdrum Prairie in Idaho and into the Spokane Valley. The early floods encountered Glacial Lake Columbia, and so overtopped its rim, spilling southward over low divides into the Palouse area. It is likely that Glacial Lake Columbia emptied during a flood of its own sometime during the later stages of the Ice Age. After Glacial Lake Columbia had drained once and for all, the Missoula floodwaters rushed through the Spokane Valley relatively unimpeded and down the drainage of the Columbia.

Today, floodwater gravels fill the Spokane Valley, reaching 800 feet (240 m) in thickness in some places. They form the Spokane aquifer, the city's primary

Groundwater filling the bottom of a quarry in flood gravels in Spokane.

source of water. You can get a good view of the gravels in a quarry just off I-90 at exit 286. The quarry is cut into the water table, so standing water occupies much its floor. In contrast to the gravels, finer-grained slack-water deposits from Glacial Lake Columbia were also deposited. They are best preserved along Hangman Creek near milepost 5, just south of town on US 195 (see photo in road guide for US 195 in this chapter). These deposits contain coarse sandy beds with intervening thin layers of fine sand and silt called varves. The varves reflect annual deposits in the lake, whereas the sandy beds mark individual flood events. The upper four sandy beds are not accompanied by varves, suggesting some floods occurred after Glacial Lake Columbia had drained.

Bedrock of the Spokane area that pre-dates the Columbia River Basalt Group includes Cretaceous granitic and metamorphic rock as well as Precambrian metamorphic rock. These rocks are exposed in some places along the valley margins beneath the basalt, whereas in other places they form the nearby mountains.

The Mt. Spokane Granite and the Newman Lake Gneiss, both Cretaceous age, are faulted together at Beacon Hill, just northeast of downtown. The faulting occurred deep in the crust where the rocks were hot. Elsewhere, good exposures of the granite show up at the intersection of Division Road with US 395, and at the Dishman Hills Natural Area, just south of I-90 about 5 miles (8 km) east of downtown. At the Dishman Hills, look for unusually large feldspar crystals in the rock. You can also visit some small lakes, now surrounded by forest but probably eroded as potholes by the Missoula floods. The Newman Lake Gneiss is beautifully exposed at Mirabeau Point Park on the east side of the city. In addition to interesting ribbon-like features in the deformed gneiss, you can see numerous pegmatite dikes, some of which are cut by small faults that are filled with the mineral chlorite.

Geologist inspecting the Prichard Formation on East Trent Avenue. His hands are on either side of a dark band of amphibolite.

Precambrian-age gneiss and schist of the Hauser Lake Gneiss formed through the metamorphism of still older sedimentary rocks. Most researchers think the original sedimentary rocks belonged to the Prichard Formation, the oldest rock of the Belt Supergroup, but some argue they are even older. East of town, a roadcut just west of the intersection of North Barker Road and East Trent Avenue consists of west-dipping gneiss, schist, and amphibolite of the Prichard Formation. The interlayering of schist and gneiss likely reflects the original sedimentary rock, in which sandier layers metamorphosed to gneiss and the muddier ones to schist. A close look at the schist reveals fibrous clusters of sillimanite, an aluminum-bearing silicate mineral that forms only at very high temperatures of metamorphism. The amphibolite, which contains garnets, likely originated as basaltic sills that intruded the Prichard Formation prior to metamorphism.

INTERSTATE 82
ELLENSBURG—OREGON BORDER
134 miles (216 km)

Beginning in the synclinal Kittitas Valley and finishing in the Pasco Basin, I-82 provides the best look at the Yakima fold belt of any major highway. Because the folds are in the erosionally resistant Columbia River Basalt Group, the anticlines form ridges and the synclines form large, flat-floored valleys. From

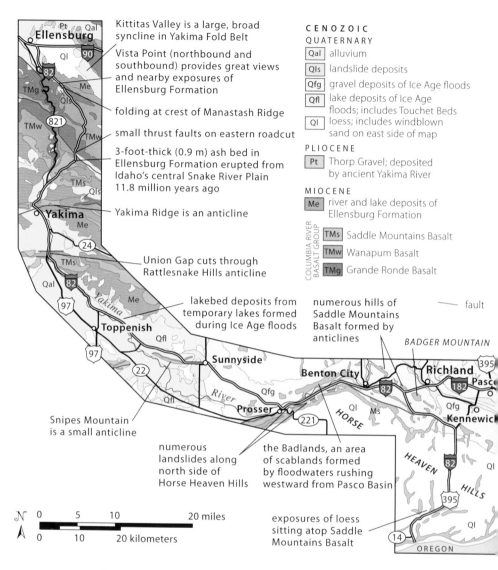

CENOZOIC

QUATERNARY

Qal	alluvium
Qls	landslide deposits
Qfg	gravel deposits of Ice Age floods
Qfl	lake deposits of Ice Age floods; includes Touchet Beds
Ql	loess; includes windblown sand on east side of map

PLIOCENE

Pt	Thorp Gravel; deposited by ancient Yakima River

MIOCENE

Me	river and lake deposits of Ellensburg Formation

COLUMBIA RIVER BASALT GROUP

TMs	Saddle Mountains Basalt
TMw	Wanapum Basalt
TMg	Grande Ronde Basalt

— fault

Kittitas Valley is a large, broad syncline in Yakima Fold Belt

Vista Point (northbound and southbound) provides great views and nearby exposures of Ellensburg Formation

folding at crest of Manastash Ridge

small thrust faults on eastern roadcut

3-foot-thick (0.9 m) ash bed in Ellensburg Formation erupted from Idaho's central Snake River Plain 11.8 million years ago

Yakima Ridge is an anticline

Union Gap cuts through Rattlesnake Hills anticline

lakebed deposits from temporary lakes formed during Ice Age floods

numerous hills of Saddle Mountains Basalt formed by anticlines

BADGER MOUNTAIN

Snipes Mountain is a small anticline

numerous landslides along north side of Horse Heaven Hills

the Badlands, an area of scablands formed by floodwaters rushing westward from Pasco Basin

exposures of loess sitting atop Saddle Mountains Basalt

Geology along I-82 between Ellensburg and the Oregon border.

Ellensburg, I-82 heads due south toward the Yakima River Canyon, but just south of milepost 2 the interstate veers southeastward, skipping the canyon. (See the road guide for WA 821 for a trip through the canyon.)

I-82 approaches Manastash Ridge at milepost 3. Bedrock exposures of the Wanapum Basalt appear at and south of milepost 6. Vista Point, near milepost 8 for both southbound and northbound lanes, features spectacular views of Kittitas Valley to the north and the Stuart Range in the Cascades to the west, and offers the chance to inspect steeply north-dipping Grande Ronde Basalt and interlayered Ellensburg Formation. Much of the Ellensburg here consists of crossbedded sandstone deposited in streams. A close look at the rock shows the sand grains consist mostly of tiny grains of quartz and basalt. At milepost 8, at the crest of Manastash Ridge, you can look to the eastern roadcut to see the Ellensburg Formation and basalt rolled over into steep northward dips from gentle ones. More Grande Ronde Basalt and Ellensburg exposures show up in roadcuts south of the crest; southbound lanes can see pillow basalt and Ellensburg Formation just south of milepost 9.

South of milepost 14, I-82 passes back into the Wanapum Basalt, which it follows to the crest of North Umtanum Ridge, a quarter mile (400 m) south of milepost 17. Pillows of the Wanapum Basalt show up along northbound lanes just north of milepost 15.

On the north side of the crest of South Umtanum Ridge, just south of milepost 20, I-82 reaches the crest of another anticline as defined by sandstone beds of the Ellensburg Formation. Those beds south of the crest dip gently southward, whereas those to the north dip gently northward. Two small thrust faults, visible in the eastern roadcuts, cut the sandstone and offset it toward the south. The top of the ridge at milepost 21 contains south-dipping Ellensburg Formation beneath Saddle Mountains Basalt. As reported by Barbara Nash of the University of Utah, the Ellensburg Formation here contains an ash bed that reaches 3 feet (0.9 m) in thickness. It erupted 11.8 million years ago from the central Snake River Plain in Idaho.

Steeply dipping beds of Ellensburg Formation between basalt flows at the crest of Manastash Ridge. At the far south end of the roadcut (right), rocks dip less steeply.

Selah Creek valley viewed to the west from the Selah Creek Rest Area. Dipping Saddle Mountains Basalt controls the slope of the land as the road descends to river level.

As I-82 descends to the level of the Yakima River, it follows a sloping surface developed on the south-dipping Saddle Mountains Basalt, from which you gain wonderful views of Mt. Rainier and Mt. Adams. As you cross Selah Creek near milepost 24, you can see the Saddle Mountains Basalt forming a prominent cliff down the length of the Selah Creek valley, with slopes of the Wanapum Basalt below. Called the Selah Cliffs, the area is home to a variety of wildlife and native plants, including the endemic basalt daisy (*Erigeron basalticus*).

A quarry at milepost 29 exposes Ellensburg Formation and Wanapum Basalt on the north side of Yakima Ridge, another anticline. The highway passes through the ridge at the Yakima River gap and crosses both the Yakima and Naches Rivers, with some outstanding exposures of the Wanapum Basalt along the northbound lanes between mileposts 30 and 31.

Between Yakima and Union Gap, I-82 crosses the virtually flat Ahtanum Valley, a broad syncline between the Yakima Ridge anticline to the north and the Rattlesnake Hills anticline to the south. The Yakima River flows through the ridges at water gaps. We normally expect water to go around an obstacle rather than right through it, but the Yakima River drainage likely established its path before the development of the folds. It was able to cut down through the basalt ridges at the same pace as the basalt rose across the river's path. A quarter mile (400 m) south of milepost 38, you can look eastward to glimpse the steep, north-dipping limb of the Rattlesnake Hills anticline.

South of Union Gap, I-82 skirts the northeastern edge of the much larger Yakima Valley, another broad syncline. The Rattlesnake Hills, with occasional exposures of the Saddle Mountains Basalt and Ellensburg Formation, lie to the north. Beginning near milepost 45 and exposed intermittently to the southeast for the next 40 miles (64 km), lakebeds north of the highway have an altogether different origin than the Ellensburg. They formed in a temporary lake created when the Ice Age floods pouring into the Pasco Basin backed up behind Wallula Gap, a dramatic constriction in the flood's channel. The lake drained and filled repeatedly with each successive flood, leaving behind thin, rhythmically bedded deposits. The enormous scale of these floods becomes clear when you consider that Wallula Gap is nearly 80 miles (128 km) away as the crow flies. Some especially clear examples of the lakebeds form scattered low bluffs just north of the highway from near milepost 51 to east of milepost 53.

Snipes Mountain, another anticline of the Yakima fold belt, forms the low ridge south of the highway between Granger and Sunnyside. The gentle, even slope reflects the gentle dip of the bedrock, which consists of the Saddle Mountains Basalt and Ellensburg Formation on the north limb of the fold.

Prosser lies at the foot of the Horse Heaven Hills, one of the larger anticlinal ridges of the fold belt. It extends more than 40 miles (64 km) to the southwest from here, nearly all the way to Status Pass on US 97, and about 10 miles (16 km) to the northeast, where it turns southeastward and continues all the way to Wallula Gap. South of Prosser, the Horse Heaven Hills rise more than 1,200 feet (370 m) above the Yakima River and are cored by the Wanapum Basalt, with the Ellensburg Formation and Saddle Mountains Basalt on its limbs. The most notable features, however, are the landslides that form hummocky surfaces along its north flank. You can see the slides all the way from Prosser to about a half mile (0.8 km) east of milepost 92.

Bluffs made of lakebeds deposited during the Ice Age floods.

Hummocky landslide deposits along the north flank of the Horse Heaven Hills.

Between mileposts 88 and 92 on the north side of the road, you can see a long stretch of scablands referred to locally as the Badlands. This area funneled Ice Age floodwater from the Pasco Basin into the Yakima Valley and so was subject to faster water and more erosion than the wide basins on either side. You can access the Badlands by driving west on Old Inland Empire Highway from Benton City.

Between Benton City and Kennewick, I-82 crosses a gentle surface made of flood deposits punctuated by some impressive cliffs of the Saddle Mountains Basalt, especially near mileposts 96 and 97. Near milepost 100, the highway passes a series of hills that mark small anticlines aligned in a northwest direction, including Badger Mountain. These anticlines mark the southwestern edge of the Pasco Basin, a wide syncline in the Yakima fold belt. The basin's formation most certainly started early in the history of the Columbia River Basalt Group, as geophysical studies indicate the basin holds an inordinate thickness of basalt, greater than 12,000 feet (3,660 m).

South of the I-82 interchange, a couple of interesting roadcuts of the Saddle Mountains Basalt border I-82 for nearly the entire distance between mileposts 109 and 110. Look for some brecciated parts of the flows as well as colonnade. South of the interchange with US 395, I-82 crests the Horse Heaven Hills near milepost 118. South of there, it's a long downgrade to the Columbia River past scattered exposures of the Saddle Mountains Basalt. Probably the most instructive stretch runs south from milepost 125 for about a half mile (0.8 km), where you can see Pleistocene loess on top of the basalt. In addition, a pull-out just north of milepost 127 offers a good view into a canyon with basaltic bedrock capped by loess. The prominent hills east of the highway near milepost 131 also consist of the Saddle Mountains Basalt overlain by loess.

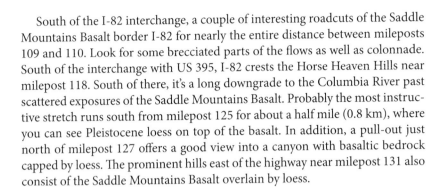

INTERSTATE 90
ELLENSBURG—SPOKANE—IDAHO BORDER
190 miles (306 km)

East of Ellensburg, you can really sense that the Kittitas Valley is a large bowl, a broad downwarp in the Yakima fold belt, with ridges of Grande Ronde Basalt on all sides sloping toward the valley floor. I-90 climbs east out of the valley along a gentle grade with few outcrops. The valley bottom is covered by loess, windblown dust from the glacial period.

The Grande Ronde Basalt shows up between mileposts 118 and 119. Thorp Gravel, deposited by an ancestral Yakima River 3.7 million years ago, caps the hillsides on either side of milepost 120. Just west of milepost 121 you can see west-tilted Grande Ronde Basalt on the south side of the highway beneath the railroad trestle; a quarter mile (400 m) east, a long exposure with colonnade lies on the north side. Amid windmills and a rolling landscape underlain by Wanapum Basalt, you cross Ryegrass Summit near milepost 126.

On the east side of Ryegrass Summit, I-90 descends to the Columbia River, first dropping into the older Grande Ronde Basalt, with decent exposures beginning halfway between mileposts 128 and 129. Roadcuts of the Wanapum Basalt begin just west of milepost 130. Even though the highway descends eastward, it passes into younger rocks because they dip eastward more steeply than the highway. Look for reddish paleosols in between flows and some outstanding examples of colonnade. Pillow basalt, formed when the lava flows poured into lakes, shows up on the north side of the highway near milepost 136.

You can also see a number of good exposures of the light-colored Ellensburg Formation, which was deposited in rivers and lakes during quiet times between major eruptions of the basalt. Look for it on the north side of the highway beginning a quarter mile (400 m) east of milepost 132, and then in several places until about milepost 135. The light-colored deposits at milepost 136, however, were deposited in temporary lakes created when the Ice Age floods backed up behind constrictions in the channel.

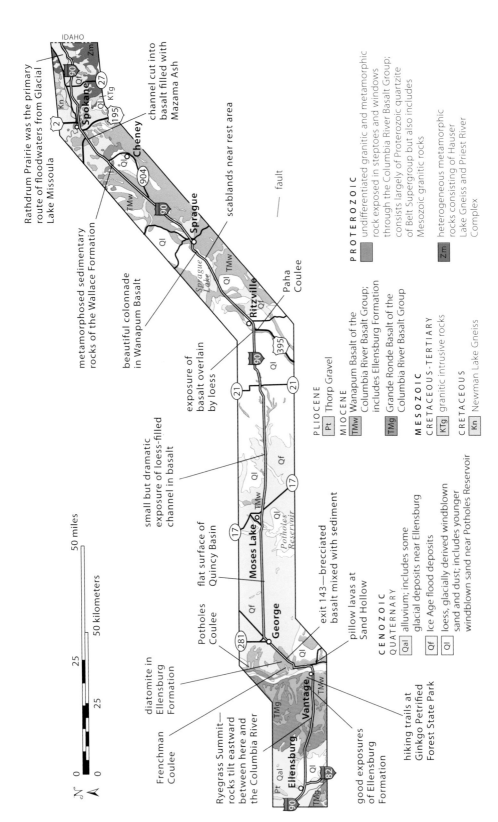

Geology along I-90 between Ellensburg and the Idaho border.

IDAHO

Rathdrum Prairie was the primary route of floodwaters from Glacial Lake Missoula

channel cut into basalt filled with Mazama Ash

Spokane

Cheney

scablands near rest area

— fault

metamorphosed sedimentary rocks of the Wallace Formation

beautiful colonnade in Wanapum Basalt

Sprague

Sprague Lake

Ritzville

Paha Coulee

exposure of basalt overlain by loess

small but dramatic exposure of loess-filled channel in basalt

Moses Lake

Potholes Reservoir

George

flat surface of Quincy Basin

Potholes Coulee

diatomite in Ellensburg Formation

exit 143—brecciated basalt mixed with sediment

Frenchman Coulee

Ryegrass Summit— rocks tilt eastward between here and the Columbia River

pillow lavas at Sand Hollow

Vantage

hiking trails at Ginkgo Petrified Forest State Park

Ellensburg

good exposures of Ellensburg Formation

PROTEROZOIC

undifferentiated granitic and metamorphic rock exposed in steptoes and windows through the Columbia River Basalt Group; consists largely of Proterozoic quartzite of Belt Supergroup but also includes Mesozoic granitic rocks

Zm   heterogeneous metamorphic rocks consisting of Hauser Lake Gneiss and Priest River Complex

PLIOCENE
Pt   Thorp Gravel

MIOCENE
TMw   Wanapum Basalt of the Columbia River Basalt Group; includes Ellensburg Formation

TMg   Grande Ronde Basalt of the Columbia River Basalt Group

MESOZOIC

CRETACEOUS-TERTIARY
KTg   granitic intrusive rocks

CRETACEOUS
Kn   Newman Lake Gneiss

CENOZOIC
QUATERNARY
Qal   alluvium; includes some glacial deposits near Ellensburg

Qf   Ice Age flood deposits

Ql   loess, glacially derived windblown sand and dust; includes younger windblown sand near Potholes Reservoir

50 miles

50 kilometers

N

Ginkgo Petrified Forest State Park

At Vantage, exit 136, you can visit Ginkgo Petrified Forest State Park. Drive north 0.7 mile (1.1 km) through town to Ginkgo Avenue, turn right and continue 0.4 mile (640 m) to visit the museum and gain some spectacular views of the Columbia River and surrounding area. Driving 2 miles (3.2 km) farther west on Vantage Highway leads to a parking lot and trailhead at the contact of the Grande Ronde and Wanapum Basalts. A quarter-mile-long (400 m) interpretive trail through the Wanapum Basalt passes a variety of fossil tree stumps in their growth positions, partially excavated and protected by steel grating. Most of the trees consist of elm, spruce, ginkgo, and walnut.

Petrified stump at the visitor center for Ginkgo Petrified Forest State Park. Sentinel Gap, where the Columbia River cuts through the Saddle Mountains, lies in the background to the right.

As I-90 crosses the Columbia River, look southward to see the floodplain narrow considerably at Sentinel Gap. There, the river cuts across the Saddle Mountains, an east-trending anticline that's part of the Yakima fold belt. Because the river cuts directly across this structure instead of finding an easier route around the edges, we know the river was flowing in this spot before the anticline started to form. As the anticline grew, the river continued to erode downward into the rising rock.

Sand Hollow Pillow Lavas

To see some outstanding pillow lavas, take exit 137. Head south on WA 26 along the eastern shore of the Columbia River, passing below cliffs of the Wanapum Basalt, for just over 1 mile (1.6 km), to where the road turns east up a canyon that cuts through the cliffs. The first roadcut in the canyon exposes a fantastic array of basalt pillows and yellowish palagonite. (The road shoulder is unusually rough for parking.) The pillows formed as the lava poured into a lake, and the palagonite is an alteration mineral of the volcanic glass that formed when the lava was instantly chilled. The lava eventually filled the water body, so the upper part of the flow does not contain pillows. If you stand back from the rocks you can see that the pillows form inclined layers: as the lava spilled into the lake these layers formed along the original sloping shoreline.

Pillow lavas at Sand Hollow.

On the west side of the Columbia River, the interstate climbs along a sloping bench of the Grande Ronde Basalt between mileposts 138 and 141. Called the Babcock Bench, it's a flood-eroded terrace, perched about 400 feet (120 m) above river level, that extends some 20 miles (32 km) upriver. The cliffs above consist of the Wanapum Basalt. About a half mile (0.8 km) north of milepost 141, the highway rises into the Wanapum Basalt. Over the next couple miles, you pass into the Quincy Basin, a depression in the plateau that collected waters from many of the Ice Age floods. At exit 143, you can see outstanding exposures of brecciated basalt and sediment mixed together at the north end of the exit ramp; similarly good exposures lie at the south end of the entrance ramp. These mixtures generally form when lava flows invade soft lake sediments. The explosions resulting from the lava-water interactions cause the fragmentation and mixing of both materials.

The landscape along I-90 flattens into the Quincy Basin just east of exit 143. If you look westward toward the exit, you can see a subtle slope funneling into the head of Frenchman Coulee and imagine the immense volumes of water that

Frenchman Coulee

A 2-mile (3.2 km) detour at exit 143 leads to breathtaking views down Frenchman Coulee, a 300-foot-deep (90 m) canyon carved into the Wanapum Basalt by the Ice Age floods. As the floods poured out of the Quincy Basin and into the Columbia River 600 feet (180 m) below, they exploited the existing drainages, tearing bedrock from the valley walls and greatly enlarging and deepening the valleys in the process. Several other coulees empty into the Columbia in the vicinity, the closest being another arm of Frenchman Coulee immediately to the south, and Potholes Coulee about 10 miles (16 km) to the north.

Ice Age floods flowing from the Quincy Basin toward the Columbia River, visible in the distance, carved Frenchman Coulee.

To reach the canyon, follow Silica Road northward about 0.75 mile (1.2 km) to Vantage Road and turn left. Within another mile, the road follows a ledge between cliffs of the Wanapum Basalt. Notice the flow contact between colonnade and an underlying paleosol approximately at eye level. At the parking lot, you see the deep coulee descending toward the Babcock Bench and the Columbia River, as well as more colonnade and prominent horizontal fracturing in the basalt, both of which are artifacts of the cooling of the flow. Across the canyon is a waterfall, fed by surface water as well as springs issuing from between individual lava flows. The rounded, white hills sitting on top of the basalt consist of diatomite, deposited in lakes as part of the Ellensburg Formation during a quiet period between eruptions of the Wanapum Basalt. Diatomite consists of the silica shells of diatoms, single-celled organisms that thrive in some lakes and oceans. Several now-defunct mines quarried the diatomite for its silica. You can access the deposits on the way back to the highway by turning northward on Silica Road and driving over scablands for fewer than 2 miles (3.2 km).

---

eroded it. At George, you can see how the land rises to the north and south to enclose the basin, but overall the land is flat. This flatness reflects the underlying basaltic bedrock, which is nearly horizontal, and the blanket of flood gravels and loess over the top. The gravels are visible from place to place in low hillsides, and the loess becomes increasingly common east of Moses Lake.

This part of the Columbia Basin is arid, with generally less than 8 inches (20 cm) of rainfall each year. Still, significant wetlands show up along the highway, especially near Moses Lake. Potholes Reservoir, impounded by O'Sullivan Dam just south of Moses Lake, is a well-known locality for recreation and birding. The interstate crosses the west and east arms of Moses Lake on either side of milepost 177. East of the reservoir, I-90 passes some exposures of Pliocene lakebed deposits. About halfway between mileposts 186 and 187 you can see a dramatic but unnatural exposure of flood gravel in a levee embankment.

At the Adams County line, 0.1 mile (160 m) west of milepost 192, you can see the Wanapum Basalt on top of a red paleosol on the north side of the road; windblown loess sits on top of the outcrop. Directly across from milepost 192, the loess fills a prominent channel in the basalt. Between mileposts 209 and 213, the interstate follows the north edge of Bauer Coulee, a shallow floodwater channel, and then crosses it.

Just west of the US 395 interchange at Ritzville, I-90 drops into the north-south Paha Coulee. Look for a good exposure of Wanapum Basalt overlain by loess on the north side of the road just west of milepost 220.

Loess filling a channel eroded into the Wanapum Basalt at milepost 192.

Scablands near Sprague Lake Rest Area.

Between Ritzville and the rest area at Sprague Lake, at milepost 242, the landscape remains mostly blanketed by loess but punctuated in places by exposures of the Wanapum Basalt. The basalt forms scablands where the Ice Age floods eroded the bedrock. Some interesting scablands lie on both sides of the highway between mileposts 225 and 230, but the most extensive scablands lie along both sides of Sprague Lake. You can drive through and see more of the scablands by driving a half mile (0.8 km) south on WA 23 at exit 245 for

Sprague. Turn west on First Street and drive to where it becomes Max Harder Road, and then follow it for about 4 miles (6.4 km) back to the southeast shore of the lake. The floods probably scoured out the shallow depression of the lake. At the rest area, look for the beautiful colonnade in the Wanapum lava.

Between Sprague and exit 270 for Cheney, I-90 rises about 500 feet (150 m) into a forested area marked by numerous scablands. The bedrock everywhere along this 25-mile (40 km) stretch consists of the Wanapum Basalt, and many of the scoured low areas are filled with lakes or marshlands. At milepost 269, however, a roadcut exposes metamorphosed sedimentary rocks of the Wallace Formation. These rocks are part of the Proterozoic Belt Supergroup, which lies beneath the basalt and forms the bedrock of the hills in the immediate area.

At the exit ramp for Garden Springs (exit 277A), and less than a quarter mile (400 km) east of milepost 277, look for an ash-filled channel in the basalt on both sides of the highway. This reworked Mazama Ash erupted 7,700 years ago during the catastrophic eruption that formed Crater Lake in Oregon.

Mazama Ash filling a channel eroded into the Wanapum Basalt just east of milepost 277.

Driving through Spokane and its valley, I-90 passes numerous bedrock exposures of the Grande Ronde Basalt, which lies beneath the Wanapum Basalt. The best exposures are in the downtown area near milepost 282. (See also the guide for Spokane in this chapter.) The 20 miles (32 km) between Spokane and the Idaho state line follow Rathdrum Prairie, the primary route for floodwaters coming from Glacial Lake Missoula to the Spokane area and the Channeled Scabland. Flood gravels cover this flat, mostly developed area.

US 2
US 97—Spokane
140 miles (225 km)

West of Orondo on the Columbia River, US 2 climbs more than 1,600 feet (490 m) in Corbaley (Pine) Canyon to the basalt-capped Waterville Plateau. The canyon cuts deeply into cliffs of biotite gneiss of the Swakane terrane, the deepest exposed part of the North Cascades. The gneiss formed through the metamorphism of sedimentary rock deposited about 72 million years ago. It was metamorphosed just 4 million years later (68 million years ago), implying a geologically rapid rate of deposition and burial. Numerous light-colored granitic dikes and dark basaltic dikes cut through the gneiss. The hills north of the road in the first quarter mile (400 m) nicely display the metamorphic layering, and the pull-out just up from milepost 143 offers a good chance to see the gneiss and some dikes. At milepost 144, you pass the contact between the gneiss and the overlying basalt, but it is covered by landslide debris, which is exposed in the roadcut. Above, the basalt forms the skyline. You pass some erosionally resistant entablature of the Grande Ronde a quarter mile (400 m) west of milepost 145.

Lakebeds of the Ellensburg Formation crop out at the tight switchback near milepost 146. The Ellensburg Formation, which was deposited during volcanically quiet times between eruptions of the Columbia River Basalt Group, can be

View up Corbaley (Pine) Canyon. Some of the many dikes in the canyon hold up the end of the graffiti-decorated ridge in the middle of the photo.

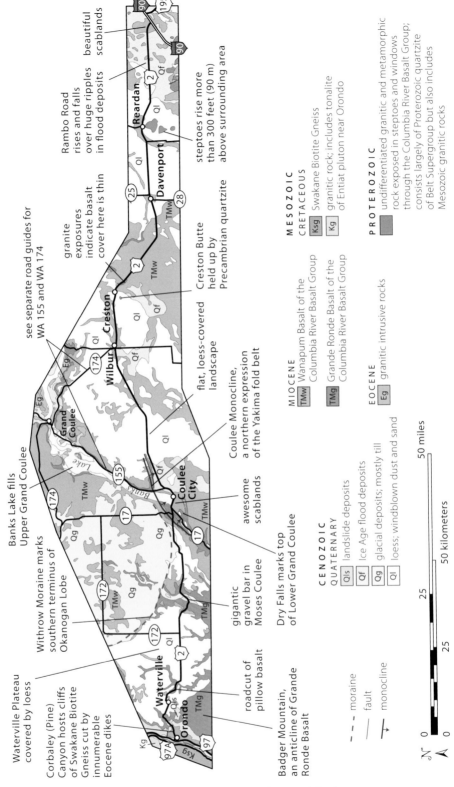

Waterville Plateau covered by loess

Corbaley (Pine) Canyon hosts cliffs of Swakane Biotite Gneiss cut by innumerable Eocene dikes

Banks Lake fills Upper Grand Coulee

Withrow Moraine marks southern terminus of Okanogan Lobe

see separate road guides for WA 155 and WA 174

granite exposures indicate basalt cover here is thin

Rambo Road rises and falls over huge ripples in flood deposits

beautiful scablands

steptoes rise more than 300 feet (90 m) above surrounding area

Creston Butte held up by Precambrian quartzite

flat, loess-covered landscape

Coulee Monocline, a northern expression of the Yakima fold belt

awesome scablands

Dry Falls marks top of Lower Grand Coulee

gigantic gravel bar in Moses Coulee

roadcut of pillow basalt

Badger Mountain, an anticline of Grande Ronde Basalt

MESOZOIC
CRETACEOUS
Ksg Swakane Biotite Gneiss
Kg granitic rock; includes tonalite of Entiat pluton near Orondo

PROTEROZOIC
undifferentiated granitic and metamorphic rock exposed in steptoes and windows through the Columbia River Basalt Group; consists largely of Proterozoic quartzite of Belt Supergroup but also includes Mesozoic granitic rocks

MIOCENE
TMw Wanapum Basalt of the Columbia River Basalt Group
TMg Grande Ronde Basalt of the Columbia River Basalt Group

EOCENE
Eg granitic intrusive rocks

CENOZOIC
QUATERNARY
Qls landslide deposits
Qf Ice Age flood deposits
Qg glacial deposits; mostly till
Ql loess; windblown dust and sand

- - - moraine
——— fault
⊤ monocline

N

0   25   50 miles
0   25   50 kilometers

Geology along US 2 between US 97 and Spokane.

*Close-up view of the contact of a granitic dike (left)
and the Swakane Biotite Gneiss (right).*

found interbedded with all the members of the group. Here, the Ellensburg is interbedded with the Grande Ronde Basalt. The roadcut immediately below the milepost sign shows the basalt cutting downward into the lakebeds. A double roadcut a quarter mile (400 m) higher offers nice examples of colonnade in the basalt and a wide pull-out to take in the view.

The Waterville Plateau is mostly covered by loess, predominantly wind-blown dust from glacial erosion, but bedrock shows up as local high areas and in river channels cut through the loess. The bedrock consists of the Grande Ronde Basalt overlain by the Wanapum Basalt, with a layer of the Ellensburg Formation sandwiched between the basalts. West of Waterville near milepost 149, you can look southward to the rounded crest of Badger Mountain, an anticline of Grande Ronde Basalt.

Southeast from Waterville, US 2 passes an outstanding exposure of pillow basalt in the Grande Ronde, just east the curve at milepost 152. The pillows display a crude layering that dips westward to indicate the direction the lava poured into a lake. Individual basalt pillows are surrounded by brown

Withrow Moraine

A half mile (0.8 km) east of milepost 163, WA 172 leads north 9 miles (14 km) to the Withrow Moraine, the terminus of the Okanogan ice lobe. The moraine extends southeastward as a distinct topographic feature nearly to Banks Lake in Grand Coulee. You start seeing the moraine as a low ridge after driving about 3 miles (4.8 km) to the north. After 7.5 miles (12 km), the road is deflected around a gigantic haystack boulder; it crosses a narrow gully and then climbs 250 feet (76 m) to the top of the sagebrush-covered moraine. The haystack boulder is a glacial erratic carried here by an earlier ice advance.

palagonite, formed by the alteration of volcanic glass, a typical by-product of the flash-freezing of lava as it contacts cold water. Several more outcrops of the Grande Ronde Basalt frame the edge of this small valley until you climb out the other side of it onto a 9-mile (14 km) stretch of gently rolling landscape covered in loess.

At milepost 166, US 2 starts down the western grade into Moses Coulee, cut into the Wanapum and Grande Ronde Basalts by the Ice Age floods. You can see loess sitting on top of basalt in a couple of places on either side of milepost 167, and there is an impressive exposure of the contact between two lava flows of the Grande Ronde Basalt near the bottom of the grade, about halfway between mileposts 169 and 170. A beautiful colonnade forms the base of the upper flow, and a rubbly zone forms the top of the lower flow. If you look closely, you can also see that the rock near the top of the lower flow is filled with the holes left by gas bubbles.

Moses Coulee measures more than 1 mile (1.6 km) wide in some places with cliffs of the Grande Ronde Basalt reaching up to 500 feet (150 m) high. From US 2, the coulee continues southwestward for about 30 miles (48 km) to join the Columbia River south of Wenatchee. Moses Coulee was carved by Ice Age floodwaters after the Okanogan Lobe advanced across the Columbia River drainage and diverted floods southward. As the glacial advance continued, the ice eventually filled the northern reaches of the newly formed Moses Coulee and diverted the floods farther east down what is now Grand Coulee. The floor of Moses Coulee is covered in deposits of these floods, some of which are piled in a gigantic gravel bar that the road descends on the west side of the coulee bottom. The bar lies on the downcurrent side of a bedrock protrusion into the coulee and at an inside bend, where currents would be expected to be slower and thus deposit material.

Jameson Lake Road intersects US 20 on the east side of the coulee. If you follow it north for 4.5 miles (7.2 km), you'll cross the Withrow Moraine within the coulee. Because the moraine lies within the coulee, the flooding that created the coulee must be older than the moraine. Some large boulders, including one unusually large haystack rock, lie on the moraine surface.

Coauthor Darrel Cowan standing by the lower of two lava flows of the Grande Ronde Basalt near the bottom of Moses Coulee. Note the colonnade of the upper flow and the rubbly zone between the flows.

View northward up Moses Coulee.

On the east side of the coulee, you gain a nice view to the west of the large gravel bar on the west side and pass more outstanding cross-sectional views of the Grande Ronde Basalt. Directly across from milepost 173, look for some interesting contact relations between different flows.

Between Moses Coulee and Grand Coulee are 10 miles (16 km) or so of gently rolling farmland. Look for some haystack rocks south of the highway a quarter mile (400 m) east of milepost 182 and at milepost 184. These giant boulders, which seem to appear out of nowhere, were deposited during a glacial advance that pre-dated the Withrow Moraine.

US 2 starts descending through the loess on the west side of Grand Coulee at milepost 185, passing scattered, generally poor, underlying exposures of basalt for the next few miles. At milepost 188, south-dipping basalt crops out on the south side of the road: the dip is easier to discern when looking west. The dipping basalt is part of the Coulee Monocline, a fold that helped determine the location of both the upper and lower stretches of Grand Coulee. The fold also explains why the basalt at lower elevations to the southeast consists of the younger Wanapum Basalt. East of the turnoff to Dry Falls State Park is a deposit of flood gravels.

The Coulee Monocline, which runs parallel to the Lower Grand Coulee and crosses the Upper Grand Coulee just north of Coulee City, channelized the Ice Age floods. South of Coulee City, the fold's shape caused the land to rise immediately to the west, thereby providing a barrier to contain the flows on that side. Simultaneously, folding of the rigid basaltic rock caused it to break within the hinge, making it easier to erode, thereby localizing the flow and erosion. North of Coulee City, the fold's shape caused the land to drop southward, increasing the velocity of the floodwaters and forming a cataract at that location. This cataract retreated some 20 miles (32 km) upstream. Its only remnant is the south face of Steamboat Rock.

Sun Lakes–Dry Falls State Park

Less than 2 miles (3.2 km) south of US 2 on WA 17, the overlook and visitor center for the Sun Lakes–Dry Falls State Park provides a stunning view of a 400-foot (120 m) cliff over which the Ice Age floods poured. Called the Dry Falls cataract, the arcuate cliffs stretch more than 1 mile (1.6 km) eastward at the head of Dry Falls Coulee. Monument Coulee lies immediately east with a similar cliff, and Deep Lake Coulee lies 1.5 miles (2.4 km) east of there with another cliff. Together, the three cliffs form the top of the Lower Grand Coulee.

At the bottom of each cliff lies a lake-filled depression, called a plunge pool, that formed due to the intense erosion at the base of the now-dry waterfall. Over multiple flood events, this erosion caused the cataract to retreat upstream. The three dry cataracts here are termed *recessional cataracts* because they actually formed about 20 miles (32 km) to the south, near Soap Lake. As they retreated, eroding the cliff over which they flowed, they left behind the narrow canyon we call Lower Grand Coulee. From the visitor center, you gain a clear view of Dry Falls Lake, the plunge pool beneath Dry Falls, and a more distant view of Red Alkali Lake, beneath the head of Monument Coulee. A rim of white alkali deposits develops around Red Alkali Lake as it slowly evaporates. The lakes farther south down Grand Coulee become increasingly alkaline until you reach Soap Lake,

Looking north over Dry Falls Lake to Dry Falls. Red Alkali Lake rests at the bottom of Monument cataract, at the far right (east) side of the photo.

which is the most alkaline of all. At the end of the Pleistocene Epoch, the lakes were all connected as one. Called Lake Bretz, the surface of this lake reached an elevation of 1,160 feet (354 m), as recorded by silty and alkaline deposits at that elevation along the margin of the coulee. As the climate warmed and dried, Lake Bretz began to dry up, leaving behind today's isolated lakes.

If you drive an additional 2 miles (3.2 km) southward to the turnoff to the Sun Lakes campground, you gain an outstanding view of the Coulee Monocline, which forms the edge of the Waterville Plateau. The Wanapum Basalt forms the top of the plateau at an elevation of 2,200 feet (670 m), but it drops to below 1,500 feet (460 m) at the roadcut near milepost 93. Across the road from the turnoff sits a landslide deposit with groundwater seeping out of its base.

Scablands cover the area just south of US 2 as it crosses Dry Falls Dam, which backs up Banks Lake to the north. Banks Lake continues northward nearly as far as the Grand Coulee Dam, where it receives water pumped from the Columbia River. The lake is the primary storage reservoir for the Columbia Basin Project, which supplies irrigation water to more than 950 square miles (2,500 km²) of central Washington farmland. For a wonderful detour, take WA 155 along the shores of Banks Lake to the town of Grand Coulee; from there you can head southeastward on WA 174 to rejoin US 2 at Wilbur (see road guides in this chapter). Even if you choose to continue on US 2, a 2-mile (3.2 km) side trip up WA 155 gives a good view of the Coulee Monocline.

Some well-exposed flood gravels lie on both sides of the road near milepost 194 east of Coulee City. Just east of the intersection with WA 155, the road rises up a gentle incline of flood gravels at milepost 196. Between milepost 196 and Wilbur, a distance of 25 miles (40 km), US 2 crosses a gentle, mostly loess-covered surface that varies less than 400 feet (120 m) in elevation. Some exposures of the Wanapum Basalt and the overlying loess do show up along this flat stretch of road, illustrating how the underlying bedrock controls the shape of the land: flat basalt flows generally mean a flat land surface, here mostly covered by farmland.

Between Wilbur and Creston, US 2 crosses over scattered exposures of the Wanapum Basalt along the south edge of Goose Creek, a shallow channel eroded by the Ice Age floods. West of Creston, near milepost 230, eastbound travelers gain a nice view of Creston Butte rising nearly 400 feet (120 m) above the nearly flat, loess- and basalt-covered plain. Creston Butte is made of Precambrian-age quartzite. Without any faults or folds nearby to explain why such old rock would rise above the much younger basalt flows, we can only surmise that it was a prominent hill at the time of the basaltic eruptions about 15 million years ago. The flowing lava encircled but did not overtop the hill. Similar but smaller steptoe buttes rise from the surrounding lava farther east along this route.

For a short distance at milepost 233, US 2 passes from basalt bedrock into granite of Cretaceous or early Tertiary age before rising back into the overlying basalt. The presence of the granite suggests that the Wanapum Basalt here is

Scablands eroded into granitic bedrock east of Creston. Note the flat-lying basalt in the distance.

relatively thin. If so, then the steptoe butte at Creston probably did not rise substantially higher over the surrounding area 15 million years ago than it does today.

Between Creston and a little east of milepost 244, US 2 crosses scablands in the Wanapum Basalt. These scablands appear as numerous small hilltops, narrow ridges with intervening gullies, and closed depressions filled by small lakes where they intersect groundwater. From west of Davenport to Reardan, scablands show up only intermittently because most of the land is covered by loess. Geologists have used the outcrop patterns of the loess and exposed basalt to map out the flood channels. Places covered by loess escaped the flooding, whereas areas of scablands experienced it intensively.

The hills just south of Reardan consist of Precambrian quartzite as well as granitic rocks of Cretaceous or early Tertiary age. These steptoes, each of which rises more than 300 feet (90 m) above the surrounding landscape, form a nearly straight line that heads southward for about 7 miles (11 km).

Between Reardan and Deep Creek, US 2 crosses a mostly flat surface covered by loess. You can see more steptoes about 4 miles (6.4 km) to the south near milepost 271. At Deep Creek, the road passes eastward from loess into flood gravels, some of which formed giant ripples as they were being deposited. At the east edge of Fairchild Air Force Base, Rambo Road leads northward into the ripple field, noticeably rising and falling as it passes over the ripples about 1.1 miles (1.8 km) north of US 2.

From just east of milepost 281 to I-90, US 2 passes through scablands of the Wanapum Basalt, including some beautiful exposures of colonnade. If you take West Sunset Highway about 1 mile (1.6 km) east and then head north on South Basalt Street for about 1.5 miles (2.4 km), you reach Palisades Park, where you can walk over the Wanapum Basalt with a beautiful view of the Spokane Valley.

US 12
Pasco—Clarkston
143 miles (230 km)

Southeast of Pasco on the east edge of the Pasco Basin, US 12 follows the north bank of the Columbia River. The Pasco Basin was a downwarp in the crust during the eruptions of the Columbia River Basalt Group about 16 to 15 million years ago, and according to recent geophysical surveys, it is now filled with more than 12,000 feet (3,660 m) of the lava flows.

US 12 crosses the Snake River between mileposts 294 and 295. Its mouth is only 1 mile (1.6 km) to the south, at the Columbia. The Snake River rises in northwestern Wyoming, more than 1,000 river miles (1,600 km) away. It's the Columbia River's largest tributary, draining an area of more than 100,000 square miles (260,000 km²), including the Teton Range in Wyoming, most of southern Idaho and eastern Oregon, and much of southeastern Washington. The only parts of southeastern Washington that do not drain into the Snake

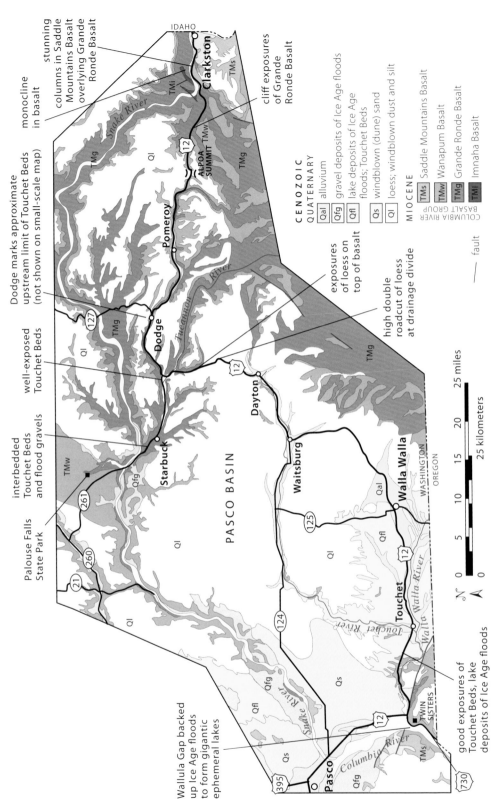

Geology along US 12 between Pasco and Clarkston.

River are those crossed by US 12 between milepost 296 and the drainage divide a few miles north of Dayton, at milepost 367. Rivers in this area feed south and westward directly into the Columbia.

Near milepost 300, look south to Wallula Gap, which forms a break in the prominent high ridge of basalt. Known as the Horse Heaven Hills, the ridge is a northwest-trending anticline in the Yakima fold belt and forms the southern margin of the Pasco Basin. Wallula Gap forms a distinct constriction in the otherwise wide floodplain of the Columbia River. At the time of the Ice Age floods, floodwaters poured into the Pasco Basin and backed up behind the constriction to form a large temporary lake that inundated all the low-lying parts of the region. Fine-grained sediments accumulated on the lake bottom. These sediments, called Touchet Beds, for the town of Touchet (pronounced TOO-shee) at milepost 320, consist of multiple beds of sand, silt, and clay.

Looking south across the Columbia River toward Wallula Gap, a water gap through the Horse Heaven Hills.

Wallula Gap

Head 2 miles (3.2 km) south on US 730, which intersects US 12 between mileposts 307 and 308, to see soaring cliffs of Wanapum Basalt in Wallula Gap. Here, the Columbia River flows in a narrow channel through the Horse Heaven Hills, a west-northwest-trending anticline. Numerous pull-outs allow easy access to the rocks as well as photo opportunities. Directly across the river, you can see nearly flat lava flows of the Wanapum Basalt dipping north in a broad, open anticline.

Wallula Gap, only 1 mile (1.6 km) wide, formed an important constriction in the path of the Ice Age floods, the largest of which overtopped the cliffs on either side of the river. The constriction caused the floodwaters to back up and temporarily fill the Pasco Basin, as well as the Walla Walla Valley and parts of Yakima Valley.

Wallula Gap also demonstrates the Columbia River's antiquity. As water always takes the least resistant path downhill, the only reasonable explanation for the Columbia cutting across this anticline is that the river pre-dated the anticline and continued to flow across it as the anticline grew. The anticline, part of the Yakima fold belt, likely started its period of growth during the eruption of the flood basalts, probably about 15 million years ago. Various researchers have found numerous examples of flows that become noticeably thinner or pinch out altogether over the crests of some of the folds, or become thicker in the basins.

Beautiful cliffs of the Wanapum and overlying Saddle Mountains Basalt along US 730 exhibit many examples of colonnade as well as entablature. Some striking examples of red paleosols (soil layers) lie between the lava flows. Perhaps most important, the continuity of individual flows over great distances becomes obvious. Some lava flows of the Columbia River Basalt Group are among the largest known on Earth. The Twin Sisters wayside near milepost 4 (a couple miles south of US 12) provides an especially good spot to take in the view and inspect the Wanapum Basalt. The presence of scablands in the basalt more than 200 (60 m) feet above the river indicates a minimum depth of the floodwaters. You can see other outstanding examples of scablands if you climb the short steep trail up to the rocks and look eastward.

The Twin Sisters, scablands eroded by the enormous Ice Age floods.

In many cases the beds gradually become finer grained upward because the finer material settled from suspension. Successive floods inundated the area, forming successive temporary lakes and depositing more beds of fine material.

Between milepost 312 and Touchet are the best exposures of Touchet Beds, although numerous good exposures show up all the way to Walla Walla. Between Touchet and Walla Walla, the road crosses mostly alluvium, deposited by the Walla Walla River. The hill fronts south of milepost 308 exhibit triangular facets eroded by the Ice Age floods. An especially dramatic double roadcut of Touchet Beds resting atop Wanapum Basalt is just east of milepost 315. A large pull-out on the east side offers the chance to inspect the deposits.

Only a few miles north of Walla Walla, US 12 passes into Palouse country, although this area south of the Snake River is sometimes excluded from the region. Like the Palouse Hills around Pullman, rolling wheat fields grow on thick deposits of loess, windblown dust derived largely from glacial abrasion to the north. Beneath the loess lies bedrock of the Wanapum Basalt, which you can also see in some of the roadcuts, a good example being in the little canyon at milepost 349. The light-colored, fine-grained loess resembles the Touchet Beds, but having been transported by wind, the loess is by definition very fine grained, whereas the Touchet Beds, having been transported by water, contain sandy and even cobble-rich zones.

Touchet Beds sitting on top of basalt just east of milepost 315.

Author Marli Miller standing in front of a thick deposit of loess at the drainage divide north of Dayton.

Between Waitsburg and Dayton, US 12 follows the floodplain of the Touchet River, bordered by hills of the Palouse. Bedrock exposures of Grande Ronde and Wanapum Basalts are scattered near the bases of some of the hills. North of Dayton, however, the road passes some better exposures as it climbs toward a drainage divide just north of milepost 373, marked by a high double roadcut of loess. This divide separates streams flowing northward into the Snake River from those flowing south and westward into the Columbia.

Between mileposts 375 and 376, US 12 descends past some great roadcuts of loess overlying the Wanapum Basalt. The exposures are clear enough that you can see the basal contact of the loess, which reflects the topography at the time it was deposited. Northward to milepost 377, US 12 passes through the older Grande Ronde Basalt, which forms prominent cliffs farther down valley to the north. Where the road bends sharply east into the Pataha Creek drainage near milepost 382, you can see some beautifully exposed Touchet Beds that indicate

WA 261 to Lyons Ferry Bridge

Between US 12 and Starbuck, WA 261 follows the beautiful Tucannon River past walls of Grande Ronde Basalt and exposures of lake deposits. West of Starbuck, and about a quarter mile (400 m) east of milepost 9, a great exposure of flood gravels interlayered with lakebeds shows up on the north side of the road. The lakebeds were deposited when the floods backed up behind Wallula Gap near the Washington-Oregon border to form Lake Lewis, which filled during each major flood and emptied soon afterward. Near milepost 11, you can see some scablands eroded into entablature of the basalt. Immediately south of the Lyons Ferry Bridge that crosses the Snake River, you can see portions of a giant gravel bar deposited by the Ice Age floods that flowed down the Palouse River and entered the Snake River here. The low cliffs east of the highway by milepost 16 belong to the upper part of the Grande Ronde Basalt, but you rise westward into the younger Wanapum Basalt. All the basalt exposures from here west to US 395 consist of the Wanapum Basalt. Between mileposts 20 and 19, just south of the turnoff for Palouse Falls State Park, the road passes hills of loess streamlined in a northeast-southwest direction by the floods. (See the discussion of Palouse Falls on page 319 in the US 395 road guide.)

this canyon was inundated by Ice Age floodwater that temporarily collected in the Pasco Basin. WA 261 leads west to Palouse Falls State Park. (See the discussion of the falls in the US 395 road guide.)

East of the WA 261 junction, US 12 follows the Pataha Valley upstream for nearly 30 miles (48 km) to Alpoa Summit. Although unexposed, the Touchet Beds are mapped by US Geological Survey geologists as far east as milepost 390, just west of Dodge. Their elevation of just below 1,300 feet (400 m) suggests an upper limit of the water that filled the Pasco Basin and back-flooded the valley. The Pataha Valley is framed by cliffs of the Grande Ronde and Wanapum Basalts. Look for a good exposure of Grande Ronde Basalt at milepost 391, at the intersection with WA 127.

Many good outcrops of the Grande Ronde and Wanapum Basalts border the highway between Dodge and Alpoa Summit. Of special note is the flow that makes a prominent ledge approximately halfway up the hillsides. This lava flow belongs to the upper part of the Grande Ronde Basalt; the exposed rock above it belongs to the Wanapum. You can see the ledge disappear beneath river gravel north of the road just east of milepost 401, although there are scattered exposures of it all the way to Pomeroy. For the several miles east of Pomeroy, roadside exposures consist of the Wanapum Basalt. Some minor folding must affect the rocks because the Grande Ronde Basalt reappears farther east along the road, even though the road rises in elevation. A great exposure of two flows of the Grande Ronde Basalt separated by a red paleosol lies at milepost 411, just as the road bends eastward toward Alpoa Summit. The paleosol is an ancient soil that formed on the surface of the lower flow before the upper flow covered it.

More Grande Ronde Basalt forms cliffs in the canyon east of Alpoa Summit, all the way to the Snake River at milepost 425. Look for multiple lava flows stacked on top of each other, many of which erode into distinct ledges or are separated by red paleosols. You can also find good examples of colonnade, entablature, and rubbly flow bases.

Monoclinal fold along the Snake River in units of Columbia River Basalt Group. Note the flat layers at the far left and right and the dipping layers in the middle of the photo.

Columns in the Saddle Mountains Basalt west of milepost 430. The Ellensburg Formation lies directly beneath the columns.

The 9 miles (14 km) along the Snake River west of Clarkston show off more lava flows on the south side of the road and a suite of folded lava flows immediately north of the river. The fold defines a large monocline in which the upper and lower limbs are approximately horizontal but connected by a steeply dipping limb. You pass the side of the steeply dipping limb at and east of milepost 426. The dipping rock layers erode into large, triangular-shaped features called flatirons.

Although basalt flows of the Wanapum overlie the Grande Ronde immediately west of milepost 426, the younger Saddle Mountains Basalt directly overlies the Grande Ronde eastward as far as Clarkston. This stretch of highway traverses an area where the Wanapum was never deposited or where it was eroded before the eruption of the Saddle Mountains Basalt. The beautiful toothpick-like columns immediately west of milepost 430 belong to the Saddle Mountains Basalt. Its contact with the underlying Grande Ronde Basalt, marked by a thin zone of sedimentary rocks of the Ellensburg Formation, slopes upward to the west.

The cliffs of the Saddle Mountains Basalt end abruptly at Elm Street as you enter Clarkston. A glance downriver to the west reveals an excellent view of the monocline.

US 97
Columbia River—Yakima
78 miles (126 km)

North of the Columbia River crossing, US 97 climbs through spectacular exposures of the Wanapum Basalt to the intersection with WA 14. The lava flows dip gently southward. You are driving up the south flank of the Columbia Hills anticline, the crest of which lies near Davies Pass between mileposts 6 and 7, more than 1,500 feet (460 m) above the river. This anticline is part of the Yakima fold belt, which extends across southern Washington from the Cascades to the Washington-Idaho border.

North of the short jog on WA 14, US 97 continues its climb through the Wanapum Basalt until just south of milepost 4, at which point it crosses a thrust fault, north of which is the older Grande Ronde Basalt. The long roadcut at milepost 4 shows the Grande Ronde Basalt in a highly fractured and altered state; the rounded ridgetop just to the southwest is also capped by the Grande Ronde. Like the fold, the fault trends roughly east-west, and the two appear to have formed together. Most of the large anticlines of the Yakima fold belt are faulted along one or both of their limbs.

The Wanapum Basalt, which overlies the Grande Ronde Basalt, shows up again in the roadcut a quarter mile (400 m) above milepost 4 and continues all the way to Davies Pass. It becomes noticeably more intact the farther you move up the road, away from the fault. From the scenic overlook between mileposts 7 and 8, you gain an outstanding view to the west of Mt. Adams, a stratovolcano in the South Cascades. To the north, down the highway, you can see Lorena Butte, next to Goldendale, and the Indian Rock shield volcano, another 10 miles (16 km) to the north.

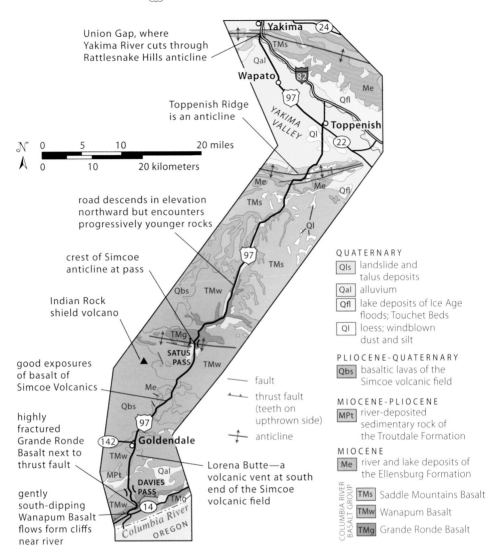

Union Gap, where
Yakima River cuts through
Rattlesnake Hills anticline

Toppenish Ridge
is an anticline

road descends in elevation
northward but encounters
progressively younger rocks

crest of Simcoe
anticline at pass

Indian Rock
shield volcano

good exposures
of basalt of
Simcoe Volcanics

highly
fractured
Grande Ronde
Basalt next to
thrust fault

gently
south-dipping
Wanapum Basalt
flows form cliffs
near river

Lorena Butte—a
volcanic vent at south
end of the Simcoe
volcanic field

QUATERNARY
Qls landslide and
talus deposits
Qal alluvium
Qfl lake deposits of Ice Age
floods; Touchet Beds
Ql loess; windblown
dust and silt

PLIOCENE-QUATERNARY
Qbs basaltic lavas of the
Simcoe volcanic field

MIOCENE-PLIOCENE
MPt river-deposited
sedimentary rock of
the Troutdale Formation

MIOCENE
Me river and lake deposits of
the Ellensburg Formation
TMs Saddle Mountains Basalt
TMw Wanapum Basalt
TMg Grande Ronde Basalt

COLUMBIA RIVER
BASALT GROUP

— fault
⊣⊣⊣ thrust fault
(teeth on
upthrown side)
⊣ anticline

Geology along US 97 between the Columbia River and Yakima.

Approaching Goldendale near milepost 11, you pass the western edge of Lorena Butte, a basaltic vent that erupted during the Pleistocene Epoch. It's one of the southernmost expressions of the Simcoe volcanic field, which extends more than 30 miles (48 km) northward from Goldendale. The field consists of more than two hundred vents, most of which are basaltic, that erupted from 4 million years ago until about 600,000 years ago. The Simcoe Volcanics are noteworthy because they formed east of the Cascades and so reflect processes that don't have a clear link to subduction.

Cliffs of south-dipping Wanapum Basalt line the Columbia River valley. Mt. Hood in Oregon looms in the background.

Looking north to the Indian Rock shield volcano from north of the overlook.

Goldendale lies on a narrow exposed stretch of the Wanapum Basalt between Lorena Butte and the main body of the Simcoe volcanic field. The Indian Rock shield volcano, part of the main body, erupted between 4.2 and 3.4 million years ago.

The highway passes along the eastern edge of the Simcoe volcanic field north of Goldendale, with Simcoe lavas sitting on top of the older Wanapum Basalt in many places. About a quarter mile (400 m) north of milepost 13, look for a well-developed paleosol, an ancient soil, at the base of a flow of the Wanapum Basalt. Only 0.1 mile (160 m) north of the paleosol are good exposures of the

Simcoe lava. An especially prominent one shows up at the corner of US 97 and Hanging Rock Road.

About halfway between mileposts 17 and 18, US 97 passes through the narrow Jenkins Creek valley, rimmed by a flow of Simcoe lava and floored by the Wanapum Basalt. A good exposure opposite milepost 18 shows a red paleosol within the Wanapum. US 97 encounters a long, low roadcut of Simcoe basalt halfway between mileposts 19 and 20. At a low pass, a quarter mile (400 m) north of milepost 20, US 97 descends back into the Wanapum Basalt.

Just north of milepost 27, Satus Pass marks the crest of the Simcoe anticline, another fold of the Yakima fold belt. Look for some beautiful colonnade in the Grande Ronde Basalt between mileposts 30 and 31 as you descend into the Satus Creek Canyon. US 97 follows this canyon downstream for about the next 20 miles (32 km), passing from the Grande Ronde into the successively younger flows of the Wanapum, at milepost 40, and then the Saddle Mountains Basalt, at about milepost 50. Although the lava flows appear relatively flat-lying, they must dip more steeply northward than the highway because you keep passing younger rocks as you descend the canyon. Near milepost 39 you gain a good view to the Saddle Mountains Basalt that forms the canyon rim, with slopes of the Wanapum Basalt underneath.

About halfway between mileposts 49 and 50, the road crosses to the west side of Satus Creek and climbs into the Saddle Mountains Basalt, which is covered by loess and grass in many places. A poor exposure of lake deposits of the Ellensburg Formation lies on the west side of the road a quarter mile (400 m) north of milepost 52, here interlayered with the Saddle Mountains Basalt.

The road crests Toppenish Ridge, another anticline, between mileposts 53 and 54. Looking eastward from just below the north side of the pass, you can see basalt dipping gently southward, but farther north, at milepost 56, you can see north-dipping flows. A prominent exposure of the basalt lies right at milepost 57, where the road hits the flat floor of the Yakima Valley.

US 97 crosses over alluvium in the wide Yakima Valley, one of the typically wide synclines between the typically narrow anticlines of the Yakima fold belt.

The smooth, sloping surface on the north side of Toppenish Ridge has formed on basalt that slopes gently down to the north.

View north along US 97 into Union Gap. The Wanapum Basalt forms the main bedrock, with light-colored beds of the Ellensburg Formation above. US 97 shares this narrow space with a railroad, the Yakima River, and I-82.

Union Gap, 17 miles (27 km) to the north, cuts through the Rattlesnake Hills anticline. You can't stop along this busy section of US 97 to inspect the roadcuts in the gap, but if you look across the river, you can see the Wanapum Basalt dipping toward the south. You can glimpse where the basalt folds over and dips steeply northward at the very far north edge of the fold just before the turnoff for Main Street. Southbound travelers get an even better view of the north-dipping flows as you enter US 97 from the interstate.

US 195
Spokane— Idaho Border
near Lewiston
94 miles (151 km)

After climbing out of the geologically rich watershed of the Spokane River, US 195 heads almost straight south across Washington's Palouse country, a seemingly endless sea of low, rolling hills covered with farmland, most of which grows wheat. The Palouse escaped most of the ravages of the Ice Age floods and remains covered by Pleistocene loess, windblown dust and fine sand created by glacial abrasion farther north. The loess rests on flows of the Columbia River Basalt Group, which crop out where streams carved valleys through the loess. The lava flows buried a hilly landscape of mostly Proterozoic-age bedrock, which now poke through the basalt in places.

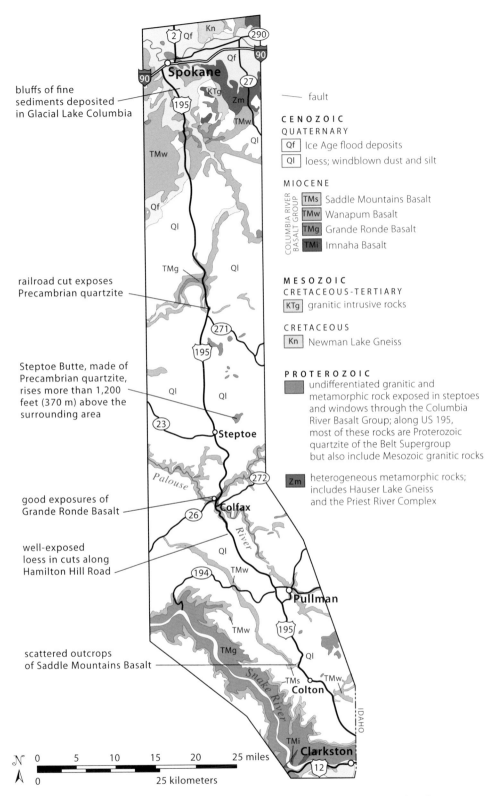

bluffs of fine
sediments deposited
in Glacial Lake Columbia

railroad cut exposes
Precambrian quartzite

Steptoe Butte, made of
Precambrian quartzite,
rises more than 1,200
feet (370 m) above the
surrounding area

good exposures of
Grande Ronde Basalt

well-exposed
loess in cuts along
Hamilton Hill Road

scattered outcrops
of Saddle Mountains Basalt

—— fault

CENOZOIC
QUATERNARY
Qf Ice Age flood deposits
Ql loess; windblown dust and silt

MIOCENE
COLUMBIA RIVER BASALT GROUP
TMs Saddle Mountains Basalt
TMw Wanapum Basalt
TMg Grande Ronde Basalt
TMi Imnaha Basalt

MESOZOIC
CRETACEOUS-TERTIARY
KTg granitic intrusive rocks

CRETACEOUS
Kn Newman Lake Gneiss

PROTEROZOIC
undifferentiated granitic and
metamorphic rock exposed in steptoes
and windows through the Columbia
River Basalt Group; along US 195,
most of these rocks are Proterozoic
quartzite of the Belt Supergroup
but also include Mesozoic granitic rocks

Zm heterogeneous metamorphic rocks;
includes Hauser Lake Gneiss
and the Priest River Complex

N

0   5   10   15   20   25 miles

0   25 kilometers

Geology along US 195 between Spokane and the Idaho border near Lewiston.

South of I-90, US 195 follows Hangman Creek, which flows north into the Spokane River. Flows of mostly Wanapum Basalt line the valley in some places. The valley bottom consists of fine-grained sediments deposited in Glacial Lake Columbia, a temporary lake filled by the Ice Age floods. Originating in Glacial Lake Missoula, the floods poured into the Spokane and Columbia Valleys but encountered another ice dam farther downstream in the vicinity of Grand Coulee. Some of the best exposures of these lake deposits show up on the east side of the highway on both sides of milepost 93. About a quarter mile (400 m) north of milepost 91, you can see a well-defined thrust fault cutting the sediments. It is not clear exactly what this fault represents. Some researchers argue that if formed on the edge of a landslide block in the lake sediments, whereas others suggest it may be related to a poorly understood bedrock fault zone.

Deposits of Glacial Lake Columbia along Hangman Creek.

A good exposure of the Wanapum Basalt exists at milepost 91, but at milepost 89 even better exposures continue for a half mile (0.8 km). Northbound lanes get a fleeting view of pillows within these basalts. A half mile (0.8 km) south of milepost 87, some poorly exposed, light-colored rocks belong to the Belt Supergroup of Proterozoic age. Their presence means the lava flows of the Columbia River Basalt Group must be relatively thin here. The old rocks formed the topography when the basalt erupted, and some are sitting at higher elevations than the nearby lava flows.

It's hard to say exactly where the northern boundary of the Palouse is, but the railroad tracks near milepost 81 seem a good approximation. There, loess laps onto the Wanapum Basalt, whereas to the north Ice Age flood gravels blanket the landscape. Northbound lanes, which dip beneath the tracks, pass through a basaltic roadcut, whereas southbound lanes rise above the tracks and miss the bedrock.

US 195 passes numerous exposures of the Wanapum Basalt beneath the loess from about milepost 71 to 64. One exposure between mileposts 68 and 69 is a half mile (0.8 km) long.

About 0.75 mile (1.2 km) south of milepost 64, you can look eastward up an old railroad grade to an exposure of Proterozoic quartzite. Then, south of the rest area between mileposts 61 and 60, you gain occasional glimpses to the east of a chain of hills rising 100 to 200 feet (30–60 m) above the surrounding Palouse. These hills, called steptoes, consist of more Proterozoic bedrock of the Belt Supergroup. These old rocks lie beneath the Columbia River Basalt Group but protrude above the younger rocks in places. They must have been highlands prior to the basalt eruptions. The highest of these buttes, named Steptoe Butte, is clearly visible between mileposts 53 and 42. You can drive to the top of the butte on a good but steep road that passes numerous outcrops of the quartzite, described in the sidebar.

Steptoe Butte (background) rises about 1,200 feet (370 m) above loess-covered basalt.

Steptoe Butte State Park

Named after Colonel Edward Steptoe (1815–1865) of the US Army, Steptoe Butte rises some 1,200 feet (370 m) above the surrounding Palouse. No faults surround the butte, yet it is made of 1.1-billion-year-old quartzite of the Belt Supergroup, whereas the bedrock around and below it consists of the much younger Wanapum and Grande Ronde Basalts. These relations tell us that the butte must have been a topographically high area during the basalt eruptions, which began 16.7 million years ago. What's more, the US Geological Survey estimates the thickness of the basalt at the nearby town of Steptoe to be more than 1,000 feet thick (300 m). Steptoe Butte must have soared more than 2,200 feet (670 m) above the surrounding area at the start of the eruptions. The name *steptoe* has been applied worldwide to any butte, hill, or mountain that rises above surrounding lava. In eastern Washington, numerous smaller steptoes punctuate the horizon to the north and south of Steptoe Butte and along US 2.

View over the Palouse from near the top of Steptoe Butte. Precambrian quartzite crops out on the left.

To visit Steptoe Butte State Park, drive Scholtz Road 2 miles (3.2 km) southeastward from the town of Steptoe to Hume Road, and follow that east and north 4 miles (6.4 km) to the park entrance road. From here, the road climbs another 4 miles (6.4 km) as it spirals to the top of the butte. Numerous good outcrops of the quartzite line the road.

Red paleosol, an ancient soil, between flows of the Grande Ronde Basalt at the north end of Colfax.

Between Steptoe and Colfax, US 195 crosses more of the Palouse and occasional good exposures of loess and the underlying Wanapum Basalt. At Colfax the highway descends to the Palouse River, dropping down through scattered poor exposures of the Wanapum Basalt into much more prominent outcrops of the underlying Grande Ronde Basalt.

Between Colfax and the Idaho border, US 195 crosses more of the Palouse, although bare exposures of the loess tend to be rare. One good exposure shows up on Hamilton Hill Road that leads west from just south of milepost 31. Near milepost 20, WA 270 leads to Washington State University, in Pullman, and exposures of Wanapum Basalt. Some of the younger Saddle Mountains Basalt shows up intermittently between mileposts 12 and 8 near Colton. You gain spectacular views of Lewiston and the Snake River at the Lewiston Summit overlook, just over 1 mile (1.6 km) southeast of the Idaho border on US 95.

US 395
PASCO—RITZVILLE
73 miles (118 km)

For the first 12 miles (19 km) north of Pasco, US 395 gradually rises out of the Pasco Basin, a large downwarp that's not particularly visible in today's landscape but accumulated more than 12,000 feet (3,660 m) of the Columbia River Basalt Group during the Miocene Epoch. The basalt is mostly covered by lakebeds or gravels related to the Ice Age floods, and numerous vineyards and orchards take advantage of the region's hot, dry summers.

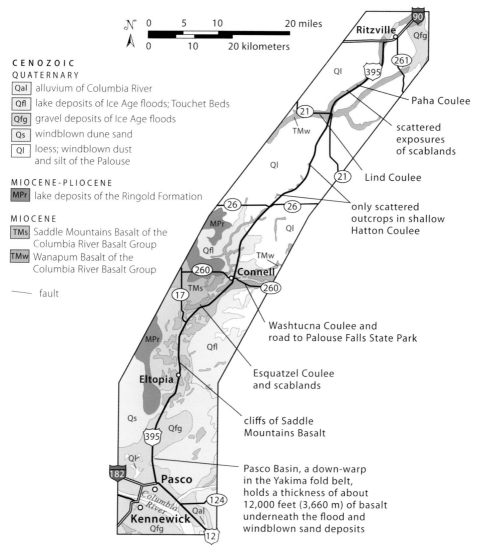

N
0   5   10   20 miles
0   10   20 kilometers

CENOZOIC
QUATERNARY
Qal alluvium of Columbia River
Qfl lake deposits of Ice Age floods; Touchet Beds
Qfg gravel deposits of Ice Age floods
Qs   windblown dune sand
Ql   loess; windblown dust
     and silt of the Palouse

MIOCENE-PLIOCENE
MPr lake deposits of the Ringold Formation

MIOCENE
TMs Saddle Mountains Basalt of the
     Columbia River Basalt Group
TMw Wanapum Basalt of the
     Columbia River Basalt Group

――― fault

Ritzville

Paha Coulee

scattered
exposures
of scablands

Lind Coulee

only scattered
outcrops in shallow
Hatton Coulee

Connell

Washtucna Coulee and
road to Palouse Falls State Park

Esquatzel Coulee
and scablands

Eltopia

cliffs of Saddle
Mountains Basalt

Pasco

Columbia River

Kennewick

Pasco Basin, a down-warp
in the Yakima fold belt,
holds a thickness of about
12,000 feet (3,660 m) of basalt
underneath the flood and
windblown sand deposits

Geology along US 395 between Pasco and Ritzville.

Near and north of milepost 34, you start seeing the first hints of basaltic bedrock, but exposures don't show up until about 1 mile (1.6 km) north of Eltopia. A prominent roadcut near milepost 39 shows several flows of the Saddle Mountains Basalt on the east side of the road. You can see some beautiful flow bases and paleosols, and, higher in the flows, entablature and colonnade. For the next 6 miles or so (10 km), you pass by scattered roadcuts and outcrops of Saddle Mountains Basalt; in some places you can see younger gravels deposited by the Ice Age floods. The flat-lying basalt is resistant to erosion and dictates the flatness of the landscape.

Esquatzel Coulee, a large valley cut by Ice Age floods and now lacking a significant stream.

At and north of milepost 48 are scablands carved into the basalt by the Ice Age floods. Between milepost 49 and just south of Connell, near milepost 57, the highway follows Esquatzel Coulee. The best views of the coulee are from mileposts 49 and 50. The Esquatzel Coulee, some 200 feet (61 m) deep, carried water of the Ice Age floods from Washtucna Coulee to the east and Providence Coulee to the north and emptied southward into the Pasco Basin. The Wanapum Basalt, which underlies the Saddle Mountains Basalt, is exposed in the bottom of the canyon. Look for some especially well-developed scablands near milepost 55. At Connell, WA 260 heads east to Palouse Falls State Park, following Washtucna Coulee.

North of Connell, exposed lava (all Wanapum Basalt) shows up only sporadically. Most of the landscape is covered by windblown loess, most of which is plowed for agriculture. Some good examples of loess include the low cuts on both sides of the road at milepost 58 and just north of milepost 75, where you can see loess sitting directly on basalt.

US 395 follows Hatton Coulee between milepost 65 and about milepost 78, and then Paha Coulee between milepost 83 and Ritzville. Hatton Coulee is the shallowest, exposing basaltic bedrock in only a few places. It was an overflow channel of the larger west-draining Lind Coulee, which you cross at milepost 82. Paha Coulee, which appears to be a tributary of Lind Coulee, is better defined than Hatton Coulee and displays much more basalt and even some scablands along the highway. Just north of milepost 89, look for some basalt overlain by loess.

Palouse Falls State Park

Palouse Falls, a comparatively small stream of water, plunges 186 feet (57 m) into a huge punchbowl-like amphitheater. It seems unlikely that, even during the highest spring flows, today's Palouse River could have eroded such a large space. As is the case with so many grand features in eastern Washington, the formation of Palouse Falls stems from the Ice Age floods, whose flows were typically more than 100,000 times the peak spring flows of today's river.

Exposed in the deep canyon walls are some 300 feet (90 m) of lava flows of the Grande Ronde and overlying Wanapum Basalts. Between them lies a thin zone of sedimentary rock, a part of the Ellensburg Formation called the Vantage Horizon. It was deposited during the volcanically quiet period between the eruption of the two basalts and shows up nearly as far west as the Pacific Ocean.

Downstream of the falls, the Palouse River canyon follows a zigzag course as it descends nearly 6 miles (10 km) to its confluence with the Snake River. The confluence marks the waterfall's original location. During the Ice Age floods, the waterfall retreated upriver as the incredible energy in its plunge pool continually undercut the lip of the falls and caused it to quickly erode. The zigzag pattern of its canyon formed as the water eroded a prominent series of

Palouse Falls. Its canyon and oversized plunge pool was eroded by Ice Age floods.

northwest-trending fractures that break the basalt. It is not entirely clear what formed the fractures, but some researchers suggest they are related to strike-slip movements along faults in the basement rock beneath the basalt.

To reach Palouse Falls, head east on WA 260 from US 395 at Connell. WA 260 follows Washtucna Coulee all the way to the intersection with WA 261, which heads east toward the falls. WA 261 crests a pass near milepost 22, where you can see hills of loess southwest of the road that were streamlined in a northeast-southwest direction by the Ice Age floods.

WA 14
MARYHILL (US 97)—PLYMOUTH (I-82)
79 miles (127 km)

WA 14 follows the north side of the Columbia River through open country east of the Cascades. For the first few miles east from US 97, the highway passes a few roadcuts of the Wanapum Basalt. Look for some pillows at the top of the prominent roadcut less than a quarter mile (400 m) west of milepost 105. The pillows formed when the lava flow poured into a lake. The orange material around the pillows consists of the mineral palagonite, an alteration product of the volcanic glass, which formed when the lava encountered water. About a quarter mile (400 m) east of milepost 105, you can look northeast and see the rounded hills of a landslide deposit.

Some large pull-outs near milepost 106, at the top of a long downhill grade, offer a chance to rest and take in the view of the John Day Dam stretching across the river. Completed in 1971 for hydroelectric power, it is the youngest dam on the Columbia. Just east of the pull-outs, the road descends to a river terrace alongside a large deposit of talus, loose rock fragments that have fallen from the basalt cliffs above. River terraces mark former positions of the floodplain that were stranded after the river cut deeper into its channel. The terrace is decorated by scablands, where erosion by the Ice Age floods cut down into the bedrock. Between the bottom of the grade and milepost 117, the bedrock consists of the Grande Ronde Basalt. Farther east, more prominent exposures of the Wanapum Basalt frame the highway. Look for gravels left by the Ice Age floods in hillsides along the highway.

The views southward across the Columbia River show long cliffs, formed by multiple lava flows of the Wanapum Basalt. The continuity of individual flows is striking, and what you can see here is only a tiny fraction of their true extent: although the specific flows are different, this part of the Wanapum Basalt, called the Frenchman Springs Member, also forms the cliffs of Palouse Falls north of Walla Walla, and many of the waterfalls at Silver Falls State Park, near Salem, Oregon. At and east of milepost 119 you can also see notches in the cliffs across

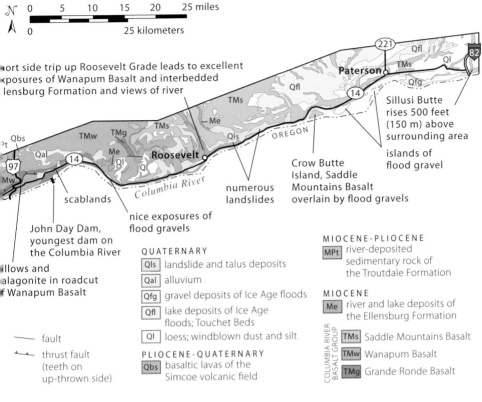

N 0   5   10   15   20   25 miles

0   25 kilometers

short side trip up Roosevelt Grade leads to excellent
exposures of Wanapum Basalt and interbedded
Ellensburg Formation and views of river

Paterson

Sillusi Butte
rises 500 feet
(150 m) above
surrounding area

islands of
flood gravel

Crow Butte
Island, Saddle
Mountains Basalt
overlain by flood gravels

numerous
landslides

Roosevelt

Columbia River

scablands

nice exposures of
flood gravels

John Day Dam,
youngest dam on
the Columbia River

pillows and
palagonite in roadcut
of Wanapum Basalt

—— fault

▲▲— thrust fault
(teeth on
up-thrown side)

QUATERNARY

Qls	landslide and talus deposits
Qal	alluvium
Qfg	gravel deposits of Ice Age floods
Qfl	lake deposits of Ice Age floods; Touchet Beds
Ql	loess; windblown dust and silt

PLIOCENE-QUATERNARY

| Qbs | basaltic lavas of the Simcoe volcanic field |

MIOCENE-PLIOCENE

| MPt | river-deposited sedimentary rock of the Troutdale Formation |

MIOCENE

| Me | river and lake deposits of the Ellensburg Formation |

COLUMBIA RIVER BASALT GROUP

TMs	Saddle Mountains Basalt
TMw	Wanapum Basalt
TMg	Grande Ronde Basalt

Geology along WA 14 between US 97 and I-82.

Scablands along WA 14.

the river, as well as the occasional gravel deposit. The Ice Age floods filled the gorge to these elevations, so the notches acted as local drains through which the water flowed into the valleys behind. Philippi Canyon, directly across the river about a half mile (0.8 km) east of milepost 119, forms one of the lowest notches and displays a prominent gravel bar near its base.

Between mileposts 117 and 142, WA 14 passes through numerous road-cuts of the Wanapum Basalt. Along the way, look for gravel left by the Ice Age floods. It is particularly well-exposed in the hillsides halfway between mile-posts 128 and 129 and at milepost 130, where it's being quarried. You can also see the occasional exposure of the Ellensburg Formation, deposited in rivers and lakes between eruptions of the Columbia River Basalt Group. These sedi-ments appear as light-colored stratified deposits at the top of the hills near the small town of Roosevelt. More Ellensburg Formation appears high on the slopes across from milepost 144.

A 2.7-mile (4.3 km) side trip up the Roosevelt Grade leads to a pull-out with a wonderful view of the river and exposures of lakebeds of the Ellensburg Formation. The first exposure, 1.9 miles (3.1 km) up the grade, lies between basalt flows but does not offer a safe pull-out. The upper and larger exposure sits on top of the basalt about 0.1 mile (160 m) below the pull-out.

At and east of milepost 142, you pass into the Saddle Mountains Basalt, which you follow all the way to I-82. The bedrock is well exposed until mile-post 163, east of which it shows up only intermittently from beneath a cover of farmland developed on loess and gravel deposits. The Saddle Mountains Basalt reflects the long final waning of the eruptions of the Columbia River Basalt Group. Although its eruptions continued from 13 to 5.5 million years ago, the basalt represents just over 1 percent of the total volume of the flood basalts. The lavas along this stretch of road erupted between about 12 and 10.5 million years ago.

Islands of flood gravel in the Columbia River.

Look for landslides between mileposts 142 and 163 and elongate islands of flood gravel in the Columbia River from about milepost 163 to milepost 170. The landslide across from milepost 142 displays the characteristic hummocky topography especially well. Between mileposts 152 and 155, the road follows a side channel in the Columbia on the north side of Crow Butte Island, which consists of the Saddle Mountains Basalt overlain by flood gravels. Sillusi Butte, which rises nearly 500 feet (150 m) directly east of I-82, and the smaller buttes immediately to the north, consist of the Saddle Mountains Basalt overlain by loess.

WA 28/WA 281
WENATCHEE—GEORGE (I-90)
40 miles (64 km)

At Wenatchee are several flat benches perched at different levels on either side of the Columbia River. The benches consist of gravel and boulders carried by Ice Age floods; the higher ones having been formed by earlier, larger floods, and the lower benches by later, smaller ones. The city center of Wenatchee sits on one of the lower benches, here called the Wenatchee-level bench for clarity. WA 28 follows this same bench for most of the next 23 miles (37 km) on the east side of the river. Studies of the gravels in this deposit indicate that they were probably deposited by floodwaters that backfilled the Columbia Valley northward after being diverted by the advancing Okanogan ice lobe down Moses Coulee, which joins the Columbia about 20 miles (32 km) downstream to the south.

About one hundred to several hundred feet above the Wenatchee-level bench sits an older flood deposit, best seen from between mileposts 3 and 6. It forms the surface beneath the main airport at Pangborn Field and reaches elevations more than 600 feet (180 m) above the river. It was deposited by a flood that came directly down the Columbia before the river was blocked by the Okanogan ice lobe. An even higher and older bench was once occupied by the airstrip at Fancher Field on the northeast side of town, but it is now subdivided. This surface is mantled by up to 6 feet (2 m) of calcium carbonate, which accumulated in the sediment through time. Its thickness, as well as deposits of loess, indicate that this oldest bench pre-dates the most recent glacial advance.

Near milepost 1, you can look across the river to cliffs of tilted Chumstick Formation that rise behind some railroad tracks just above the river level. The Chumstick Formation was deposited in alluvial fans and river channels during the Eocene Epoch and is described more fully in the road guides for US 2 west of Wenatchee and US 97 north of Ellensburg. Nearing milepost 2, you dip onto a lower and still younger bench, just above the river level. As you rise back onto the Wenatchee-level bench, the view opens up and you can see the higher and older terrace that lies beneath Pangborn Field rising to the east.

You pass the first exposure of Grande Ronde Basalt at milepost 6; it is part of a landslide deposit. Cliffs of the basalt form the horizon about 2 miles

(3.2 km) to the east. A small lake, about halfway between mileposts 6 and 7, connects to a string of lakes that sweep north and east in an arc that recrosses the highway at milepost 9. These lakes mark the outer edge of a surface scoured by Ice Age floods. This surface formed more recently than the bar that makes the Wenatchee-level bench.

Near milepost 10, you can look up Rock Island Creek, a narrow coulee that cuts northeastward through the basalt. Rock Island Dam, at milepost 11, was built between 1929 and 1933 for hydropower and was the first dam to span the Columbia River. You can see fine-grained sediments across the river upstream from the dam. These sediments were deposited in a temporary lake that backed up behind gravel deposited at the mouth of Moses Coulee by floodwaters racing down the coulee. The road here rests on bedrock of the Grande Ronde Basalt, which rises above as a series of cliffs.

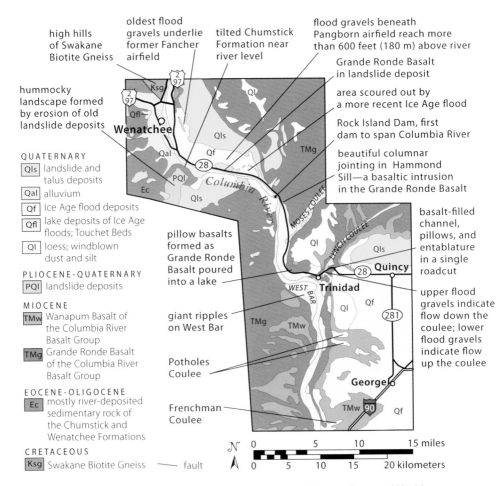

Geology along WA 28 between Wenatchee and I-90 at George, Washington.

The tilted Chumstick Formation exposed near river level extends nearly to the top of the ridge, which consists of Pliocene to early Quaternary landslide deposits.

As the road rises back onto the Wenatchee-level bench, you can see some spectacular columnar jointing in the basalt on either side of milepost 12; the columns fan into near horizontal orientations just south of the milepost. Such fanning results from temperature variations in the cooling of the lava. This rock originated as a shallow intrusion that invaded the surrounding lava flows along their layering, and the temperature variations probably occurred because of variations in the host rock. Called the Hammond Sill, this feature crops out over an area of more than 100 square miles (260 km²). If you look at it closely, you'll see that the rock is devoid of air bubbles and that it sits on top a zone of highly fragmented rock. Across the river you can see a beautiful view of the Wenatchee-level bench.

About a half mile (0.8 km) south of milepost 12, you can see more light-colored, fine-grained deposits on the east side of the highway. Like the deposits just upriver, these were deposited in a lake that formed after the Columbia River became dammed by gravel deposited at the mouth of Moses Coulee. Part of this gravel deposit is visible beneath the power lines just north of milepost 15. As you drive south from here, you can look northward up Moses Coulee. More than 700 feet (210 m) deep and more than a half mile (0.8 km) across, the coulee extends northeastward as a well-defined feature for more than 40 miles (64 km). It was carved by multiple floods that spilled from Glacial Lake Columbia, which formed after the Okanogan Lobe of the ice sheet advanced across a western part of the Columbia River drainage. The southward advance of the lobe forced later floodwaters to divert farther east and carve Grand Coulee.

Between Moses Coulee and Trinidad, the highway continues along the Wenatchee-level bench beneath cliffs of the Grande Ronde Basalt. If you look high on the cliffs south of milepost 18, you can see a flow of pillow basalt, formed when the lava poured into a lake. The pillows form layers that dip northward, indicating the direction of flow. Just south of milepost 20, you gain

Pillow basalt formed when lavas of the Grande Ronde Basalt poured into a lake. Note pillow layers dip to the north (left).

an outstanding view of West Bar on the river's inside bend. The bar is covered with giant ripples from a late Ice Age flood down the Columbia, probably from the failure of the Okanogan Lobe's ice dam across Glacial Lake Columbia. Richard Waitt of the US Geological Survey, who has studied this bar in detail, has found evidence that most of the gravel of the huge bar likely came from an earlier flood that originated from Glacial Lake Missoula. A short stroll up the hill from the pull-out on the south side of the road a quarter mile (400 m) west of milepost 22 allows another view of the bar.

At Trinidad, WA 28 turns northward up Lynch Coulee. You can turn right and follow Crescent Bar Road 0.9 mile (1.5 km) to a spectacular view southward along the Columbia and of West Bar. The gravels that form the walls of Lynch Coulee show large crossbeds inclined in a down-coulee sense, to indicating deposition by floodwaters flowing down the coulee from the Quincy Basin. About a half mile (0.8 km) farther north, where the road crosses the coulee, you can see two layers of flood gravel: the upper one shows beds inclined down-coulee, whereas the lower, older one shows beds inclined up-coulee. These older gravels reflect an older flood down the Columbia that backed up into the coulee.

Pull-outs on either side of the roadcut just north of milepost 23 allow easy access to the Wanapum Basalt. You can see a nicely exposed basalt-filled channel that cuts into an older flow with a well-developed zone of pillow basalt. Note also the beautifully developed entablature in the younger flow and the reddish paleosol, an ancient soil, at the top of the older flow. Another half mile (0.8 km) up the road, a roadcut on the north side of the highway displays some beautiful columns and a thin paleosol within the Grande Ronde Basalt.

At the rest area near milepost 25, at the top of the grade above the Columbia Valley, are flat agricultural fields taking advantage of loess. Just west of milepost 27, you cross the shallow Crater Coulee, which drains part of the Quincy

West Bar, covered by generations of flood debris, was shaped into huge ripples by a late flood down the Columbia River. The bar was originally deposited on this inside meander bend by an earlier flood from Glacial Lake Missoula.

The lowest flood gravels in Lynch Coulee are inclined to the right, having been deposited by floodwaters that flowed up the coulee. The beds above it are inclined to the left, having been deposited by younger floods that came down the coulee from the Quincy Basin.

Basin westward into Lynch Coulee. About 20 miles (32 km) to the west across the Columbia Valley, you can see the crest of the southern length of Naneum Ridge, made of the Grande Ronde Basalt and some overlying Wanapum Basalt.

WA 155
COULEE CITY—GRAND COULEE
26 miles (42 km)

WA 155 follows the east side of Banks Lake, which fills Upper Grand Coulee. Banks Lake gets its water from the Columbia River, where a pumping station at the Grand Coulee Dam elevates it some 100 feet (30 m) to a canal that feeds the lake. The lake serves as the initial storage reservoir for the vast Columbia Basin Project, which provides irrigation water to more than 950 square miles (2,500 km²) of central Washington farmland.

For the first 2 miles (3.2 km) north of US 2, you cross a flat surface of outburst flood gravels. At milepost 2, however, the road crosses the hinge of the Coulee Monocline. The narrow basaltic ridges on the east side of the road consist of moderately dipping Wanapum Basalt, whereas immediately to the north are cliffs

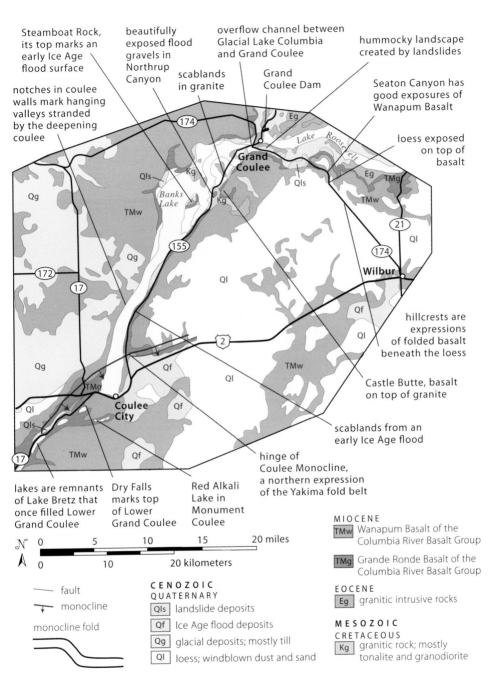

Steamboat Rock, its top marks an early Ice Age flood surface

beautifully exposed flood gravels in Northrup Canyon

scablands in granite

overflow channel between Glacial Lake Columbia and Grand Coulee

Grand Coulee Dam

hummocky landscape created by landslides

Seaton Canyon has good exposures of Wanapum Basalt

notches in coulee walls mark hanging valleys stranded by the deepening coulee

loess exposed on top of basalt

hillcrests are expressions of folded basalt beneath the loess

Castle Butte, basalt on top of granite

scablands from an early Ice Age flood

hinge of Coulee Monocline, a northern expression of the Yakima fold belt

lakes are remnants of Lake Bretz that once filled Lower Grand Coulee

Dry Falls marks top of Lower Grand Coulee

Red Alkali Lake in Monument Coulee

Grand Coulee

Banks Lake

Wilbur

Coulee City

Lake Roosevelt

174 155 172 17 2 21 174

Qls Kg Qg TMw Eg TMg Ql Qf

MIOCENE
TMw Wanapum Basalt of the Columbia River Basalt Group

TMg Grande Ronde Basalt of the Columbia River Basalt Group

EOCENE
Eg granitic intrusive rocks

MESOZOIC
CRETACEOUS
Kg granitic rock; mostly tonalite and granodiorite

CENOZOIC
QUATERNARY
Qls landslide deposits
Qf Ice Age flood deposits
Qg glacial deposits; mostly till
Ql loess; windblown dust and sand

—— fault
⊤— monocline
monocline fold

N
0   5   10   15   20 miles
0   10   20 kilometers

Geology along WA 155 between Coulee City and Grand Coulee.

of flat-lying Grande Ronde and overlying Wanapum Basalt. Now mostly eroded, the two exposures were once connected to form the hinge of the monocline. As the road follows a bench of flood gravels near milepost 3, you can see how the Grande Ronde Basalt flattens out and consists of continuous individual lava flows.

The Coulee Monocline played a critical role in the formation of Grand Coulee. After the Okanogan ice lobe advanced across the Columbia River near the site of today's dam, the floodwaters were forced to spill southward over a flat open landscape. At the monocline, the land abruptly dropped to a lower elevation, causing the floodwaters to accelerate, forming a wide cascade with

Inclined lava flows of the south side of the Coulee Monocline. Note that flows at the left side of the photo (north) are flat-lying.

immeasurable turbulence and erosive power where the water hit the base of the fold. In addition, the curved bedrock in the fold's hinge created a weak, fractured zone that was more easily eroded. With continued erosion, the flood-waters concentrated into a narrower stream, and the cascade turned into a waterfall. Subsequent floods caused the waterfall to retreat upstream through the flat-lying basalt, leaving the Grand Coulee in its wake.

Near milepost 7, the road crosses over some scablands eroded by the floods into the Grande Ronde Basalt. These scablands likely formed during one of the earlier floods when the margin of the ice lobe had advanced eastward almost across today's coulee, thereby restricting the flow to the far side. Across Banks Lake, you can see notches at the top of the coulee with steep gullies, or water-falls, below. These notches mark valleys that were left hanging by the abrupt deepening of the coulee during flooding. These hanging valleys exist on both rims but tend to be easier to see from a distance. On the east side, many are filled with loess that was not washed away in the floods.

A spectacular double roadcut of the Grande Ronde Basalt lies just south of milepost 8. A big pull-out just north of the roadcut offers the chance to stop. Northward, the road follows a long bar of outburst gravel for about the next 5 miles (8 km), where at milepost 13, you pass directly beneath an amphithe-ater-like hanging valley. Directly across Grand Coulee to the west you can see a basalt flow filling a channel cut into an older basalt flow.

The road to Steamboat Rock State Park meets the highway between mile-posts 15 and 16. Rising some 900 feet (270 m) in the middle of Grand Coulee, Steamboat Rock remains a relic of an early Ice Age flood. The flat bench at its top approximately coincides in elevation with the adjacent rims of Grand Coulee to indicate that they were once continuous. The bench also hosts some deposits of glacial till, much of which was washed away during the floods. The

The light-colored rocks at water level at the base of Steamboat Rock consist of Creta-ceous granodiorite, but the well-layered rock above is the Grande Ronde Basalt overlain by the Wanapum Basalt.

remaining till shows that the continental ice sheet covered the Steamboat Rock area before Grand Coulee was eroded. Near milepost 17, you can see outcrops of Cretaceous granodiorite beneath the basalt. As the waterfall retreated upstream from the monocline, it split where it reached the granodiorite, thus preserving Steamboat Rock.

If you turn east up Northrup Canyon near milepost 19, you can see a thick accumulation of gravel, deposited where floodwaters formed a large eddy in the granitic bedrock. The deposit lies less than 0.1 mile (160 m) up the road, but if you continue another quarter mile (400 m) you reach a beautiful area surrounded by numerous outcrops of granodiorite, quite unlike the typical basalt bedrock of the area.

Look for some pegmatite dikes with unusually large feldspar and quartz crystals within the granodiorite on the east side of the road just south of milepost 20. About 100 feet (30 m) north of the milepost, the road crosses a fault and passes into younger, lighter-colored granite of Eocene age. Castle Butte rises above the

Pegmatite dike in Cretaceous granodiorite.

road. Its base consists of both Eocene granite and Cretaceous granodiorite, and its flat top consists of the Grande Ronde and overlying Wanapum Basalt.

Between mileposts 21 and 22, you pass through scablands in the Eocene granite, which erodes into rounder, less jagged scablands than the basalt. Granite lacks the pervasive fracturing and flow surfaces of basalt that create angular shapes when they break. To the south, the view nicely shows how Castle Rock is capped by flat-lying basalt.

The double roadcut immediately north of Osborn Bay Lake reveals both granitic rocks: the Eocene granite lies on the west side and the Cretaceous granodiorite on the east. The Eocene granite contains fewer dark-colored minerals and more quartz than the darker Cretaceous granodiorite. In addition, the Cretaceous granodiorite displays numerous fracture surfaces coated with the fine-grained, dark-green mineral chlorite.

At the far northeast edge of Banks Lake, you pass between basalt ridges that each rise about 900 feet (270 m) above the road to about the same elevation

Grand Coulee Dam

To reach Grand Coulee Dam and the visitor center, continue northeast on WA 155 from the town of Grand Coulee. A couple large blocks of basalt in the first 0.1 mile (160 m) north of the intersection with WA 174 sit randomly on the north side of the road to highlight the prevalence of landslides in the area. A good exposure of Latah Formation, deposited in rivers and lakes between eruptions of the Columbia River Basalt Group, shows up in about a half mile (0.8 km) on the north side of the road. If you look southeastward, you can see a large landslide deposit decorated with houses. Shortly thereafter, you pass by a narrow exposure of the Grande Ronde Basalt and then cross into Eocene granite, which forms the prominent cliffs on the way to the visitor center and the dam's abutments. A large parking area and overlook just before (southwest of) the cliffs allows a clear view to the dam. The flat surface above and to the right of the far end of the dam consists of more deposits of Glacial Lake Columbia.

Grand Coulee Dam is immense. It stretches nearly 1 mile (1.6 km) across the Columbia River and contains more than 11.9 million cubic yards (9.1 million m³) of concrete—enough to pave a highway from Seattle to Miami, Florida! It rises 550 feet (168 m) above its base of Eocene granite and generates enough electricity for 2.3 million households. Initially constructed from 1933 to 1941, mainly to promote irrigation, the dam was converted to electricity generation at the beginning of World War II. As a result, the Columbia Basin Project was put on hold until the early 1950s. An additional powerhouse and increased electrical-generating capacity came on line in 1967. Lake Roosevelt, backed up behind the dam, has some 600 miles (970 km) of shoreline.

as the top of Steamboat Rock. This low area, now mostly urbanized, was the overflow channel between Glacial Lake Columbia and Grand Coulee. During the early Ice Age floods, this channel bottom didn't exist. Instead there was a divide 900 feet (270 m) higher that connected the ridges and separated the Grand Coulee from Glacial Lake Columbia, which was 900 feet (270 m) deeper than the modern Lake Roosevelt. The floods spilled over the divide to create an enormous waterfall that plunged into the growing Grand Coulee. Continued erosion of the waterfall, however, breached the divide and let loose an even larger flood that practically drained Glacial Lake Columbia.

At the intersection with WA 174 in the town of Grand Coulee, you can turn westward on WA 174 and drive for 2 miles (3.2) to the turnoff for Crown Point Vista; another 1.3 miles (2.1 km) leads to a spectacular viewpoint above Grand Coulee Dam. Along the way, you cross the canal that feeds water from the Columbia River to Banks Lake. You can also see granite bedrock and a spectacular array of power lines. Close to the turnoff, the road passes beneath cliffs of the Grande Ronde Basalt.

WA 174
GRAND COULEE—WILBUR
20 miles (32 km)
See map on page 329.

For the first few miles east of Grand Coulee, WA 174 travels eastward over landslide deposits marked by irregular hummocky surfaces and occasional large blocks of basalt that fell from the cliffs above. This area is especially prone to sliding because lakebeds of the Latah Formation lie beneath the Grande Ronde Basalt in some places. Because it's so fine grained, the Latah Formation tends to trap water at the base of the basalt. The elevated fluid pressures and slippery base promote sliding.

Just above milepost 23, you see some entablature and fanning columns in the Grande Ronde Basalt followed by exposures of Eocene granite along the edge of Lake Roosevelt. The granite provides the foundation for the abutments of Grand Coulee Dam. From mileposts 24 to 26, you get good views of the Grande Ronde Basalt resting on top the granite on the north shore of the lake.

Seaton Canyon, between mileposts 27 and 30, is cut into the Grande Ronde Basalt and overlying Wanapum Basalt, although the Grande Ronde is mostly covered by grass in the canyon. A pull-out on the north side of the highway about halfway between mileposts 28 and 29 marks the base of the well-exposed Wanapum Basalt. Look for pillow basalt across from a large pull-out about a quarter mile (400 m) east of milepost 29. At a pull-out a quarter mile (400 m) above there loess rests on top of the basalt. By the time you reach milepost 30, the road has leveled off onto the loess-covered plateau. The flatness is only temporary, however, as the road passes over several folds in the basalt between the top of Seaton Canyon and Wilbur. Look for some well-exposed loess in

View looking north across Lake Roosevelt to the skyline of flat-lying Grande Ronde Basalt resting on top of Eocene granite. The granite had been uplifted and exposed at Earth's surface by the time the basalt erupted.

hillsides near the crests of some of these folds, such as at milepost 32, where their high elevations allowed the loess to escape the Ice Age floods.

WA 821 (YAKIMA RIVER CANYON)
ELLENSBURG—EAST SELAH
28 miles (45 km)

WA 821, a scenic route between Ellensburg and Yakima, is about the same distance as I-82 but takes a little longer because it follows the tightly meandering Yakima River through its deep canyon cut into the Grande Ronde and Wanapum Basalts. The canyon cuts through the Manastash and Umtanum Ridges, as well as the Yakima Ridge farther south, all of which are anticlines of the Yakima fold belt. Reaching depths of nearly 2,000 feet (610 m) in places, the canyon poses an enigmatic question: Why does the river cut through the high basaltic ridges rather than flow around them? Because meandering rivers generally exhibit lower erosive power than braided ones, it seems even stranger

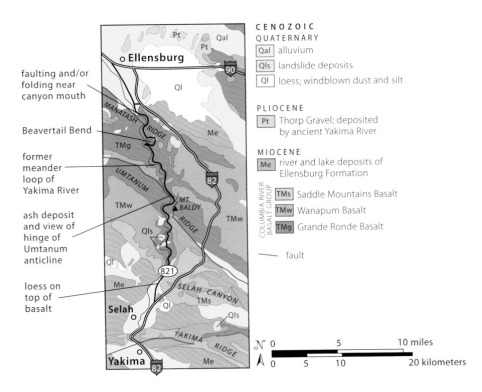

CENOZOIC
QUATERNARY
Qal	alluvium
Qls	landslide deposits
Ql	loess; windblown dust and silt

PLIOCENE
| Pt | Thorp Gravel; deposited by ancient Yakima River |

MIOCENE
| Me | river and lake deposits of Ellensburg Formation |

COLUMBIA RIVER BASALT GROUP
TMs	Saddle Mountains Basalt
TMw	Wanapum Basalt
TMg	Grande Ronde Basalt

— fault

faulting and/or folding near canyon mouth

Beavertail Bend

former meander loop of Yakima River

ash deposit and view of hinge of Umtanum anticline

loess on top of basalt

0   5   10 miles
0   5   10   20 kilometers

Geology along WA 821 between Ellensburg and East Selah.

that the river would meander the way that it does. The reason is because the Yakima River meandered over a flat landscape before the ridges started to rise, between 15 and 10 million years ago. Then, as the flat landscape was folded into ridges, the river cut downward, eventually forming a canyon.

To reach the canyon, follow WA 821 south approximately 4 miles (6.4 km) from its intersection with I-90, or from I-82 take Thrall Road (exit 3) west a half mile (0.8 km). For those heading north on I-82, take exit 26 for Selah and turn left. For those traveling downstream (south) through the canyon, milepost 24 marks the first bedrock exposure—rubbly Grande Ronde Basalt, which lies beneath steeply south-dipping Ellensburg Formation. Above the Ellensburg are fragmented pillow lavas of the Wanapum Basalt. Just 0.1 mile (160 m) down the road you see north-dipping Ellensburg atop pillow basalt. The opposing dips may be the result of folding, although it's likely more complicated than that because the position of the rock types does not match up across the fold.

The road curves sharply around the inside bend of the river at milepost 22, passing over a comparatively low-relief surface covered in alluvium. Because rivers flow more slowly around their inside bends, they typically deposit alluvium in those locations. By contrast, the opposing outer bends of rivers, called cutbanks, tend to be sites of erosion. The road passes many examples of this phenomenon throughout the canyon. A half mile (0.8 km) south of milepost 21, the road passes into a loop called Beavertail Bend, where the river meanders far enough back on itself that two cutbanks lie on opposite sides of the same narrow ridge. Through time, the river will likely erode through the ridge and cut off the large outer loop.

On either side of milepost 20 within Beavertail Bend, you can see a number of deposits left by debris flows that poured down narrow channels from the cliffs above. The gravel deposits extend as lobes into the river. The debris flows formed July 3, 1998, during a brief but unusually heavy rainstorm. Elsewhere throughout the canyon, notice the large gravel aprons blanketing the steep slopes above the road. These deposits consist of material eroded from the cliffs in a more piecemeal fashion, such as during occasional rockfalls and small avalanches. During heavy rains, sections of these gravel aprons can become unstable and mobilize as debris flows.

Between mileposts 16 and 15, the road passes a large amphitheater-like area, with arcuate cliffs set back as much as a half mile (0.8 km) from the road. It formed as a meander loop that has since been abandoned by the river. The river gravels along the highway near each edge of the amphitheater attest to the river's former position. Just downstream, the road rises past cliffs of the Grande Ronde Basalt to milepost 14. Cliffs frame the outer, erosional bend of the river, whereas alluvium and a large debris flow deposit cover the inside bend.

Incised meander bend along the Yakima River across from milepost 14. The bar on the inside consists of river gravels and debris flow deposits that flowed down the valley near the center of the photo.

From the heritage marker above the Lmuma Creek Campground, halfway between mileposts 13 and 12, you can see a prominent white ash layer on the other side of the river, as well as the anticline that defines Umtanum Ridge to the south. The ash has long been considered a deposit of the Mt. Mazama eruption in Oregon 7,700 years ago. However, an independent confirmation of the ash's age has so far been unsuccessful. The ash is interlayered with river deposits, the top of which forms a low bench—a river terrace that marks a former position of the floodplain. Look for the ash exposed along the east side of the road a quarter mile (400 m) farther south.

Looking southeastward from the heritage marker, you can see Mt. Baldy, nearly 2,000 feet (610 m) above the river level. Its top sits directly on the hinge of the Umtanum anticline: basalt flows on its southwest (right) side dip gently to the southwest, while those on the northeast side dip steeply northeastward and are even overturned near their base. The Umtanum anticline, part of the Yakima fold belt, continues for more than 20 miles (32 km) to the northwest, while to the southeast it continues across I-82 all the way to Hanford, more than 30 miles (48 km) away. As you pass beneath Baldy's cliffs, you can see the lava flows dipping to the southwest, although you can get a clearer, more distant view from milepost 10.

Looking southward to Mt. Baldy. The hinge of the overturned anticline lies east of the powerline tower on the horizon. The white ash bed on the left may be derived from the Mt. Mazama eruption in Oregon.

View southward down the lower stretches of Yakima River Canyon. High cliffs consist of Wanapum Basalt. The steep slopes below and continuing to river level are Grande Ronde Basalt covered in places by rockfall and landslide deposits.

South of milepost 6, the canyon becomes impressively deep and narrow, with bedrock exposures of the Grande Ronde Basalt interrupted by landslide deposits. The road crosses an especially large landslide at milepost 4. Across the river are cliffs of south-dipping Grande Ronde Basalt and overlying Wanapum Basalt. The Wanapum Basalt slopes down to the river and marks the canyon mouth at milepost 3.

For the remaining 3 miles (4.8 km) to I-82, the road passes over a much more subdued landscape. You gain a view northward up Selah Creek canyon and the interstate bridge and then climb along a cliff of the Saddle Mountains Basalt, with some beautiful colonnade within the first mile. Look for a nice exposure of loess overlying the basalt as the road bends near the top of the hill at milepost 2. On clear days, you can see Mt. Rainier and Mt. Adams to the west and southwest, respectively, as you pass by the fruit orchards. As you approach I-82, you can see the narrow gap where the Yakima River cuts across Yakima Ridge, another anticline of the Yakima fold belt.

Glossary

accreted. Something that has been added onto, such as an accreted terrane, which has been added to the North American continent through plate motions.

alluvial fan. A gently sloping, fan-shaped accumulation of sediments deposited by a stream where it flows out of a narrow valley onto a wider, flatter area.

alluvium. Water-transported sedimentary material.

amphibolite. A type of rock formed by regional metamorphism of basalt or gabbro under medium to high pressures and temperatures. It consists largely of the minerals amphibole and plagioclase.

andesite. A medium- to dark-colored volcanic rock that is between basalt and rhyolite in silica content. It was named after the Andes Mountains in South America, where it is common.

anticline. A fold with the oldest rock in the core; most anticlines have limbs that dip away from the core.

ash. Tiny particles of volcanic glass blown into the air during volcanic eruptions.

ash-flow tuff. The rock formed from the consolidation and compaction of an ash-flow deposit.

asthenosphere. The zone of somewhat malleable rock beneath the lithosphere. It is the zone over which the lithospheric plates move.

basalt. A dark-colored volcanic rock that contains less than 52 percent silica.

basement. The deepest crustal rocks of a given area. They are typically igneous or metamorphic rock, but some accreted terranes of Washington are sedimentary in places.

bedding. The layering as seen in a sedimentary rock. A single layer is called a bed. When different rock types are interlayered with each other, they are described as **interbedded**.

bedrock. Rock that remains in its place of origin and has not been moved by erosional processes.

biotite. Dark mica, a platy mineral. It is a minor but common mineral in igneous and metamorphic rocks.

blueschist. A metamorphic rock with the blue-colored minerals glaucophane and lawsonite, which form from basaltic rocks under high pressure–low temperature conditions. Blueschists most commonly form in subduction zones.

breccia. A rock consisting of angular fragments.

caldera. A steep-walled, subcircular depression in a volcano, at least 1 mile (1.6 km) across, that formed by collapsing into an emptied or partially emptied magma chamber below.

carbonates. Rocks, such as limestone or dolomite, that are formed by the combination of atoms of calcium or magnesium with carbon and oxygen.

chert. An extremely fine-grained sedimentary rock made of silica.

cinder. A volcanic rock, typically of basaltic composition, that contains enough air bubbles as to be noticeably less dense than one without air bubbles.

cinder cone. A steep-sided, cone-shaped accumulation of cinders that surrounds a basaltic vent.

cirque. A bowl-shaped basin on a mountain, usually where the head of a glacier once existed.

clast. A grain or fragment of a rock. Clastic rock is sedimentary rock composed of broken fragments, such as sand grains, derived from preexisting rocks.

clay. A sedimentary particle with a grain size less than 0.004 millimeter in diameter.

cleavage. The tendency of a rock (or mineral grain) to split, or cleave, in a preferred orientation.

coal. An organic-rich, dark-colored to black rock that burns. Coal forms from the compaction and long-term, low-temperature heating of plant material. Coal typically is founds in layers called seams.

coarse grained. A term used to describe a rock with large particles or crystals about 1 millimeter in diameter or larger and typically visible to the naked eye.

coastal plain. A low, gently sloping region on the margin of an ocean.

colonnade. The part of a basaltic lava flow that shows near-parallel, typically steep, fractures that break the rock into columns.

columnar jointing. The fracturing in a lava flow that causes the flow to break into columns. This is the prevalent style of fracturing in the colonnade.

concretion. A subspherical to irregularly shaped body of sedimentary rock that is more resistant to weathering than the rest of the rock. This resistance usually occurs because it is unusually well cemented.

conglomerate. A sedimentary rock composed of particles that exceed 2 millimeters in diameter.

continental shelf. The gently inclined part of the continental landmass between the shoreline and the more steeply inclined continental slope.

crossbedding. Layering in a sedimentary rock that forms at an angle to horizontal.

crust. The uppermost layer of Earth. Continental crust consists mainly of an igneous and/or metamorphic basement overlain by sedimentary and volcanic rock. Oceanic crust consists of basalt and gabbro.

dacite. A volcanic rock that is intermediate in silica content between andesite and rhyolite.

debris avalanche. A chaotic mixture of differently sized material that flows rapidly downward in response to gravity.

delta. A nearly flat accumulation of clay, sand, and gravel deposited in a lake or ocean at the mouth of a river.

diatomite. A sedimentary rock that consists mostly of diatoms, single-celled freshwater or marine algae composed of silica.

dike. A tabular intrusive body that cuts across layering in the host rock.

diorite. An intrusive igneous rock that is between gabbro and granite in silica content. It is the intrusive equivalent of andesite.

drainage basin. The land area drained by a stream and all of its tributaries.

drainage divide. A line of high elevation, typically a ridge, that separates one drainage basin from another.

drumlin. An elongate hill made of glacial till that was streamlined by glacial ice.

entablature. The part of a basaltic lava flow that shows numerous closely spaced, typically steep fractures that break the rock into thin, irregular columns.

erosion. Movement or transport of weathered material by water, ice, wind, or gravity.

erratic. A block of rock transported by glacial ice and deposited at some distance from the bedrock outcrop it came from.

fault. A fracture or zone of fractures in Earth's crust along which blocks of rock on either side have shifted.

fault scarp. An abrupt cliff or steep section in an otherwise continuous landscape caused by offset along a fault.

feeder dike. A dike of igneous rock that once fed a lava flow.

feldspar. The most abundant rock-forming mineral group. Makes up 60 percent of Earth's crust and contains calcium, sodium, or potassium with aluminum silicate. Includes plagioclase feldspars and alkali feldspars.

felsic. Said of an igneous rock with abundant light-colored minerals, such as quartz and feldspar.

fine grained. A term used to describe a rock with small particles or crystals less than about 1 millimeter in diameter and typically not visible to the naked eye.

fissure. An open crack.

floodplain. The portion of a river valley adjacent to the river that is built of sediments deposited when the river overflows its banks during flooding.

foliation. Layering in rock caused by metamorphism or deformation.

footwall. The block of rock that lies beneath a fault zone.

fore-arc basin. A sedimentary basin located between a magmatic arc and a subduction zone.

formation. A body of sedimentary, igneous, or metamorphic rock that can be recognized over a large area. It is the basic stratigraphic unit in geologic mapping. A formation may be part of a larger group and may be broken into members.

fossils. Remains, imprints, or traces of plants or animals preserved in rock.

fragmental basalt. Basaltic rock that was broken into myriad small fragments, typically during explosive interactions with water. Also referred to as brecciated basalt or basaltic breccia.

gabbro. Dark-colored intrusive igneous rock that is less than 52 percent silica. When magma of the same composition erupts at the surface, it forms basalt.

glacial. A term pertaining to a glacier.

glacial outwash. Sediment deposited by the large quantities of meltwater emerging from the terminus of a glacier.

glacial till. An unsorted mixture of silt, sand, and gravel left by a melting glacier.

glacier. A large and long-lasting mass of ice on land that flows downhill in response to gravity.

gneiss. A metamorphic rock that has a banded appearance caused by the segregation of minerals at high temperatures.

graben. A crustal block that is down-dropped between two inwardly dipping normal faults.

graded beds. Sedimentary layers in which the grain size is coarsest at the bottom and becomes finer toward the top.

granite. A light-colored, coarse-grained igneous rock with a silica content that exceeds 66 percent. It is the intrusive equivalent of rhyolite.

granitic. A term pertaining to an igneous rock that resembles and approximates the chemical composition of a granite.

granodiorite. An intrusive igneous rock with a silica content between diorite and granite.

graywacke. A type of sandstone in which a large proportion of the grains are sand-sized rock particles.

greenstone. Volcanic rocks, typically basalt, that were metamorphosed and developed green metamorphic minerals, such as chlorite and epidote, as a result.

groundwater. Subsurface water contained in fractures and pores of rock and soil.

group. Two or more formations that occur together. Two or more groups can be lumped together into a **supergroup.**

hanging wall. The block of rock that lies above a fault zone.

hinge. The point of maximum curvature on a fold. It separates the two limbs.

horst. A crustal block that is uplifted between two outwardly dipping normal faults.

igneous rock. Rock that solidified from the cooling of molten magma.

impermeable. Having a texture that does not permit water to move through. Clay is often considered a relatively impermeable sediment.

inclusion. A general term for either rocks of different composition within larger bodies of igneous rock or small crystals of a different mineral inside larger crystals.

interglacial. The period of time between glacial advances.

intrusive igneous rocks. Rocks that cool from magma beneath the surface of Earth. The body of rock is called an **intrusion.**

island arc. An offshore volcanic arc or linear chain of volcanoes formed along a convergent plate margin.

lahar. A volcanic mudflow deposit.

lava. Molten rock erupted on the surface of Earth.

limb. The flank on either side of a fold.

limestone. A sedimentary rock composed of calcium carbonate formed by precipitation in warm water, usually aided by biological activity.

lithosphere. The outer rigid shell of Earth that is broken into the tectonic plates. On average, continental lithosphere is about 100 miles (160 km) thick and old oceanic lithosphere is about 60 miles (100 km) thick.

loess. Silt and dust picked up from glacial streambeds and redeposited by the wind.

mafic. Said of an igneous rock with the approximate silica content of basalt; typically rich in dark-colored, iron- and magnesium-rich minerals.

magma. Molten rock within Earth.

magmatic arc. The zone of magma production that stretches in an arc-like pattern above and parallel to a subduction zone.

mantle. The part of Earth between the interior core and the outer crust.

marble. Metamorphosed limestone.

marine. Pertaining to the sea.

mélange. A mixture of rocks that may not have formed together. Mélanges can form at a variety of scales and most frequently indicate subduction zone settings.

metamorphic core complex. A mountain-scale complex of metamorphic rocks that is separated from overlying less- or nonmetamorphosed rocks by a low-angle normal fault. Most metamorphic core complexes are domal in shape and contain mylonites indicative of fault movements at high temperatures.

metamorphic rock. Rock derived from preexisting rock that has changed mineralogically or texturally, or both, in response to changes in temperature and/or pressure, usually deep within Earth. The prefix *meta*, such as in the terms **metasedimentary** and **metavolcanic**, indicates that the original rock has been changed by heat and pressure.

metamorphism. Recrystallization of an existing rock. Metamorphism typically occurs at high temperatures and often high pressures. The grade of metamorphism refers to the temperature affecting the original rock. **Low-grade** and **high-grade** are qualitative terms indicating relatively low and high temperatures of metamorphism, respectively.

mica. A family of silicate minerals, including biotite (black) and muscovite (white), that breaks easily into thin flakes. Micas are common in many kinds of igneous and metamorphic rocks.

migmatite. A high-temperature metamorphic rock that contains a large fraction of igneous material. In some cases, the igneous material originated from partial melting during metamorphism.

monocline. A type of fold in which two nearly horizontal limbs at different elevations are separated by a steeply dipping limb.

moraine. A mound or ridge of an unsorted mixture of silt, sand, and gravel (glacial till) left by a melting glacier.

mountain building. An event in which mountains rise. During these events, rocks are typically folded, faulted, and/or metamorphosed. Intrusive and extrusive igneous activity often accompanies these events.

mouth. The end of a stream or river where it discharges into a larger body of water, such as a larger river, a bay, or the ocean.

mudflow. A mixture of water, mud, and assorted particles that flows downhill in response to gravity.

mudstone. A sedimentary rock composed of mud.

mylonite. A metamorphic rock characterized by strong foliations, lineations, and fine grain sizes that formed through high-temperature deformation within a shear zone.

normal fault. A fault in which rocks on the hanging wall side move down relative to rocks on the footwall side and results in extension.

obsidian. Volcanic glass, typically high in silica and dark gray to black. Impurities may give rise to brown or red colors in obsidian.

olivine. An iron and magnesium silicate mineral that typically forms glassy green crystals. A common mineral in gabbro, basalt, and peridotite.

ophiolite. The sequence of rocks that makes up the oceanic lithosphere.

orthogneiss. The metamorphic rock gneiss, formed from a rock that was originally igneous.

outwash. Sediment deposited by the large quantities of meltwater emerging from the terminus of a glacier.

paleosol. A fossil soil.

paragneiss. The metamorphic rock gneiss, formed from a rock that was originally sedimentary.

peat. A deposit of semicoalified plant remains in a bog.

pebble. A rounded rock particle 0.16 to 2.5 inches (4–64 mm) in diameter.

pegmatite. An unusually coarse-grained igneous rock, usually of a granitic composition.

peridotite. Low-silica igneous rock that makes up Earth's mantle.

phyllite. A metamorphic rock that has been heated more than slate but not as much as schist; typically starts as a fine-grained sedimentary rock.

pillow basalt. Basalt that takes on a bulbous pillow shape due to the lava's interaction with water, either from erupting underwater or flowing into it.

plagioclase. A feldspar mineral rich in sodium and calcium. One of the most common rock-forming minerals in igneous and metamorphic rocks.

plagiogranite. An igneous rock that resembles granite except that nearly all of its feldspars are plagioclase.

plate tectonics. The theory that Earth's lithosphere is broken into large fragments, or plates, that move slowly over the somewhat malleable asthenosphere, with intense geological activity at plate boundaries.

Pleistocene. The last 2 million years of geologic time, during which periods of extensive continental glaciation alternated with warmer interglacial periods of glacial retreat.

pluton. A body of intrusive igneous rock. A **stitching pluton** intrudes across the boundaries of two or more terranes.

pumice. A pyroclastic rock that consists of volcanic glass with a frothy texture because of an abundance of air holes.

pyroclastic. A term used to describe fragmental volcanic particles broken during explosive eruptions.

pyroxene. An iron-and-magnesium-bearing silicate mineral abundant in basaltic rock.

quartz. A mineral composed entirely of silica; one of the most common rock-forming minerals.

quartz monzonite. A granitic rock that differs from a true granite in that it contains less than 20 percent but more than 5 percent quartz.

radiometric dating. The calculation of age based on the rate of time it takes for radioactive elements to decay.

reverse fault. A fault in which one side is pushed up and over the other side. It forms under compressional stresses.

rhyolite. A typically light-colored volcanic rock with more than 66 percent silica. It is the volcanic equivalent of granite.

ribbon chert. A hardened sedimentary rock, composed of microcrystalline silica, that features a sequence of relatively thin, ribbon-like layers.

rift zone. A strip-like area characterized by crustal extension and normal faulting. Rift zones may evolve into new ocean basins if the region stretches and thins enough to tap magma from the underlying asthenosphere.

sand. Weathered mineral grains, most commonly quartz, between 0.06 and 2 millimeters in diameter.

sandstone. A sedimentary rock made primarily of sand.

schist. A metamorphic rock that is strongly layered due to an abundance of visible, platy minerals.

sea stack. A tall outcropping of bedrock on a beach or offshore, left as a remnant of the coastline as erosion causes the coastline to retreat.

sedimentary rock. A rock formed from the compaction and cementation of sediment.

serpentinite. A rock made of minerals of the serpentine group that formed by low-grade metamorphism of iron- and magnesium-rich rocks, usually of the oceanic lithosphere.

shale. A thinly layered rock made of sedimentary particles less than 0.004 millimeter in diameter.

shearing. The action and deformation caused by two bodies of rock sliding past each other. The sheared part of a rock is called a **shear zone**.

shield volcano. A gently sloped volcano typically made of basalt. In profile, shield volcanoes resemble shields.

silica. The compound silicon dioxide. The most common mineral made entirely of silica is quartz.

sill. An igneous intrusion that parallels the planar structure or bedding of the host rock.

silt. Sedimentary particles larger than clay but smaller than sand (between 0.004 and 0.06 millimeter in diameter).

siltstone. A sedimentary rock made primarily of silt.

slate. Slightly metamorphosed shale or mudstone that breaks easily along parallel surfaces.

spit. A long, narrow, fingerlike ridge of sand extending into the water from the shore.

sorting. A term used to describe the uniformity of grain size in a sedimentary rock. A well-sorted rock has a uniform grain size.

spreading ridge. The location on a divergent margin where plates spread apart and new oceanic crust is formed in the opening rift.

steptoe. A hill or butte that is made of rock much older than the surrounding low-elevation rock, even though no folds or fault zones explain the relation. Steptoes are remnants of ancient topography.

stratification/stratified. Sequentially layered.

stratovolcano. A steep-sided volcano, typically made of andesite.

striations. Scratches on rock that result from friction of rocks moving on a fault plane or by abrasion of bedrock beneath a moving glacier.

strike-slip fault. A fault showing sideways movement or offset of the adjacent rock.

subduction zone. A long, narrow zone where an oceanic plate descends beneath another plate at a convergent boundary.

suspended sediment. Sediment in a stream that remains lifted or transported due to the turbulence and energy of the water. When the currents slow, suspended sediment falls to the bottom.

syncline. A fold with the youngest rock in the core; most synclines have limbs that dip toward the core.

talus. An accumulation of rock fragments resting at the base of the cliff it's derived from.

tectonics. The study of regional-scale deformation of Earth's crust.

tephra. A general term for the material ejected from a volcano.

terrace. An erosional remnant of a former floodplain or coastline standing above the present river or coast.

terrane. A fault-bounded crustal fragment with a geologic history that differs from adjacent fragments.

thrust fault. A fault dipping less than 45 degrees, in which the rock above the fault has moved upward and over the rock below the fault. Thrust faults typically form by horizontal compression. The hanging wall above a thrust fault is termed a **thrust sheet,** where it is relatively thin relative to its areal extent.

till. An unlayered and unsorted mixture of clay, silt, sand, gravel, and boulders deposited directly by a glacier.

tonalite. A granitic rock, common to the North Cascades, composed mostly of plagioclase feldspar and quartz.

transform fault. A special type of strike-slip fault along a midocean spreading ridge or plate boundary.

tuff. A volcanic rock made mostly of consolidated pyroclastic material, chiefly ash and pumice, derived from ash falls or pyroclastic flows. A **welded tuff** is distinctly harder because heat of its particles caused them to weld together.

turbidite. Sands and muds that settle on the seafloor from clouds of sandy, muddy water that flow as submarine density currents. They form alternating layers of sandstone and shale. Some of the sandstone layers grade upward from coarse to fine particles.

ultramafic. An igneous rock with a silica content below 50 percent. Ultramafic rocks typically contain a great deal of iron and magnesium.

unconformity. A depositional contact across which the rock record is missing. As a result, the rocks beneath the unconformity are typically significantly older than the ones above it.

vein. A deposit of minerals that fills a fracture in rock.

vent. The actual place where volcanic materials erupt. Vents are either eruptive localities on large volcanoes or mark much smaller volcanoes.

volcanic arc. A chain of volcanoes that formed above an ocean-floor subduction zone.

volcaniclastic. A sedimentary or volcanic rock made principally of volcanic particles that were transported and deposited after their initial formation.

weathering. The physical disintegration and chemical decomposition of rock at Earth's surface.

window. A hole eroded in the hanging wall of a thrust fault that exposes the footwall beneath the fault.

zeolite. A family of aluminum-silicate minerals, containing sodium, potassium, or calcium, that commonly form where volcanic rocks or ash interact with groundwater or volcanic fluids.

FURTHER READING AND REFERENCES

Washington contains such a wealth of spectacular geology that a book like this can only scratch its surface. However, there is a trove of detailed literature for those who want to dig deeper. The following books, articles, and maps reflect only a sampling of this literature, but they are the ones we found most helpful in our research. The technical readings are organized according to chapter, but we also included the special topic Ice Age floods because it did not fit into a single chapter category. A list of useful geologic maps appears at the end.

Unusually Useful Websites

Washington State Department of Natural Resources
Geologic Information Portal
http://www.dnr.wa.gov/geologyportal

Northwest Geology Field Trips (blog by Dave Tucker)
https://nwgeology.wordpress.com/

Geology Lecture Series (delivered by Nick Zentner)
http://www.nickzentner.com/

Ice Age Floods Institute
http://iafi.org/

Nontechnical Reading

Allen, J. E., Burns, M., and S. Burns. 2009. *Cataclysms on the Columbia: The Great Missoula Floods*, 2nd ed. Ooligan Press.

Atwater, B. F., Satoko, M., Kenji, S., Yoshinobu, T., Kazue, U., and D. K. Yamaguchi. 2005. *The Orphan Tsunami of 1700: Japanese Clues to a Parent Earthquake in North America*. USGS Professional Paper 1707.

Bishop, E. M. 2014. *Living with Thunder: Exploring the Geologic Past, Present, and Future of the Pacific Northwest*. Oregon State University Press.

Bjornstad, B. 2006. *On the Trail of the Ice Age Floods: A Geological Guide to the Mid-Columbia Basin*. Keokee Books.

Bjornstad, B., and E. Kiver. 2006. *On the Trail of the Ice Age Floods—Northern Reaches: A Geological Guide to Northern Idaho and the Channeled Scabland*. Keokee Books.

Brown, N. 2014. *Geology of the San Juan Islands*. Chuckanut Editions.

Carson, B., and S. Babcock. 2000. *Hiking Guide to Washington Geology*. Keokee Books.

Jennings, A., and T. Jennings. 2003. *Estuary Management in the Pacific Northwest*. Oregon State University.

Kiver, E., Pritchard, C., and R. Orndorff. 2016. *Washington Rocks! A Guide to Geologic Sites in the Evergreen State.* Mountain Press.

Miller, M. B. 2014. *Roadside Geology of Oregon.* Mountain Press.

Norman, D. K., and J. M. Roloff. 2004. *A Self-Guided Tour of the Geology of the Columbia River Gorge—Portland Airport to Skamania Lodge, Stevenson, Washington.* Washington Division of Geology and Earth Resources Open File Report 2004-7.

Orr, W. N., and E. L. Orr. 2002. *Geology of the Pacific Northwest,* 2nd ed. Waveland Press.

Pringle, P. 2002. *Roadside Geology of Mount St. Helens National Volcanic Monument and Vicinity,* rev. ed. Washington Department of Natural Resources, Division of Geology and Earth Resources Information Circular 88.

Tabor, R. W. 1975. *Guide to the Geology of Olympic National Park.* University of Washington Press.

Tabor, R. W., and R. A. Haugerud. 1999. *Geology of the North Cascades.* The Mountaineers.

Tucker, D. 2015. *Geology Underfoot in Western Washington.* Mountain Press.

Technical Reading

Plate Tectonics and Regional Geology

Burchfiel, B. C., Cowan, D. S., and G. A. Davis. 1992. Tectonic overview of the Cordilleran orogeny in the western United States. In *The Cordilleran Orogeny, Conterminous US,* The Geology of North America Volume G-3, GSA, eds. B. C. Burchfiel, P. W. Lipman, and M. L. Zoback, p. 407–79.

Christiansen, R. L., and R. S. Yeats, with contributions by Graham, S. A., Niem, W. A., Niem, A. R., and P. D. Snavely. 1992. Post-Laramide geology of the US Cordilleran region. In *The Cordilleran Orogeny, Conterminous US,* The Geology of North America Volume G-3, GSA, eds. B. C. Burchfiel, P. W. Lipman, and M. L. Zoback, p. 261–406.

Goodge, J. W., Vervoort, J. D., Fanning, C. M., Brecke, D. M., Farmer, G. L., Williams, I. S., Myrow, P. M., and D. J. DePaolo. 2008. A positive test of East Antarctica–Laurentia juxtaposition within the Rodinia supercontinent. *Science* 321 (5886): 235–40.

Irving, E., Woodsworth, G. J., Wynne, P. J., and A. Morrison. 1985. Paleomagnetic evidence for displacement from the south of the Coast Plutonic Complex, British Columbia. *Canadian Journal of Earth Sciences* 22: 584–98.

Umhoefer, P. J. 1987. Northward translation of "Baja British Columbia" along the Late Cretaceous to Paleocene margin of western North America. *Tectonics* 6: 377–94.

Wells, R. E., Engebretson, D. C., Snavely, P. D., Jr., and R. S. Coe. 1984. Cenozoic plate motions and the volcano-tectonic evolution of western Oregon and Washington. *Tectonics* 3: 275–94.

Wells, R. E., and P. L. Heller. 1988. The relative contribution of accretion, shear, and extension to Cenozoic tectonic rotations in the Pacific Northwest. *GSA Bulletin* 100: 324–38.

Ice Age Floods

Atwater, B. F. 1987. Status of Glacial Lake Columbia during the last floods from Glacial Lake Missoula. *Quaternary Research* 27: 182–201.

Baker, V. R., Bjornstad, B. N., Gaylord, D. R., et al. 2016. Pleistocene megaflood landscapes of the Channeled Scabland. In *Exploring the Geology of the Inland Northwest*, GSA Field Guide 41, eds. R. S. Lewis and K. L. Schmidt, p. 1–73.

Benito, G., and J. E. O'Connor. 2003. Number and size of last-glacial Missoula floods in the Columbia River valley between the Pasco Basin, Washington, and Portland, Oregon. *GSA Bulletin* 115 (5): 624–38.

O'Connor, J. E., and S. F. Burns. 2009. Cataclysms and controversy: Aspects of the geomorphology of the Columbia River Gorge. In *Volcanoes to Vineyards: Geologic Field Trips Through the Dynamic Landscape of the Pacific Northwest*, GSA Field Guide 15, eds. J. E. O'Connor, R. J. Dorsey, and I. P. Madin, p. 237–51.

O'Connor, J. E., and R. B. Waitt. 1995. Beyond the Channeled Scabland: A field trip to the Missoula flood features in the Columbia, Yakima, and Walla Walla Valleys of Washington and Oregon—Part 2: Field trip, day one. *Oregon Geology* 57 (4): 75–86.

Pritchard, C. J., and L. Cebula. 2016. Geologic and anthropogenic history of the Palouse Falls area: Floods, fractures, clastic dikes, and the receding falls. In *Exploring the Geology of the Inland Northwest*, GSA Field Guide 41, eds. R. S. Lewis and K. L. Schmidt, p. 75–92.

Waitt, R. B. 1985. Case for periodic, colossal jökulhlaups from Pleistocene Glacial Lake Missoula. *GSA Bulletin* 96: 1271–86.

Waitt, R. B. 2016. Megafloods and Clovis cache at Wenatchee, Washington. *Quaternary Research* 85: 430–44.

Waitt, R. B., Breckenridge, R. M., Kiver, E. P., and D. F. Stradling. 2016. Late Wisconsin Cordilleran Ice Sheet and colossal floods in northeast Washington and north Idaho. In *The Geology of Washington and Beyond*, University of Washington Press, ed. E. S. Cheney, p. 233–56.

Waitt, R. B., Denlinger, R. P., and J. E. O'Connor. 2009. Many monstrous Missoula floods down Channeled Scabland and Columbia Valley. In *Volcanoes to Vineyards: Geologic Field Trips through the Dynamic Landscape of the Pacific Northwest*, GSA Field Guide 15, eds. J. E. O'Connor, R. J. Dorsey, and I. P. Madin, p. 775–844.

Coast Range

Armentrout, J. M. 1987. Cenozoic stratigraphy, unconformity-bounded sequences, and tectonic history of southwestern Washington. In *Selected Papers on the Geology of Washington*, Washington Division of Geology and Earth Resources Bulletin 77, ed. J. E. Schuster, p. 291–320.

Beeson, M. H., Perttu, R., and J. Perttu. 1979. The Origin of the Miocene basalts of coastal Oregon and Washington: An alternative hypothesis. *Oregon Geology* 41 (10): 159–65.

Brandon, M. T., Roden-Tice, M. K., and J. I. Garver. 1998. Late Cenozoic exhumation of the Cascadia accretionary wedge in the Olympic Mountains, northwest Washington State. *GSA Bulletin* 110: 985–1009.

Chan, C. F., Tepper, J. H., and B. K. Nelson. 2012. Petrology of the Grays River Volcanics, southwest Washington: Plume-influenced slab window magmatism in the Cascadia forearc. *GSA Bulletin* 124: 1324–38.

Eddy, M. P., Bowring, S. A., Umhoefer, P. J., Miller, R. B., McLean, N. M., and E. E. Donaghy. 2016. High-resolution temporal and stratigraphic record of Siletzia's accretion and triple junction migration from nonmarine sedimentary basins in central and western Washington. *GSA Bulletin* 128: 425–41.

Evarts, R. C., Conrey, R. M., Fleck, R. J., and J. T. Hagstrum. 2009. The Boring volcanic field of the Portland-Vancouver area, Oregon and Washington: Tectonically anomalous forearc volcanism in an urban setting. In *Volcanoes to Vineyards: Geologic Field Trips through the Dynamic Landscape of the Pacific Northwest*, GSA Field Guide 15, eds. J. E. O'Connor, R. J. Dorsey, and I. P. Madin, p. 253–70.

Rau, W. W. 1973. *Geology of the Washington Coast between Point Grenville and the Hoh River*. Washington Department of Natural Resources, Geology and Earth Resources Division Bulletin 66.

Rau, W. W. 1980. *Washington Coastal Geology between the Hoh and Quillayute Rivers*. Washington Department of Natural Resources, Geology and Earth Resources Division Bulletin 72.

Snavely, P. D., Jr., Wells, R. E., and D. Minasian. 1993. *The Cenozoic Geology of the Oregon and Washington Coast Range and Road Log for the Northwest Petroleum Association 9th Annual Field Trip*. USGS Open-File Report 93–189.

Suczek, C. A., Babcock, R. S., and D. C. Engebretson. 1994. Tectonostratigraphy of the Crescent terrane and related rocks, Olympic Peninsula, Washington. In *Geologic Field Trips in the Pacific Northwest*, vol. 1, Department of Geological Sciences, University of Washington, eds. D. A. Swanson and R. A. Haugerud, p. 1H-11–20.

Tabor, R. W., and W. M. Cady. 1978. *Geologic Map of the Olympic Peninsula, Washington*. USGS Map I-994.

Wells, R., Bukry, D., Friedman, R., et al. 2014. Geologic history of Siletzia, a large igneous province in the Oregon and Washington Coast Range: Correlation to the geomagnetic polarity time scale and implications for a long-lived Yellowstone hotspot. *Geosphere* 10: 692–719.

Puget Lowland

Booth, D. B. 1994. Glaciofluvial infilling and scour of the Puget Lowland, Washington, during ice-sheet glaciation. *Geology* 22: 695–98.

Booth, D. B., and B. Goldstein. 1994. Patterns and processes of landscape development by the Puget Lobe ice sheet. In *Regional Geology of Washington State*, Washington Division of Geology and Earth Resources Bulletin 80, eds. R. Lasmanis and E. S. Cheney, p. 207–18.

Booth, D. B., and B. Hallet. 1993. Channel networks carved by subglacial water: Observations and reconstruction in the eastern Puget Lowland of Washington. *GSA Bulletin* 105: 671–83.

Booth, D. B., Troost, K. G., and J. T. Hagstrum. 2004. Deformation of Quaternary strata and its relationship to crustal folds and faults, south-central Puget Lowland, Washington State. *Geology* 32: 505–8.

Bourgeois, J., and S. Y. Johnson. 2001. Geologic evidence of earthquakes at the Snohomish delta, Washington, in the past 1,200 years. *GSA Bulletin* 113: 482–94.

Brandon, M. T., Cowan, D. S., and J. A. Vance. 1988. *The Late Cretaceous San Juan Thrust System, San Juan Islands, Washington*. GSA Special Paper 221.

Brown, E. H., Bradshaw, J. Y., and G. E. Mustoe. 1979. Plagiogranite and keratophyre in ophiolite on Fidalgo Island, Washington, part I. *GSA Bulletin* 90: 493–507.

Brown, E. H., Housen, B. A., and E. R. Schermer. 2007. Tectonic evolution of the San Juan Islands thrust system, Washington. In *Floods, Faults, and Fire: Geological Field Trips in Washington State and Southwest British Columbia*, GSA Field Guide 9, eds. P. Stelling and D. S. Tucker, p. 143–77.

Bucknam, R. C., Hemphill-Haley, E., and E. B. Leopold. 1992. Abrupt uplift within the past 1,700 years at southern Puget Sound, Washington. *Science* 258: 1611–14.

Cramer, M. D., and N. N. Barger. 2014. Are Mima-like mounds the consequence of long-term stability of vegetation spatial patterning? *Palaeogeography, Palaeoclimatology, Palaeoecology* 409: 72–83.

Garver, J. 1988. Fragment of the Coast Range ophiolite and the Great Valley sequence in the San Juan Islands, Washington. *Geology* 16: 948–51.

Garver, J. 1988. Stratigraphy, depositional setting, and tectonic significance of the clastic cover to the Fidalgo ophiolite, San Juan Islands, Washington. *Canadian Journal of Earth Sciences* 25: 417–32.

Gusey, D., and E. H. Brown. 1987. The Fidalgo ophiolite, Washington. In *GSA Centennial Field Guide: Cordilleran Section*, ed. M. Hill, p. 389–92.

Johnson, S. Y., Potter, C. J., and J. M. Armentrout. 1994. Origin and evolution of the Seattle basin and Seattle fault. *Geology* 22: 71–74.

Mustoe, G. E., Dillhoff, R. M., and T. A. Dillhoff. 2007. Geology and paleontology of the early Tertiary Chuckanut Formation. In *Floods, Faults, and Fire: Geological Field Trips in Washington State and Southwest British Columbia*, GSA Field Guide 9, eds. P. Stelling and D. S. Tucker, p. 121–35.

Thorson, R. M. 1989. Glacio-isostatic response of the Puget Sound area, Washington. *GSA Bulletin* 101: 1163–74.

Washburn, A. L. 1988. *Mima Mounds: An Evaluation of Proposed Origins with Special Reference to the Puget Lowland*. Washington Division of Geology and Earth Resources Report of Investigations 29.

Whetten, J. T., Carroll, P. I., Gower, H. D., Brown, E. H., and F. Pessl Jr. 1989. *Bedrock Geologic Map of the Port Townsend 30- by 60-Minute Quadrangle, Puget Sound Region, Washington*. USGS Map I-1198-G.

North Cascades

Barksdale, J. D. 1985. *Geology of the Methow Valley, Okanogan County, Washington*. Washington Department of Natural Resources Bulletin 68.

Beeson, M. H., and T. L. Tolan. 1987. Columbia River Gorge: The geologic evolution of the Columbia River in northwestern Oregon and southwestern Washington. In *GSA Centennial Field Guide: Cordilleran Section*, ed. M. Hill, p. 321–26.

Brown, E. H. 1987. Structural geology and accretionary history of the Northwest Cascades system, Washington and British Columbia. *GSA Bulletin* 99: 201–14.

Cheney, E. S. 2014. Tertiary stratigraphy and structure of the eastern flank of the Cascade Range, Washington. In *Trials and Tribulations of Life on an Active Subduction Zone:*

Field Trips in and around Vancouver, Canada, GSA Field Guide 38, eds. S. Dashgard and B. Ward, p. 193–226.

Cheney, E. S., and N. W. Hayman. 2009. The Chiwaukum structural low, eastern Cascade Range, Washington. In *Volcanoes to Vineyards: Geologic Field Trips through the Dynamic Landscape of the Pacific Northwest*, GSA Field Guide 15, eds. J. E. O'Connor, R. J. Dorsey, and I. P. Madin, p. 19–52.

Cheney, E. S., and N. W. Hayman. 2009. The Chiwaukum structural low: Cenozoic shortening of the central Cascade Range, Washington State, USA. *GSA Bulletin* 121: 1135–53.

Cheney, E. S., and N. W. Hayman. 2010. The Chiwaukum Structural Low: Cenozoic shortening of the central Cascade Range, Washington State, USA: Reply. *GSA Bulletin* 122: 2103–8.

Evans, J. E. 1994. Depositional history of the Eocene Chumstick Formation: Implications of tectonic partitioning for the history of the Leavenworth and Entiat–Eagle Creek fault system, Washington. *Tectonics* 13: 1425–44.

Evans, J. E. 2010. The Chiwaukum structural low: Cenozoic shortening of the central Cascade Range, Washington State, USA: Comment. *GSA Bulletin* 122: 2097–102.

Haugerud, R. A., and R. W. Tabor. 2009. *Geologic Map of the North Cascade Range, Washington*. USGS Map SIM 2940.

Hildreth, W., Fierstein, J., and M. Lanphere. 2003. Eruptive history and geochronology of the Mount Baker volcanic field, Washington. *GSA Bulletin* 115: 729–64.

Magloughlin, J. F. 1994. Migmatite to fault gouge: Fault rocks and the structural and tectonic evolution of the Nason terrane, North Cascade Mountains, Washington. In *Geologic Field Trips in the Pacific Northwest*, GSA Annual Meeting, eds. D. A. Swanson and R. A Haugerud, p. 2B1–2B17.

McGroder, M. F. 1991. Reconciliation of two-sided thrusting, burial metamorphism, and diachronous uplift in the Cascades of Washington and British Columbia. *GSA Bulletin* 103: 189–209.

Miller, R. B., Gordon, S. M., Bowring, S. A., et al. 2009. Linking deep and shallow crustal processes in an exhumed continental arc, North Cascades, Washington. In *Volcanoes to Vineyards: Geologic Field Trips through the Dynamic Landscape of the Pacific Northwest*, GSA Field Guide 15, eds. J. E. O'Connor, R. J. Dorsey, and I. P. Madin, p. 373–406.

Misch, P. 1966. Tectonic evolution of the Northern Cascades of Washington State. *Canadian Institute of Mining and Metallurgy*, Special Volume 8: 101–48.

Misch, P. 1987. The type section of the Skagit Gneiss, North Cascades, Washington. In *GSA Centennial Field Guide: Cordilleran Section*, ed. M. Hill, p. 393–98.

Mustoe, G. E., and E. B. Leopold. 2014. Paleobotanical evidence for the post-Miocene uplift of the Cascade Range. *Canadian Journal of Earth Sciences* 51: 809–24.

Tabor, R. W. 1994. Late Mesozoic and possible early Tertiary accretion in western Washington State: The Helena-Haystack mélange and the Darrington-Devils Mountain fault zone. *GSA Bulletin* 106: 217–32.

Tabor, R. W., Frizzell, V. A., Jr., Booth, D. B., and R. B. Waitt. 2000. *Geologic Map of the Snoqualmie Pass 30 x 60 Minute Quadrangle, Washington*. USGS map I-2538.

Tabor, R. W., Frizzell, V. A., Jr., Booth, D. B., Waitt, R. B., Whetten, J. T., and R. E. Zartman. 1993. *Geologic Map of the Skykomish River 30 x 60 Minute Quadrangle, Washington.* USGS map I-1963.

Tabor, R. W., and R. A. Haugerud. 2016. Geology of the North Cascades: Summary and enigmas. In *The Geology of Washington and Beyond*, University of Washington Press, ed. E. S. Cheney, p. 131–55.

Tabor, R. W., Waitt, R. B., Frizzell, V. A., Jr., Swanson, D. A., Byerly, G. R., and R. D. Bentley. 1982. *Geologic Map of the Wenatchee 1:100,000 Quadrangle, Central Washington.* USGS map I-1311.

Tepper, J. H., Nelson, B. K., Bergantz, G. W., and A. J. Irving. 1993. Petrology of the Chilliwack Batholith, North Cascades, Washington: Generation of calc-alkaline granitoids by melting of mafic lower crust with variable water fugacity. *Contributions to Mineralogy and Petrology* 113: 333–51.

Tucker, D. S., Scott, K. M., and D. R. Lewis. 2007. Field guide to Mount Baker volcanic deposits in the Baker River valley: Nineteenth-century lahars, tephras, debris avalanches, and early Holocene subaqueous lava. In *Floods, Faults, and Fire: Geological Field Trips in Washington State and Southwest British Columbia*, GSA Field Guide 9, eds. P. Stelling and D. S. Tucker, p. 83–98.

South Cascades

Cameron, K. A., and P. Pringle. 1986. Post-glacial lahars of the Sandy River Basin, Mount Hood, Oregon. *Northwest Science* 60 (4): 225–37.

Campbell, N. 1975. *A Geologic Road Log over Chinook, White Pass, and Ellensburg to Yakima Highways.* Washington Division of Geology and Earth Resources Information Circular 54.

Evarts, R. C., and D. A. Swanson. 1994. Geologic transect across the Tertiary Cascade Range. In *Geologic Field Trips in the Pacific Northwest*, vol. 2, Department of Geological Sciences, University of Washington, eds. D. A. Swanson and R. A. Haugerud, p. 2H-1–31.

Fiske, R. S., Hopson, C. A., and A. C. Waters. 1964. *Geologic Map and Section of Mount Rainier National Park, Washington.* USGS Miscellaneous Geologic Investigations Map I-432.

Hammond, P. E., Brunstad, K. A., and J. F. King. 1994. Mid-Tertiary volcanism east of Mount Rainier: Fifes Peak volcano-caldera and Bumping Lake pluton-Mount Aix caldera. In *Geologic Field Trips in the Pacific Northwest*, vol. 2, Department of Geological Sciences, University of Washington, eds. D. A. Swanson and R. A. Haugerud, p. 2J-1–19.

Hildreth, W. 2007. *Quaternary Magmatism in the Cascades: Geologic Perspectives.* USGS Professional Paper 1744.

Hildreth, W., and J. Fierstein. 1995. *Geologic Map of the Mount Adams Volcanic Field, Cascade Range of Southern Washington.* USGS Map I-2460; Accompanying Report.

Jutzeler, M., McPhie, J., and S. R. Allen. 2014. Facies architecture of a continental, below-wave-base volcaniclastic basin: The Ohanapecosh Formation, ancestral Cascades arc (Washington, USA). *GSA Bulletin* 126: 352–76.

Mattinson, J. M. 1977. Emplacement history of the Tatoosh volcanic-plutonic complex, Washington: Ages of zircons. *GSA Bulletin* 88: 1509–14.

Stockstill, K. R., Vogel, T. A., and T. W. Sisson. 2003. Origin and emplacement of the andesite of Burroughs Mountain, a zoned, large-volume lava flow at Mount Rainier, Washington, USA. *Journal of Volcanology and Geothermal Research* 119: 275–96.

Tolan, T. L., and M. H. Beeson. 1984. Intracanyon flows of the Columbia River Basalt Group in the lower Columbia River Gorge and their relationship to the Troutdale Formation. *GSA Bulletin* 95: 463–77.

Tolan, T. L., Beeson, M. H., and B. F. Vogt. 1984. Exploring the Neogene history of the Columbia River: Discussion and geologic field trip guide to the Columbia River Gorge. Part I. Discussion. *Oregon Geology* 46 (8): 87–97.

Tolan, T. L., Beeson, M. H., and B. F. Vogt. 1984. Exploring the Neogene history of the Columbia River: Discussion and geologic field trip guide to the Columbia River Gorge. Part II. Road log and comments. *Oregon Geology* 46 (9): 103–12.

Vance, J. A., Clayton, G. A., Mattinson, J. M., and C. W. Naeser. 1987. Early and middle Cenozoic stratigraphy of the Mount Rainier–Tieton River area, southern Washington Cascades. In *Selected Papers on the Geology of Washington*, Washington Division of Geology and Earth Resources Bulletin 77, ed. J. E. Schuster, p. 269–90.

Wang, Y., Hofmeister, R. J., McConnell, V. S., Burns, S. F., Pringle, P. T., and G. L. Peterson. 2002. Columbia River Gorge landslides. In *Field Guide to Geologic Processes in Cascadia, Oregon*, Department of Geology and Mineral Industries Special Paper 36, ed. G. W. Moore, p. 273–88.

Okanogan Highlands

Brown, S. R., Gibson, H. D., Andrews, G. D. M., et al. 2012. New constraints on Eocene extension within the Canadian Cordillera and identification of Phanerozoic proto-liths for footwall gneisses of the Okanagan Valley shear zone. *Lithosphere* 4: 354–77.

Buddington, A. M., Wang, D., and P. T. Doughty. 2016. Pre-Belt basement tour: Late Archean–early Proterozoic rocks of the Cougar Gulch area, southern Priest River Complex, Idaho. In *Exploring the Geology of the Inland Northwest*, GSA Field Guide 41, eds. R. S. Lewis and K. L. Schmidt, p. 265–84.

Cheney, E., and A. Buddington. 2012. *Geology of the Continental Margin of Ancestral North America: Laurentia in Northeastern Washington*. National Association of Geoscience Teachers Pacific Northwest Section Summer Conference Field Trip.

Cheney, E. S., and P. T. Doughty. 2016. Overview of Laurentian rocks in the Pacific Northwest. In *The Geology of Washington and Beyond*, University of Washington Press, ed. E. S. Cheney, p. 23–27.

Cheney, E. S., and J. T. Figge. 2016. Overview of the terranes of Washington. In *The Geology of Washington and Beyond*, University of Washington Press, ed. E. S. Cheney, p. 95–100.

Cheney, E. S., Rasmussen, M. G., and M. Miller. 1994. Major faults, stratigraphy, and identity of Quesnellia in Washington and adjacent British Columbia. In *Regional Geology of Washington State*, Washington Division of Geology and Earth Resources Bulletin 80, eds. R. Lasmanis and E. S. Cheney, p. 49–71.

Cheney, E. S., and G. A. Zieg. 2016. Regional fold and thrust belts in northeastern Washington and adjacent British Columbia. In *The Geology of Washington and Beyond*, University of Washington Press, ed. E. S. Cheney, p. 43–61.

Doughty, P. T., Buddington, A. M., Cheney, E. S., and R. E. Derkey. 2016. Geology of the Priest River Complex and adjacent Paleozoic strata south of the Spokane River valley, Washington. In *The Geology of Washington and Beyond*, University of Washington Press, ed. E. S. Cheney, p. 77–92.

Dings, M. G., and D. H. Whitebread. 1965. *Geology and Ore Deposits of the Metaline Zinc-Lead District Pend Oreille County, Washington*. USGS Professional Paper 489.

Fox, K. F., Jr. 1994. Geology of metamorphic core complexes and associated extensional structures in north-central Washington. In *Regional Geology of Washington State*, Washington Division of Geology and Earth Resources Bulletin 80, eds. R. Lasmanis and E. S. Cheney, p. 21–47.

Gager, B. R. 1984. The Tiger Formation: Paleogene fluvial sediments deposited adjacent to a deforming Cordilleran metamorphic core complex, northeastern Washington. *Sedimentary Geology* 38: 393–420.

Gillespie, A. R. 2016. Overview of the Quaternary Period in Washington State. In *The Geology of Washington and Beyond*, University of Washington Press, ed. E. S. Cheney, p. 223–32.

Harms, T. A., and R. A. Price. 1992. The Newport fault: Eocene listric normal faulting, mylonitization, and crustal extension in northeast Washington and northwest Idaho. *GSA Bulletin* 104: 745–61.

Hurlow, H. A., and B. K. Nelson. 1993. U-Pb zircon and monazite ages for the Okanogan Range batholith, Washington: Implications for the magmatic and tectonic evolution of the southern Canadian and northern United States Cordillera. *GSA Bulletin* 105: 231–40.

Kruckenberg, S. C., Whitney, D. L., Teyssier, C., Fanning, C. M., and W. J. Dunlap. 2008. Paleocene-Eocene migmatite crystallization, extension, and exhumation in the hinterland of the northern Cordillera: Okanogan dome, Washington, USA. *GSA Bulletin* 120: 912–29.

Lund, K., and E. S. Cheney. 2016. Correlation of unconformity-bounded sequences of the Neoproterozoic Windermere Supergroup in Idaho, Washington, and southern British Columbia. In *The Geology of Washington and Beyond*, University of Washington Press, ed. E. S. Cheney, p. 28–42.

McCollum, L. B., McCollum, M. B., and M. M. Hamilton. 2016. Mesoproterozoic and Cambrian geology of the northeastern Columbia Plateau of Washington: A view from the steptoes. In *The Geology of Washington and Beyond*, University of Washington Press, ed. E. S. Cheney, p. 62–76.

Miller, C. F., and L. J. Bradfish. 1980. An inner Cordilleran belt of muscovite-bearing plutons. *Geology* 8: 412–16.

Miller, F. K. 1994. The Windermere Group and late Proterozoic tectonics in northeastern Washington and northern Idaho. In *Regional Geology of Washington State*, Washington Division of Geology and Earth Resources Bulletin 80, eds. R. Lasmanis and E. S. Cheney, p. 1–19.

Miller, F. K., and J. C. Engels. 1975. Distribution and trends of discordant ages of the plutonic rocks of northeastern Washington and northern Idaho. *GSA Bulletin* 86: 517–28.

Molenaar, D. 1988. *The Spokane Aquifer, Washington: Its Geologic Origin and Water-Bearing and Water-Quality Characteristics*. USGS Water-Supply Paper 2265.

Monger, J. W. H., Price, R. A., and D. J. Templeman-Kluit. 1982. Tectonic accretion and the origin of two metamorphic belts in the Canadian Cordillera. *Geology* 10: 70–75.

Orr, K. E. 1985. *Structural Features along the Margins of Okanogan and Kettle Domes, Northeastern Washington and Southern British Columbia*. University of Washington PhD dissertation.

Orr, K. E., and E. S. Cheney. 1987. Kettle and Okanogan domes, northeastern Washington and southern British Columbia. In *Selected Papers on the Geology of Washington*, Washington Division of Geology and Earth Resources Bulletin 77, ed. J. E. Schuster, p. 55–71.

Smith, M. T., and G. E. Gehrels. 1992. Stratigraphy and tectonic significance of lower Paleozoic continental margin strata in northeastern Washington. *Tectonics* 11: 607–20.

Tepper, J. H. 2016. Eocene rollback and breakoff of the Farallon slab: An explanation for the "Challis event." *GSA Abstracts with Programs* 48 (4).

Watkinson, A. J., and M. A. Ellis. 1987. Recent structural analyses of the Kootenay arc in northeastern Washington. In *Selected Papers on the Geology of Washington*, Washington Division of Geology and Earth Resources Bulletin 77, ed. J. E. Schuster, p. 41–53.

Columbia Basin

Carson, R. J., Tolan, T. L., and S. Reidel. 1987. Geology of the Vantage area, south-central Washington: An introduction to the Miocene flood basalts, Yakima fold belt, and the Channeled Scabland. In *GSA Centennial Field Guide: Cordilleran Section*, ed. M. Hill, p. 357–62.

Hooper, P. R., Camp, V. E., Reidel, S. P., and M. E. Ross. 2007. The origin of the Columbia River flood basalt province: Plume vs. nonplume models. In *Plates, Plumes, and Planetary Processes*, GSA Special Paper 430, eds. G. R. Foulger and D. M. Jurdy, p. 635–68.

Nash, B. P., and M. E. Perkins. 2012. Neogene fallout tuffs from the Yellowstone hotspot in the Columbia Plateau region, Oregon, Washington, and Idaho, USA. *PLoS ONE* 7(10): e44205.

Reidel, S. P., Camp, V. E., Martin, B. S., Tolan, T. L., and J. A. Wolff. 2016. The Columbia River Basalt Group of western Idaho and eastern Washington—Dikes, vents, flows, and tectonics along the eastern margin of the flood basalt province. In *Exploring the Geology of the Inland Northwest*, GSA Field Guide 41, eds. R. S. Lewis and K. L. Schmidt, p. 127–50.

Reidel, S. P., Camp, V. E., Tolan, T. L., and B. S. Martin. 2013. The Columbia River flood basalt province: Stratigraphy, areal extent, volume, and physical volcanology. In *The Columbia River Basalt Province*, GSA Special Paper 497, eds. S. P. Reidel, V. E. Camp, M. E. Ross, et al., p. 1–44.

Reidel, S. P., Campbell, N. P., Fecht, K. R., and K. A. Lindsey. 1994. Late Cenozoic structure and stratigraphy of south-central Washington. In *Regional Geology of Washington State*, Washington Division of Geology and Earth Resources Bulletin 80, eds. R. Lasmanis and E. S. Cheney, p. 159–80.

Reidel, S. P., Fecht, K. R., Hagood, M. C., and T. L. Tolan. 1989. The geologic evolution of the central Columbia Plateau. In *Volcanism and Tectonism in the Columbia River Flood-Basalt Province*, GSA Special Paper 239, eds. P. R. Hooper and S. P. Reidel, p. 247–64.

Reidel, S. P., Martin, B. S., and H. L. Petcovic. 2003. The Columbia River flood basalts and the Yakima fold belt. In *Western Cordillera and Adjacent Areas*, GSA Field Guide 4, ed. T. W. Swanson, p. 87–105.

Sherrod, B. L., Blakely, R. J., Lasher, J. P., et al. 2016. Active faulting on the Wallula fault zone within the Olympic-Wallowa lineament, Washington State, USA. *GSA Bulletin* 128: 1636–59.

Tolan, T. L, Martin, B. S., Reidel, S. P., Anderson, J. L., Lindsey, K. A., and W. Burt. 2009. An introduction to the stratigraphy, structural geology, and hydrogeology of the Columbia River Flood-Basalt Province: A primer for the GSA Columbia River Basalt Group field trips. In *Volcanoes to Vineyards: Geologic Field Trips through the Dynamic Landscape of the Pacific Northwest*, GSA Field Guide 15, eds. J. E. O'Connor, R. J. Dorsey, and I. P. Madin, p. 599–643.

Waitt, R. B. 1979. *Late Cenozoic Deposits, Landforms, Stratigraphy, and Tectonism in Kittitas Valley, Washington*. USGS Professional Paper 1127.

Wells, R. E., Niem, A. R., Evarts, R. C., and J. T. Hagstrum. 2009. The Columbia River Basalt Group: From the gorge to the sea. In *Volcanoes to Vineyards: Geologic Field Trips through the Dynamic Landscape of the Pacific Northwest*, GSA Field Guide 15, eds. J. E. O'Connor, R. J. Dorsey, and I. P. Madin, p. 737–74.

Wells, R. E., Simpson, R. W., Bentley, R. D., Beeson, M. H., Mangan, M. T., and T. L. Wright. 1989. Correlation of Miocene flows of the Columbia River Basalt Group from the central Columbia River Plateau to the coast of Oregon and Washington. In *Volcanism and Tectonism in the Columbia River Flood-Basalt Province*, GSA Special Paper 239, eds. P. R. Hooper and S. P. Reidel, p. 113–30.

State Geologic Maps

Dragovich, J. D., Logan, R. L., Schasse, H. W., et al. 2002. *Geologic Map of Washington—Northwest Quadrant*. Washington Division of Geology and Earth Resources, Geologic Map GM-50.

Schuster, J. E., Gulick, C. W., Reidel, S. P., Fecht, K. R., and S. Zurenko. 1997. *Geologic Map of Washington—Southeast Quadrant*. Washington Division of Geology and Earth Resources, Geologic Map GM-45.

Stoffel, K. L., Joseph, N. L., Zurenko Waggoner, S., Gulick, C. W., Korosec, M. A., and B. B. Bunning. 1991. *Geologic Map of Washington—Northeast Quadrant*. Washington Division of Geology and Earth Resources, Geologic Map GM-30.

Walsh, T. J., Korosec, M. A., Phillips, W. M., Logan, R. L., and H. W. Schasse. 1987. *Geologic Map of Washington—Southwest Quadrant*. Washington Division of Geology and Earth Resources, Geologic Map GM-34.

INDEX

Page numbers in bold indicate a photo

Upper reaches of the American River on WA 410 east of Mt. Rainier.

Marli Miller is a senior instructor and researcher at the University of Oregon. She completed her BA in geology at Colorado College in 1982 and her MS and PhD in structural geology at the University of Washington in 1987 and 1992, respectively, where coauthor Darrel Cowan was her thesis advisor. Marli teaches a variety of courses, including introductory geology, structural geology, field geology, and geophotography. In addition to numerous technical papers, she is the author of *Roadside Geology of Oregon* and *Geology of Death Valley National Park*, with coauthor Lauren A. Wright. She is the photographer for *What's So Great About Granite*, written by Jennifer Carey, the editor of this Roadside Geology book! Marli has two daughters, Lindsay and Megan.

Darrel Cowan received his BS and PhD in geology from Stanford University and served as an exploration geologist in the Alaska Division of Shell Oil Company from 1971 until 1974. He accepted a faculty position at the University of Washington in Seattle in 1974 and currently teaches in the Department of Earth and Space Sciences. Darrel's research interests include processes at ancient subduction zones and the late Mesozoic tectonic evolution of the western margin of North America. Although his field research has

—BATBAATAR JIGJIDSUREN

taken him and his students all over the world, the San Juan Islands in Washington rank high on his list of geologic puzzles worthy of study. In 2014, Darrel received the Career Contribution Award from the Division of Structural Geology and Tectonics of the Geological Society of America.